Recent Titles in This Series

(Continued in the back of this publication)

Arithmetic Geometry

CONTEMPORARY MATHEMATICS

174

Arithmetic Geometry

Conference on Arithmetic Geometry
with an Emphasis on Iwasawa Theory
March 15–18, 1993
Arizona State University

Nancy Childress
John W. Jones
Editors

American Mathematical Society
Providence, Rhode Island

Serial Math

The Conference on Algebraic Geometry with an Emphasis on Iwasawa Theory was held at Arizona State University, Tempe, Arizona, March 15–18, 1993.

1991 *Mathematics Subject Classification*. Primary 11–06; Secondary 11F33, 11R23, 11R42, 12F12.

Library of Congress Cataloging-in-Publication Data

Conference on Arithmetic Geometry with an Emphasis on Iwasawa Theory (1993: Arizona State University)
  Arithmetic geometry: Conference on Arithmetic Geometry with an Emphasis on Iwasawa Theory, March 15–18, 1993, Arizona State University/Nancy Childress, John W. Jones.
    p.   cm. — (Contemporary mathematics, ISSN 0271-4132; v. 174)
  Includes bibliographical references.
  ISBN 0-8218-5174-8
  1. Algebraic number theory—Congresses. 2. Geometry, Algebraic—Congresses. I. Childress, Nancy, 1958– . II. Jones, John W., 1961– . III. Title. IV. Series: Contemporary mathematics (American Mathematical Society); v. 174.
QA247.C64   1993
512′.74—dc20
                                                                                94-22801
                                                                                CIP

Contents

Preface

This book is the result of a research conference in arithmetic geometry which took place at Arizona State University in March 1993. The lectures at this conference were devoted to important recent advances in Iwasawa theory, elliptic curves, modular forms and modular curves, p-adic heights, L-functions, and the inverse Galois problem. In addition to expanded versions of many of the talks presented at the conference, this volume contains a previously unpublished letter of John Tate, written to J.-P. Serre in 1973. We thank Professors Tate and Serre for generously agreeing to the inclusion of it here.

We are grateful to all of the participants, and especially the speakers, for making the conference a success. We also thank Tonya Blair, Lydia Dancel, Bruce Long, Mary Sabel, and Marlene Salvato of the ASU Mathematics Department for their fine work in helping to organize the conference, and Craig Huneke, Donna Harmon and the staff at the AMS for their help with the preparation of this volume. Finally, we wish to acknowledge the financial support of the NSF and Arizona State University, which made travel to the conference possible for many of the participants.

Nancy Childress
John W. Jones

Contemporary Mathematics
Volume **174**, 1994

Real Hilbertianity and the
Field of Totally Real Numbers

MICHAEL D. FRIED[*†], DAN HARAN[*‡]
AND HELMUT VÖLKLEIN[*•]

ABSTRACT. We use moduli spaces for covers of the Riemann sphere to solve regular embedding problems, with prescribed extendability of orderings, over PRC fields. As a corollary we show that the elementary theory of $\mathbb{Q}^{tr}$ is decidable. Since the ring of integers of $\mathbb{Q}^{tr}$ is undecidable, this gives a natural undecidable ring whose quotient field is decidable.

Introduction

The theory and use in [**F**] of moduli spaces of covers of the Riemann sphere with prescribed ramification data has been further developed in [**FV1**]. There the main theme is that K-rational points of the moduli spaces correspond to covers defined over K. Furthermore, [**FV2**] notes a correspondence between existence of K-rational points on certain related spaces and the solvability of regular embedding problems over K. Thus, using moduli spaces allows us to prove solvability of regular embedding problems over fields K suitably large for such varieties to have the requisite K-rational points.

This principle appears in [**FV2**] to show that the absolute Galois group of a countable Hilbertian PAC field of characteristic 0 is free. The natural extension of this to the (larger) class of Hilbertian PRC fields appears in [**FV3**].

Recall [**FJ**, p. 129] that K is **PAC (pseudo algebraically closed)** if every absolutely irreducible variety V defined over K has a K-rational point. Furthermore [**P2**], K is **PRC (pseudo real closed)** if every absolutely irreducible

---

1991 *Mathematics Subject Classification*. Primary 12D15, 12E25, 12F12, 11G25.
[*] Support from the Institute for Advanced Study at Hebrew University, 1991–92.
[†] Supported by NSA grant MDA 14776 and BSF grant 87-00038.
[‡] Supported by Max-Planck-Institut für Mathematik, Bonn, 1992–93.
[•] Supported by NSA grant MDA 904-89-H-2028.
This paper is in final form and no version of it will be submitted for publication elsewhere.

variety V defined over K has a K-rational point, provided that V has a non-singular point over each real closure of K. The latter condition on V is equivalent to the following one: the function field $K(V)$ of V is a **totally real** extension of K, that is, every ordering on K extends to $K(V)$.

The field $\mathbb{Q}^{\mathrm{tr}}$ of all totally real algebraic numbers is the fixed field of all involutions in the absolute Galois group $G(\mathbb{Q})$ of $\mathbb{Q}$. From Pop [**P**], $\mathbb{Q}^{\mathrm{tr}}$ is PRC. By Weissauer's theorem [**FJ**, Proposition 12.4] every proper finite extension of $\mathbb{Q}^{\mathrm{tr}}$ is Hilbertian. Also, by Prestel's extension theorem [**P2**, Theorem 3.1] every algebraic extension of $\mathbb{Q}^{\mathrm{tr}}$ is PRC. Hence, the absolute Galois group of a proper finite extension of $\mathbb{Q}^{\mathrm{tr}}$ is known [**FV3**]. The field $\mathbb{Q}^{\mathrm{tr}}$, however, is not Hilbertian. For example, $Z^2 - (a^2 + 1)$ is reducible over $\mathbb{Q}^{\mathrm{tr}}$ for every $a \in \mathbb{Q}^{\mathrm{tr}}$.

In [**FHV**] we cover also the case of $\mathbb{Q}^{\mathrm{tr}}$. The key observation is that $\mathbb{Q}^{\mathrm{tr}}$ satisfies a certain weakening of the Hilbertian property. This allows specializing Galois extensions of $K(x)$ whose Galois group is generated by *real* involutions to obtain Galois extensions of K with the same Galois group. As a result we determine the absolute Galois group of $\mathbb{Q}^{\mathrm{tr}}$.

In the present paper we extend the methods and results of [**FV3**]. This solves regular embedding problems over a PRC field K so that the orderings of K extend to prescribed subfields (Theorem 5.2). Thus, Theorem 5.3 gives new information about the absolute Galois group of the field $K(x)$ of rational functions over K .

The first 5 sections setup the proof of Theorem 5.2. We need an approximation theorem for varieties over PRC fields (section 1), supplements about the moduli spaces (section 2), group-theoretic lemmas (section 3), and the determination of the real involutions in Galois groups over $\mathbb{R}(x)$ (section 4).

In section 6 we define the concept of totally real Hilbertian, and show that $\mathbb{Q}^{\mathrm{tr}}$ has this property. Section 7 combines these results to determine the absolute Galois group of a countable totally real Hilbertian PRC field satisfying these properties: it has no proper totally real algebraic extensions; and its space of orderings has no isolated points. This group is the free product of groups of order 2, indexed by the Cantor set X_ω (Theorem 7.6). In particular, $G(\mathbb{Q}^{\mathrm{tr}}) = \mathrm{Aut}(\tilde{\mathbb{Q}}/\mathbb{Q}^{\mathrm{tr}})$ is isomorphic to this group.

As a corollary we introduce the notion of *real Frobenius fields*: $\mathbb{Q}^{\mathrm{tr}}$ is an example (Corollary 8.3). Following the Galois stratification procedure of [**FJ**, Chap. 25] and [**HL**] we show that the elementary theory of real Frobenius fields allows elimination of quantifiers in the appropriate language. In particular, $\mathbb{Q}^{\mathrm{tr}}$ is primitive recursively decidable (Theorem 10.1). On the other hand the ring of integers of $\mathbb{Q}^{\mathrm{tr}}$ is undecidable (A. Prestel pointed us to Julia Robinson's proof of this [**R2**].) Thus we obtain a natural example of an undecidable ring with a decidable quotient field. Compare this with the possibility that $\mathbb{Q}$ is decidable (cf. Robinson [**R1**, p. 951]). Furthermore, we give (Corollary 10.5) a system of axioms for the theory of $\mathbb{Q}^{\mathrm{tr}}$.

Affirmations. We are grateful to Moshe Jarden for numerous suggestions that led to an improved presentation of this paper.

F. Pop told us the characterization of $G(\mathbb{Q}^{\mathrm{tr}})$ also follows from his "$\frac{1}{2}$ Riemann existence theorem." His method uses rigid analytic geometry, versus our use of the classical Riemann existence theorem. We look forward to seeing a written account.

This paper corresponds to a portion of the talk of the first author at the Tempe conference on Arithmetic Geometry, March 1993. Appropos the theme of the Tempe conference, this paper uses profinite ideas of Iwasawa for characterizing absolute Galois groups of fields. The remainder of the talk discussed regular realizations of dihedral groups. It considered the dihedral groups D_{p^n} of order $2p^n$ with p an odd prime and realization of these as Galois groups over $\mathbb{Q}(x)$ with a bounded numbers of branch points [**DF2**, section 5.2]. The talk emphasized the relation with rational points on modular curves. Extension of these ideas with a general group G replacing D_p will appear under the title **Modular stacks and the Inverse Galois Problem**.

1. Ordered fields and an approximation theorem for PRC fields

Let K be a field of characteristic 0, and let $G(K)$ be its absolute Galois group. Recall [**P1**, §6] that the set of orderings $X(K)$ of K is a boolean topological space in its natural Harrison topology. This topology is given by a subbase consisting of sets of the form $H(c) = \{P \in X(K)|\ c \in P\}$, for $c \in K^{\times}$. Here P denotes the positive elements in an ordering.

By Artin-Schreier theory [**L**, XI,§2], the real closures of K (inside a fixed algebraic closure $\tilde{K}$ of K) are the fixed fields of the involutions in $G(K)$. This identifies the set $\hat{\mathcal{X}}$ of real closures of K with a topological subspace of $G(K)$. It is a boolean space, since the set of involutions is closed in $G(K)$. Observe that $\overline{H}(z) = \{R \in \hat{\mathcal{X}}|\ z \in R\}$ is open in $\hat{\mathcal{X}}$, for each $z \in \tilde{K}^{\times}$.

For each $R \in \hat{\mathcal{X}}$ let $\pi(R)$ be the restriction of the unique ordering of R to K. Then $\pi(R_1) = \pi(R_2)$ if and only if R_1 and R_2 are conjugate by an automorphism of $\tilde{K}$ over K. The map $\pi\colon \hat{\mathcal{X}} \to X(K)$ is continuous: $\pi^{-1}(H(c)) = \overline{H}(\sqrt{c})$. Moreover, there exists a closed subset $\mathcal{X}$ of $\hat{\mathcal{X}}$ such that $\pi\colon \mathcal{X} \to X(K)$ is a homeomorphism [**HJ1**, Corollary 9.2]. The corresponding closed subset of involutions in $G(K)$ contains exactly one representative from each conjugacy class of involutions. Having fixed such $\mathcal{X}$, identify $X(K)$ with it.

REMARK 1.1. *Comments on orderings.*

(a) If K is PRC, then every clopen subset of $X(K)$ is of the form $H(c)$ for a suitable $c \in K^{\times}$ [**P2**, Proposition 1.3].

(b) Let R be a real closed field, and let $a \in R$, $c \in R^{\times}$. If either $c < 0$ or $a > 0$, then the system $\quad Y^2 + cZ^2 = a, \quad Y \neq 0 \quad$ has a solution in R.

(c) For $\mathbf{X} = (X_1, \dots, X_n)$ put $\|\mathbf{X}\|^2 = \sum_{i=1}^{n} X_i^2$. Let K be an ordered field, and let $\mathbf{a}, \mathbf{b}, \mathbf{c} \in K^n$ and $\nu \in K^{\times}$. From the triangle inequality (over the real closure of K), if $\|\mathbf{a} - \mathbf{b}\|^2, \|\mathbf{b} - \mathbf{c}\|^2 < (\frac{\nu}{2})^2$ then $\|\mathbf{a} - \mathbf{c}\|^2 < \nu^2$.

PROPOSITION 1.2. *Let K be a PRC field, and let $V \subseteq \mathbb{A}^n$ be an absolutely irreducible affine variety defined over K. Let $\mathcal{X}$ be a closed set of real closures of K, one for each ordering of K. Let $\mathcal{X}_1, \ldots, \mathcal{X}_m$ be disjoint clopen subsets of $\mathcal{X}$ that cover $\mathcal{X}$. Let $\mathbf{x}_1, \ldots, \mathbf{x}_m$ be nonsingular points on V such that $\mathbf{x}_j \in V(R)$ for every $R \in \mathcal{X}_j$, for $j = 1, \ldots, m$. Let $\nu_1, \ldots, \nu_m \in K^\times$. Then there is $\mathbf{x} \in V(K)$ such that for each $1 \leq j \leq m$*

$$(1) \qquad \|\mathbf{x} - \mathbf{x}_j\|^2 < \nu_j^2 \text{ in } R, \qquad \text{for every } R \in \mathcal{X}_j.$$

PROOF. Fix j and put $L = K(\mathbf{x}_j)$. Let $R \in \mathcal{X}_j$. Then $L \subseteq R$. As K is dense in R [**P2**, Proposition 1.4], there is $\mathbf{a}_j \in K^n$ such that

$$(2) \qquad \|\mathbf{x}_j - \mathbf{a}_j\|^2 < \left(\frac{\nu_j}{2}\right)^2 \text{ in } R.$$

This $\mathbf{a}_j$ depends on R, but if $R' \in \mathcal{X}_j$ is sufficiently close to R, then (2) holds also with R' instead of R. Indeed, the restriction $\mathcal{X}_j \to X(L)$ is continuous, and (2) describes a basic open set in $X(L)$. Use compactness of $\mathcal{X}_j$ to partition $\mathcal{X}_j$ into smaller clopen subsets (and thereby increase m). Associate with each of them the original point $\mathbf{x}_j$ such that (2) holds with suitable $\mathbf{a}_j$ for all $R \in \mathcal{X}_j$.

By Remark 1.1(a), for each j there is $c_j \in K^\times$ such that $\mathcal{X}_j = H(c_j)$. Suppose that V is defined by $f_1(\mathbf{X}), \ldots, f_r(\mathbf{X}) \in K[X_1, \ldots, X_n]$. These together with the additional polynomials

$$f_{r+j}(\mathbf{X}, \mathbf{Y}, \mathbf{Z}) = \left(\frac{\nu_j}{2}\right)^2 - \|\mathbf{X} - \mathbf{a}_j\|^2 - Y_j^2 - c_j Z_j^2, \quad j = 1, \ldots, m$$

define an absolutely irreducible variety $W \subseteq \mathbb{A}^{n+2m}$ of dimension $\dim V + m$.

Indeed, by induction on m we may assume that $m = 1$. Let $\mathbf{x}$ be the generic point of V over $\widetilde{K}$, that is, the image of $\mathbf{X}$ in in the integral domain $\widetilde{K}[V] = \widetilde{K}[\mathbf{X}]/(f_1, \ldots, f_r)$. Let $u = \left(\frac{\nu_1}{2}\right)^2 - \|\mathbf{x} - \mathbf{a}_1\|^2$, and let y_1 be transcendental over $\widetilde{K}(V)$. Observe that $u \neq 0$, since, by (2), $\left(\frac{\nu_1}{2}\right)^2 - \|\mathbf{x}_1 - \mathbf{a}_1\|^2 \neq 0$. Therefore $f_{r+1}(\mathbf{x}, y_1, Z_1) = u - y_1^2 - c_1 Z_1^2$ is irreducible over $\widetilde{K}(V)(y_1)$. Let z_1 be its root in the algebraic closure M of $\widetilde{K}(V)(y_1)$. Clearly

$$\widetilde{K}[W] = \widetilde{K}[\mathbf{X}, Y_1, Z_1]/(f_1, \ldots, f_r, f_{r+1}) \cong \widetilde{K}[V][y_1, z_1] \subseteq M.$$

It follows that $\widetilde{K}[W]$ is an integral domain, and $\mathrm{tr.deg.}(W) = \mathrm{tr.deg.}(V) + 1$. Thus W is absolutely irreducible and $\dim W = \dim V + 1$.

Let $R \in \mathcal{X}$. With no loss, assume $R \in \mathcal{X}_1$, and hence $R \notin \mathcal{X}_2, \ldots, \mathcal{X}_m$. Thus, c_1 is positive, and $c_2, \ldots, c_m$ are negative in R. Apply (2) and Remark 1.1(b) to complete the $\mathbf{x}_1$ to a point $(\mathbf{x}_1, \mathbf{y}, \mathbf{z}) \in W(R)$ with $y_j \neq 0$ for each $1 \leq j \leq m$. In particular, $\frac{\partial f_{r+j}}{\partial Y_j}(\mathbf{x}_1, \mathbf{y}, \mathbf{z}) \neq 0$: $(\mathbf{x}_1, \mathbf{y}, \mathbf{z})$ is a nonsingular point on W.

By the PRC property of K there exists a point $(\mathbf{x}, \mathbf{y}, \mathbf{z}) \in W(K)$. Clearly $\mathbf{x} \in V(K)$, and for each j we have $\|\mathbf{x} - \mathbf{a}_j\|^2 \leq (\frac{\nu_j}{2})^2$ in R, for each $R \in \mathcal{X}_j$. This and (2) imply (1), by Remark 1.1(c). □

Applying Proposition 1.2 to $V = \mathbb{A}^1$ yields the Block Approximation Lemma of [**P3**, p. 354]:

COROLLARY 1.3. *Let K be a PRC field. Let $H_1, \ldots, H_m$ be disjoint clopen subsets of $X(K)$, and let $x_1, \ldots, x_m \in K$, and $\nu_1, \ldots, \nu_m \in K^\times$. Then there is $x \in K$ such that for every j $(x - x_j)^2 \leq_P \nu_j^2$ for every $P \in H_j$.*

DEFINITION 1.4. *Let $\widetilde{K}$ be algebraically closed with $\iota \in \mathrm{Aut}(\widetilde{K})$ of order 2.*

(a) For $c \in \widetilde{K}$ let $|c|_\iota^2 = c \cdot \iota(c)$.

(b) For $\mathbf{z} \in \widetilde{K}^n$ let $\|\mathbf{z}\|_\iota^2 = \sum_{i=1}^n |z_i|_\iota^2$.

(c) For a $\widetilde{K}$-linear morphism $f\colon \mathbb{A}^n \to \mathbb{A}^m$ given by a matrix $A = (a_{ij}) \in M_{m \times n}(\widetilde{K})$ let $\|f\|_\iota^2 = \sum_{i,j} |a_{ij}|_\iota^2$.

In the above definition, the fixed field R of ι is real closed, and $\|c\|_\iota^2, \|\mathbf{z}\|_\iota^2, \|f\|_\iota^2$ are nonnegative elements of R. If $\mathbf{z} \in R^n$ then $\|\mathbf{z}\|_\iota^2 = \|\mathbf{z}\|^2$. Also, in the unique ordering of R, for all $\mathbf{z} \in \widetilde{K}^n$ the Schwartz inequality gives

$$(3) \qquad \|f(\mathbf{z})\|_\iota^2 \leq \|f\|_\iota^2 \cdot \|\mathbf{z}\|_\iota^2.$$

REMARK 1.5. The space $X(\mathbb{Q}^{\mathrm{tr}})$ is homeomorphic to $X_\omega = \{0, 1\}^{\aleph_0}$, the universal Boolean space of weight $\aleph_0$ (cf. the concluding Remark of [**FV3**]). In particular it has no isolated points.

LEMMA 1.6. *If K is a finitely generated field, then the set $X^a(K)$ of archimedian orderings on K is dense in $X(K)$.*

PROOF. By induction on the number of generators of $K/\mathbb{Q}$ it suffices to show the following. Let K/K_0 be a simple extension of countable fields, let $P \in X(K)$, and let $P_0 = \mathrm{res}_{K_0} P \in X(K_0)$. If P_0 is in the closure of $X^a(K_0)$, then P is in the closure of $X^a(K)$.

The restriction $X(K) \to X(K_0)$ is open [**ELW**, 4.bis], hence we may assume that P_0 is archimedian. If K/K_0 is algebraic, then P is also archimedian. Otherwise K is the field of rational functions in one variable t over K_0. Replace (K_0, P_0) by its real closure (cf. [**C**, Lemma 8]) to assume that K_0 is real closed.

By [**C**, Corollary 9(c)], every neighborhood U of P in $X(K)$ contains a set of the form $\{Q \in X(K)| \ a < t < b \text{ in } Q\}$, where $a, b \in K_0$ and $a < b$ in P_0. As P_0 is archimedian, we can embed K_0 into $\mathbb{R}$. Since K_0 is countable, there is $c \in \mathbb{R} \setminus K_0$ in the interval (a, b) in $\mathbb{R}$. This c is then transcendental over K_0. The K_0-embedding $K \to \mathbb{R}$ given by $t \mapsto c$ induces an archimedian ordering Q on K, and $a < t < b$ in Q. Thus $Q \in U$. $\quad\square$

For a subset I of a group G let $\mathrm{Con}_G(I) = \bigcup_{\sigma \in G} I^\sigma$. We say that I is a **conjugacy domain**, if I is closed under the conjugation, that is, $I = \mathrm{Con}_G(I)$.

DEFINITION 1.7. Let F/E be a Galois extension of fields with F not formally real. We say an involution $\epsilon \in G(F/E)$ is **real** if its fixed field $F(\epsilon)$ in F is formally real. Equivalently, ϵ is the restriction of an involution in the absolute Galois group $G(E)$ of E. Let $I(F/E)$ be the set of real involutions of $G(F/E)$.

Furthermore, assume that E is a totally real extension of a field K, and let $P \in X(K)$. Denote the involutions $\epsilon \in G(F/E)$ for which P extends to an ordering of $F(\epsilon)$ by $I_P(F/E)$. For $X \subseteq X(K)$, let $I_X(F/E) = \bigcup_{P \in X} I_P(F/E)$. If F is the algebraic closure of E, write $I_P(E)$ for $I_P(F/E)$, etc.

REMARK 1.8. (a)   If $E = K$, then $I_P(F/E)$ is a conjugacy class in $G(F/E)$. In the general case $I_P(F/E)$ is a conjugacy domain in $G(F/E)$; in fact,

$$I_P(F/E) = \bigcap_{\substack{Q \in X(E) \\ Q \supseteq P}} I_Q(F/E).$$

(b)   If M/N is a finitely generated extension of fields, then the restriction map of orderings $X(M) \to X(N)$ is closed and open [**ELW**, Theorem 4.1 and 4.bis]. In particular, let I be a set of involutions in $G(F/E)$, and assume that F/K is finitely generated. Then so is $F(\epsilon)/K$, for every involution $\epsilon \in G(F/E)$. Hence the set $\{P' \in X(K) \mid I_P(F/E) = I\}$ is closed and open in $X(K)$.

LEMMA 1.9.   Let $(K, P) \subseteq (K', P')$ be an extension of ordered fields. Let x be transcendental over K', and put $E = K(x)$ and $E' = K'(x)$. Furthermore, let F/E and F'/E' be Galois extensions with $F' = F \cdot E$. Assume that F, and hence also F', is not formally real. Then $I_P(F/E) = \mathrm{Con}_{G(F/E)} \mathrm{res}_F I_{P'}(F'/E')$.

PROOF.   We have $I_P(F/E) = \bigcap_{\substack{Q \in X(E) \\ Q \supseteq P}} I_Q(F/E)$. Also,

$$I_{P'}(F'/E') = \bigcap_{\substack{Q' \in X(E') \\ Q' \supseteq P'}} I_{Q'}(F'/E').$$

As E and K' are linearly disjoint over K, each extension Q of P to E extends to an ordering Q' of E' that extends P'. Thus it suffices to show

$$I_Q(F/E) = \mathrm{Con}_{G(F/E)} \mathrm{res}_F I_{Q'}(F'/E')$$

for each ordering Q of E and for each extension Q' of Q to E'.

Let $\epsilon' \in I_{Q'}(F'/E')$ and $\epsilon = \mathrm{res}_F \epsilon'$. There is an ordering R' of $F'(\epsilon')$ that extends Q'. Its restriction to $F(\epsilon)$ is an extension of Q, and hence $\epsilon \in I_Q(F/E)$. Since $I_Q(F/E)$ is a conjugacy class in $G(F/E)$, the assertion follows.   □

2. Moduli spaces for covers of the Riemann sphere

In this section we add remarks to the notation and results from [FV1] in the form to be used later. Let G be a finite group, and $r \geq 3$ an integer.

(2.1) Covers of the sphere. Let $\mathbb{P}^1 = \mathbb{C} \cup \{\infty\}$ denote the Riemann sphere. We consider covers $\chi \colon X \to \mathbb{P}^1$ of compact (connected) Riemann surfaces. Call two such covers $\chi \colon X \to \mathbb{P}^1$ and $\chi' \colon X' \to \mathbb{P}^1$ **equivalent** if there exists an isomorphism $\alpha \colon X \to X'$ with $\chi' \circ \alpha = \chi$. Let $\mathrm{Aut}(X/\mathbb{P}^1)$ denote the group of automorphisms α of X with $\chi \circ \alpha = \chi$. We say that χ is **Galois** if $\mathrm{Aut}(X/\mathbb{P}^1)$ is transitive on the fibers of χ. From now on χ will always denote a Galois cover. All but finitely many points of $\mathbb{P}^1$ have the same number of inverse images under χ. These finitely many exceptional points are called the **branch points** of χ.

(2.2) Punctured spheres. Let $a_1, \dots, a_r \in \mathbb{P}^1$ be the branch points of (the Galois cover) $\chi: X \to \mathbb{P}^1$, and set $\mathbf{a} = \{a_1, \dots, a_r\}$. Then χ restricts to an (unramified) topological covering of the punctured sphere $\mathbb{P}^1 \smallsetminus \mathbf{a}$. Choose a base point a_0 on this punctured sphere, and consider the (topological) fundamental group $\Gamma = \Pi_1(\mathbb{P}^1 \smallsetminus \mathbf{a}, a_0)$, based at a_0 (with the composition law: $\gamma_1 \gamma_2$ is the path γ_1 followed by γ_2).

Depending on the choice of a base point $p_0 \in \chi^{-1}(a_0)$, we get an epimorphism $\iota: \Gamma \to \operatorname{Aut}(X/\mathbb{P}^1)$ as follows. For each path γ representing an element $[\gamma]$ of Γ, let p_1 be the endpoint of the unique lift of γ to $X \smallsetminus \chi^{-1}(\mathbf{a})$ with initial point p_0. Then, ι sends $[\gamma]$ to the unique element α of $\operatorname{Aut}(X/\mathbb{P}^1)$ with $\alpha(p_0) = p_1$.

(2.3) Related equivalence classes of covers. Let $\mathcal{H}^{\mathrm{ab}} = \mathcal{H}_r(G)^{\mathrm{ab}}$ be the set of equivalence classes $[\chi]$ of all Galois covers $\chi: X \to \mathbb{P}^1$ with r branch points and with $\operatorname{Aut}(X/\mathbb{P}^1) \cong G$. Let $\mathcal{H}^{\mathrm{in}} = \mathcal{H}_r(G)^{\mathrm{in}}$ be the set of equivalence classes $[\chi, h]$ of pairs (χ, h) where $\chi: X \to \mathbb{P}^1$ is a Galois cover with r branch points, and $h: \operatorname{Aut}(X/\mathbb{P}^1) \to G$ is an isomorphism. Here (χ, h) and $(\chi': X' \to \mathbb{P}^1,\ h')$ are equivalent if there is an isomorphism $\delta: X \to X'$ with $\chi' \circ \delta = \chi$ and $h' \circ \delta_* = h$, where $\delta_*: \operatorname{Aut}(X/\mathbb{P}^1) \to \operatorname{Aut}(X'/\mathbb{P}^1)$ is the isomorphism $\alpha \mapsto \delta \circ \alpha \circ \delta^{-1}$. Let $\Lambda: \mathcal{H}^{\mathrm{in}} \to \mathcal{H}^{\mathrm{ab}}$ be the map sending $[\chi, h]$ to $[\chi]$.

(2.4) G-covers. Think of points of $\mathcal{H}^{\mathrm{in}}$ as equivalence classes $[\mathbf{a}, a_0, f]$ of triples $(\mathbf{a}, a_0, f)$. Here $\mathbf{a} = \{a_1, \dots, a_r\}$ is a set of r points of $\mathbb{P}^1$, and $a_0 \in \mathbb{P}^1 \smallsetminus \mathbf{a}$, and $f: \Gamma = \Pi_1(\mathbb{P}^1 \smallsetminus \mathbf{a}, a_0) \to G$ is an epimorphism that does not factor through the canonical map $\Gamma \to \Pi_1((\mathbb{P}^1 \smallsetminus \mathbf{a}) \cup \{a_i\}, a_0)$, for any i. (The latter condition means that the corresponding cover χ has **exactly** r branch points). Call two such triples $(\mathbf{a}, a_0, f)$ and $(\tilde{\mathbf{a}}, \tilde{a}_0, \tilde{f})$ equivalent if $\mathbf{a} = \tilde{\mathbf{a}}$ and there is a path ω from a_0 to $\tilde{a}_0$ in $\mathbb{P}^1 \smallsetminus \mathbf{a}$ such that $\tilde{f} \circ \omega^* = f$. Here $\omega^*: \Pi_1(\mathbb{P}^1 \smallsetminus \mathbf{a}, a_0) \to \Pi_1(\mathbb{P}^1 \smallsetminus \mathbf{a}, \tilde{a}_0)$ is the isomorphism $\gamma \mapsto \omega^{-1} \gamma \omega$.

(2.5) Covers versus cycle descriptions. Here is the correspondence between the above pairs and triples [**FV1**, §1.2]. Given $[\chi, h] \in \mathcal{H}^{\mathrm{in}}$, with $\chi: X \to \mathbb{P}^1$ as above, let $\mathbf{a}$ be the set of branch points of χ, and choose $a_0 \in \mathbb{P}^1 \smallsetminus \mathbf{a}$ and $p_0 \in \chi^{-1}(a_0)$. Set $\Gamma = \Pi_1(\mathbb{P}^1 \smallsetminus \mathbf{a}, a_0)$ as above, and define $f: \Gamma \to G$ as $f = h \circ \iota$, where $\iota: \Gamma \to \operatorname{Aut}(X/\mathbb{P}^1)$ is the map from (2.2). Recall that ι is canonical up to composition with inner automorphisms of $\operatorname{Aut}(X/\mathbb{P}^1)$. Thus h and f determine each other up to inner automorphisms of G. This is compatible with the equivalence of pairs (resp., triples).

(2.6) The topology on $\mathcal{H}^{\mathrm{in}}$. To specify a neighborhood $\mathcal{N}$ of the point $[\mathbf{a}, a_0, f]$ of $\mathcal{H}^{\mathrm{in}}$, where $\mathbf{a} = \{a_1, \dots, a_r\}$, choose pairwise disjoint open discs $D_1, \dots, D_r$ around $a_1, \dots, a_r$, with $a_0 \notin D_1 \cup \dots \cup D_r$. Then $\mathcal{N}$ consists of all points $[\tilde{\mathbf{a}}, a_0, \tilde{f}]$ such that $\tilde{\mathbf{a}}$ has exactly one point in each D_i, and $\tilde{f}$ is the composition of the canonical isomorphisms

$$\Pi_1(\mathbb{P}^1 \smallsetminus \tilde{\mathbf{a}}, a_0) \cong \Pi_1(\mathbb{P}^1 \smallsetminus (D_1 \cup \dots \cup D_r), a_0) \cong \Pi_1(\mathbb{P}^1 \smallsetminus \mathbf{a}, a_0)$$

with f. These $\mathcal{N}$ form a basis for the topology. They are connected. The sets $\Lambda(\mathcal{N})$ form a basis for a topology on $\mathcal{H}^{\mathrm{ab}}$, such that $\Lambda \colon \mathcal{H}^{\mathrm{in}} \to \mathcal{H}^{\mathrm{ab}}$ becomes an (unramified) covering.

(2.7) r-tuples of unordered branch points. Let $\mathcal{U}_r$ denote the space of all subsets of cardinality r of the Riemann sphere $\mathbb{P}^1$. It has a natural structure of algebraic variety defined over $\mathbb{Q}$ [**FV1**, §1.1]; it is isomorphic to $\mathbb{P}^r \setminus D$, where D, the discriminant locus, is a hypersurface in $\mathbb{P}^r$. In particular, $\mathcal{U}_r$ is an affine variety. Furthermore, if K is a subfield of $\mathbb{C}$ and $\mathbf{a} = \{a_1, \dots, a_r\} \in \mathcal{U}_r$ with $a_1, \dots, a_r \neq \infty$, then $\mathbf{a}$ is K-rational if and only if $\prod_{i=1}^r (X - a_i) \in K[X]$. As a complex manifold, the topology of $\mathcal{U}_r$ has a basis consisting of sets $\mathcal{D}$ of the following form: Given pairwise disjoint open discs $D_1, \dots, D_r$ on $\mathbb{P}^1$, let $\mathcal{D}$ be the set of all $\mathbf{a} \in \mathcal{U}_r$ with $|\mathbf{a} \cap D_i| = 1$ for $i = 1, \dots, r$.

(2.8) Maps to $\mathcal{U}_r$. Define $\Psi \colon \mathcal{H}^{\mathrm{in}} \to \mathcal{U}_r$ and $\bar{\Psi} \colon \mathcal{H}^{\mathrm{ab}} \to \mathcal{U}_r$ by sending $[\chi, h]$ and $[\chi]$, respectively, to the set of branch points of χ. These maps are (unramified) coverings and $\bar{\Psi} \circ \Lambda = \Psi$. Through these coverings the spaces $\mathcal{H}^{\mathrm{ab}}$ and $\mathcal{H}^{\mathrm{in}}$ inherit a structure of complex manifold from $\mathcal{U}_r$.

(2.9) The algebraic structure on covers. Each cover $\chi \colon X \to \mathbb{P}^1$ as above is an algebraic morphism of algebraic varieties over $\mathbb{C}$, compatible with its analytic structure (Riemann's existence theorem). An automorphism β of $\mathbb{C}$ defines an automorphism β^* of $\mathbb{P}^1$ by $(x_0 : x_1) \mapsto (\beta^{-1}(x_0) : \beta^{-1}(x_1))$. Consider the cover $\beta(\chi) \colon \beta(X) \to \mathbb{P}^1$ obtained from $\chi \colon X \to \mathbb{P}^1$ through base change with β^*. Furthermore, for each $\alpha \in \mathrm{Aut}(X/\mathbb{P}^1)$ let $\beta_*(\alpha) = \beta(\alpha) \in \mathrm{Aut}(\beta(X)/\mathbb{P}^1)$ be the morphism obtained by the same base change.

(2.10) The algebraic structure on $\mathcal{H}^{\mathrm{in}}$. The spaces $\mathcal{H}^{\mathrm{ab}}$ and $\mathcal{H}^{\mathrm{in}}$ have a unique structure of (the set of complex points of) a (reducible) algebraic variety defined over $\mathbb{Q}$ [**FV1**, Theorem 1]. This variety structure is compatible with the analytic structure of $\mathcal{H}^{\mathrm{ab}}$ and $\mathcal{H}^{\mathrm{in}}$, and it makes the maps Ψ, $\bar{\Psi}$ and Λ into algebraic morphisms defined over $\mathbb{Q}$. Also, each automorphism β of $\mathbb{C}$—in its natural action $(x_1, \dots, x_n) \mapsto (\beta(x_1), \dots, \beta(x_n))$ on the complex points of (the affine pieces of) a variety defined over $\mathbb{Q}$—sends the point $[\chi] \in \mathcal{H}^{\mathrm{ab}}$ to $[\beta(\chi)]$ and the point $[\chi, h] \in \mathcal{H}^{\mathrm{in}}$ to $[\beta(\chi), h \circ \beta_*^{-1}]$.

(2.11) Complex conjugation acting on $\mathcal{H}^{\mathrm{in}}$. We can describe the action of complex conjugation c on the triples of (2.4) that compose $\mathcal{H}^{\mathrm{in}}$. Namely, c naturally acts on paths in $\mathbb{P}^1$. Thus, it induces a map $\Pi_1(\mathbb{P}^1 \setminus \mathbf{a}, a_0) \to \Pi_1(\mathbb{P}^1 \setminus c(\mathbf{a}), c(a_0))$. Denote this map by $\gamma \mapsto c\gamma$.

LEMMA. *If $\mathbf{p} = [\mathbf{a}, a_0, f] \in \mathcal{H}^{\mathrm{in}}$, then $c(\mathbf{p}) = [c(\mathbf{a}), c(a_0), cf]$. Here $(cf)(c\gamma) = f(\gamma)$ for each $\gamma \in \Pi_1(\mathbb{P}^1 \setminus \mathbf{a}, a_0)$.*

PROOF. Write $\mathbf{p}$ as $\mathbf{p} = [\chi, h]$. Then $c(\mathbf{p}) = [c(\chi), h \circ c_*^{-1}]$. It remains to show that this point is represented by the triple $(c(\mathbf{a}), c(a_0), cf)$. This is a straightforward consequence of the definitions (cf. [**DF1**, Lemma 2.1]).   □

(2.12) r-tuples of all conjugacy classes of G. Let $\mathbf{b} = \{b_1, \dots, b_r\} \in \mathcal{U}_r$ such that $0 \notin \mathbf{b}$. We can choose generators $\gamma_1, \dots, \gamma_r$ for the fundamental group $\Pi_1(\mathbb{P}^1 \smallsetminus \mathbf{b}, 0)$ so that $\gamma_1 \cdots \gamma_r = 1$ is the only relation among them. Indeed, assume that $b_j = \zeta^j$, for $j = 1, \dots, r$, where $\zeta = e^{\frac{2\pi i}{r}}$. Otherwise apply a homeomorphism $\mathbb{P}^1 \to \mathbb{P}^1$ that maps 0 onto itself and b_j onto ζ_j. Let $\tilde{\gamma}_j$ be a path starting at 0, going up on a straight line to a neighborhood of b_j, traversing a small disk around b_j in the counterclockwise direction, and following the same straight line back to 0. Then $\tilde{\gamma}_1, \dots, \tilde{\gamma}_r$ do not intersect except at 0. Let γ_j be the homotopy class of $\tilde{\gamma}_j$. Then $\gamma_1, \dots, \gamma_r$ generate $\Pi_1(\mathbb{P}^1 \smallsetminus \mathbf{b}, 0)$ and $\gamma_1 \cdots \gamma_r = 1$.

Represent a point $\mathbf{p} \in \Psi^{-1}(\mathbf{b})$ by a triple $(\mathbf{b}, 0, f)$. The r-tuple $(\sigma_1, \dots, \sigma_r) = (f(\gamma_1), \dots, f(\gamma_r))$ determines the epimorphism $f \colon \Pi_1(\mathbb{P}^1 \smallsetminus \mathbf{b}, 0) \to G$. It has the following properties: $\sigma_1 \cdots \sigma_r = 1$, $\sigma_1, \dots, \sigma_r$ generate G, and $\sigma_j \neq 1$ for all j [**FV1**, §1.3]. Let $\mathcal{E}_r$ denote the set of such r-tuples $(\sigma_1, \dots, \sigma_r)$. Clearly, each $(\sigma_1, \dots, \sigma_r) \in \mathcal{E}_r$ arises in the above way from some $\mathbf{p} \in \Psi^{-1}(\mathbf{b})$. Let $\mathcal{L}(G)$ be the collection of conjugacy classes $\neq \{1\}$ of G, and let $\mathcal{E}^{(r)}$ be all r-tuples $(\sigma_1, \dots, \sigma_r) \in \mathcal{E}_r$ where each $\mathcal{C} \in \mathcal{L}(G)$ is represented exactly $r/|\mathcal{L}(G)|$ times among $\sigma_1, \dots, \sigma_r$.

(2.13) When commutators generate the Schur multiplier of G. For the rest of section 2 assume that r is a multiple of $|\mathcal{L}(G)|$ and suitably large [**FV1**, Appendix], and the Schur multiplier of G is generated by commutators. We explain the latter condition. Let R be a group of maximal order with the property that R has a subgroup $M \le R' \cap Z(R)$ satisfying $R/M \cong G$. Then $M \cap \{g^{-1}h^{-1}gh \mid g, h \in R\}$ generates M, the Schur multiplier of G.

Fix $\mathbf{b} \in \mathcal{U}_r$ and $\gamma_1, \dots, \gamma_r$ as above. By [**FV1**, §1.3] there is a (unique) connected component $\mathcal{H}$ of $\mathcal{H}^{\mathrm{in}}$ containing $\{[\mathbf{b}, 0, f] \mid (f(\gamma_1), \dots, f(\gamma_r)) \in \mathcal{E}^{(r)}\}$. Let $\bar{\mathcal{H}} = \Lambda(\mathcal{H})$ be its image in $\mathcal{H}^{\mathrm{ab}}$. We call $\mathcal{H}$ and $\bar{\mathcal{H}}$ **Hurwitz spaces.** By [**FV1**, Thm. 1] they are absolutely irreducible algebraic varieties defined over $\mathbb{Q}$. Moreover, since $\Psi \colon \mathcal{H} \to \mathcal{U}_r$ and $\bar{\Psi} \colon \bar{\mathcal{H}} \to \mathcal{U}_r$ are finite normal covers of an affine variety, $\mathcal{H}$ and $\bar{\mathcal{H}}$ are affine [**H**, Exc. III.4.1].

(2.14) Automorphisms of $\mathcal{H}^{\mathrm{in}} \to \mathcal{H}^{\mathrm{ab}}$. For $A \in \mathrm{Aut}(G)$ (acting from the left on G), let $\delta_A \colon \mathcal{H} \to \mathcal{H}$ be the map sending the point $[\chi, h]$ to $[\chi, A \circ h]$. Then δ_A is an automorphism of the covering $\Lambda \colon \mathcal{H} \to \bar{\mathcal{H}}$. It depends only on the class of A modulo $\mathrm{Inn}(G)$. In fact, Λ is a Galois covering, and the map $A \mapsto \delta_A$ induces an isomorphism $\delta \colon \mathrm{Out}(G) = \mathrm{Aut}(G)/\mathrm{Inn}(G) \to \mathrm{Aut}(\mathcal{H}/\bar{\mathcal{H}})$ [**FV1**, §6.1]. Furthermore, δ_A is a morphism defined over $\mathbb{Q}$ [**FV1**, §6.2]. In the description of $\mathcal{H}^{\mathrm{in}}$ in (2.4), δ_A sends the point $[\mathbf{a}, a_0, f]$ to $[\mathbf{a}, a_0, A \circ f]$. As $\Lambda \colon \mathcal{H}^{\mathrm{in}} \to \mathcal{H}^{\mathrm{ab}}$ is an unramified covering (2.8), δ_A has no fixed points.

For the rest of this section assume that G has trivial center. Accordingly, identify G with the subgroup $\mathrm{Inn}(G)$ of $\mathrm{Aut}(G)$ (acting from the left on G). Let $\mathbf{p} \in \mathcal{H}$ and let $K \subseteq L$ be subfields of $\mathbb{C}$ such that $\Lambda(\mathbf{p}) \in \bar{\mathcal{H}}(K)$ and $L = K(\mathbf{p})$.

(2.15) Fields of definition of covers. Write $\mathbf{p}$ as $\mathbf{p} = [\chi, h]$. Then, the cover $\chi \colon X \to \mathbb{P}^1$ can be defined over L (in a unique way) such that all automorphisms of χ are defined over L [**FV1**, Cor. 1]. Thus, there is a unique cover $\chi_L \colon X_L \to \mathbb{P}^1_L$ such that base change with the embedding $L \to \mathbb{C}$ gives χ from χ_L and the automorphisms of χ from the automorphisms of χ_L.

(2.16) Fields of definition of automorphisms. We recall some facts from [**FV1**, §6.3]. The function field $F = L(X_L)$ is regular over L, and the extension $F/L(x)$ induced by χ is Galois. Here, x is the identity function on $\mathbb{P}^1$. The group $G(F/L(x))$ (acting from the left on F) is canonically isomorphic to $\mathrm{Aut}(X/\mathbb{P}^1)$, via the map that sends $\alpha \in \mathrm{Aut}(X/\mathbb{P}^1)$ to the element $g \mapsto g \circ \alpha^{-1}$ of $G(F/L(x))$. Let $h_0 \colon G(F/L(x)) \to G$ be the composition of this isomorphism with $h \colon \mathrm{Aut}(X/\mathbb{P}^1) \to G$.

(2.17) Indentification of automorphisms of G. Furthermore, L/K and $F/K(x)$ are Galois extensions, and the centralizer of $G(F/L(x))$ in $G(F/K(x))$ is trivial. This implies h_0 extends to a unique embedding $h_1 \colon G(F/K(x)) \to \mathrm{Aut}(G)$. [**FV1**, Proposition 3] says: $H := h_1(G(F/K(x)))$ equals

$$\{A \in \mathrm{Aut}(G) \mid \delta_A(\mathbf{p}) \text{ is conjugate to } \mathbf{p} \text{ under } G(L/K)\}.$$

(2.18) Action by autmorphisms of $\mathbb{C}$. Let β be an automorphism of $\mathbb{C}$, and let K and K' be two subfields of $\mathbb{C}$ such that $\beta(K) \subseteq K'$. Put $\mathbf{p}' = \beta(\mathbf{p})$ and $L' = K'(\mathbf{p}')$. Then $\beta(L) \subseteq L'$, and $\Lambda(\mathbf{p}') \in \bar{\mathcal{H}}(K')$. Let $F'/L'(x)$ be the Galois extension associated to K' and the point $\mathbf{p}'$ of $\mathcal{H}$, and let $h'_1 \colon G(F'/K'(x)) \to \mathrm{Aut}(G)$ be the associated embedding. Then the following holds:
Let $\beta \colon L(x) \to L'(x)$ be the extension of β (fixing x). This map extends further to $\beta \colon F \to F'$ such that canonically

$$(1) \qquad\qquad F' \cong \beta(F) \otimes_{\beta(L)} L' \cong F \otimes_L L'.$$

Consider restriction $\beta^* \colon G(F'/K'(x)) \to G(F/K(x))$: $\sigma \in G(F'/K'(x))$ goes to $\beta^{-1}\sigma|_{\beta(F)}\beta$. It is injective and it gives an isomorphism $G(F'/L'(x)) \to G(F/L(x))$. Further, it makes the following diagram commutative:

$$(2)$$

$$\begin{array}{ccc}
G(F'/K'(x)) & \xrightarrow{\;\;\beta^*\;\;} & G(F/K(x)) \\
 & \underset{h'_1}{\searrow} \quad \underset{h_1}{\swarrow} & \\
 & \mathrm{Aut}(G) &
\end{array}$$

PROOF (2) COMMUTES. We have $\mathbf{p}' = [\beta(\chi), h \circ \beta_*^{-1}]$ by (2.10). The natural action of $\beta \in \mathrm{Aut}(\mathbb{C})$ on functions defined over L extends β to a map from $F = L(X)$ to $F' = L(\beta(X))$. Then (1) follows from the fact that F is regular over L, and $[F' \colon L'(x)] = [F \colon L(x)] \; (= \deg(\chi))$. The proof of (2) is straightforward from the definitions. $\quad \square$

(2.19) Conclusion from (2.18). In (2.18) and Lemma 1.9 we have

$$h'_1(I_{P'}(F'/E')) \subseteq h_1(I_P(F/E)) \quad \text{and} \quad \mathrm{Con}_H \, h'_1(I_{P'}(F'/E')) = h_1(I_P(F/E)),$$

where H is the image of h_1 in $\mathrm{Aut}(G)$. If the 'restriction' map $\beta^* \colon G(F'/K'(x)) \to G(F/K(x))$ is an isomorphism, then $h'_1(I_{P'}(F'/E')) = h_1(I_P(F/E))$. Indeed, without loss of generality assume that the map $\beta\colon K \to K'$ is an inclusion of fields. Hence $\beta^* = \mathrm{res}_F$. In the commutative diagram (2) we may replace $\mathrm{Aut}(G)$ by H, so that h_1 is an isomorphism. The assertions follow from the commutativity of that diagram.

3. Group-theoretic lemmas

For a group G and $r \geq 0$ put $\dot{G} = G \smallsetminus \{1\}$. Let $\mathcal{L}(G)$ be the nontrivial conjugacy classes of G, and put $l = |\mathcal{L}(G)|$. For an r-tuple $\boldsymbol{\sigma} = (\sigma_1, \ldots, \sigma_r) \in \dot{G}^r$ and $\mathcal{C} \in \mathcal{L}(G)$, let $n_{\mathcal{C}}(\boldsymbol{\sigma})$ be the number of indices, $1 \leq i \leq r$, with $\sigma_i \in \mathcal{C}$. Then $\sum_{\mathcal{C} \in \mathcal{L}(G)} n_{\mathcal{C}}(\boldsymbol{\sigma}) = r$.

LEMMA 3.1. *Let G be a finite group. Every sufficiently large multiple r of $4l$ satisfies the following. Let $\epsilon \in \mathrm{Aut}(G)$ be of order 2 and let I be involutions in $(G \rtimes \langle \epsilon \rangle) \smallsetminus G$ with $\epsilon \in I$. Let $e = 8 \cdot |G|!$ if $|I| \geq 2$, and $e = 0$ if $|I| = 1$. Put $m = r - e$. Then there are sequences $\boldsymbol{\sigma} \in \dot{G}^e$, $\boldsymbol{\tau} \in \dot{G}^m$ with these properties:*

- (a1) $\sigma_i^\epsilon = \sigma_1 \cdots \sigma_{i-1} \sigma_i^{-1} \sigma_{i-1}^{-1} \cdots \sigma_1^{-1}$, *for each* $1 \leq i \leq e$;
- (a2) $\tau_j^\epsilon = \tau_{m+1-j}^{-1}$ *for each* $1 \leq j \leq m$;
- (b) $I = \{\epsilon, \epsilon\sigma_1, \epsilon\sigma_1\sigma_2, \ldots, \epsilon\sigma_1\sigma_2 \cdots \sigma_e\}$;
- (c) $(\boldsymbol{\sigma}, \boldsymbol{\tau}) \in \mathcal{E}^{(r)}(G)$; *or* $\langle \sigma_1, \ldots, \sigma_e, \tau_1, \ldots, \tau_m \rangle = G$, $\sigma_1 \cdots \sigma_e \tau_1 \cdots \tau_m = 1$, *and* $n_{\mathcal{C}}(\boldsymbol{\sigma}, \boldsymbol{\tau}) = r/l$ *for each* $\mathcal{C} \in \mathcal{L}(G)$.

PROOF. We may consider $\epsilon \in \mathrm{Aut}(G)$, $|\epsilon| = 2$, and one set I of involutions.

A. Separation of $\boldsymbol{\sigma}$ from $\boldsymbol{\tau}$. Define an equivalence relation on $\mathcal{L}(G)$: the class $[\mathcal{C}]$ is $\{\mathcal{C}, \mathcal{C}^{-1}, \mathcal{C}^\epsilon, \mathcal{C}^{-\epsilon}\}$. Part B constructs $\boldsymbol{\sigma} \in \dot{G}^e$ with (a1), (b),

- (c1) $\sigma_1 \cdots \sigma_e = 1$; and
- (d1) for each $[\mathcal{C}]$ there is $\mu_{[\mathcal{C}]} \geq 0$ such that $n_{\mathcal{C}}(\boldsymbol{\sigma}) = 4\mu_{[\mathcal{C}]}$.

Observe that $e = \sum_{\mathcal{C} \in \mathcal{L}(G)} n_{\mathcal{C}}(\boldsymbol{\sigma}) = 4 \sum_{\mathcal{C} \in \mathcal{L}(G)} \mu_{[\mathcal{C}]}$. Also, for each $[\mathcal{C}]$ let $\nu_{[\mathcal{C}]}$ be a positive integer and $m = 4 \sum_{\mathcal{C} \in \mathcal{L}(G)} \nu_{[\mathcal{C}]}$. Part C shows there is $\boldsymbol{\tau} \in \dot{G}^m$ satisfying (a2),

- (c2) $\langle \tau_1, \ldots, \tau_m \rangle = G$ and $\tau_1 \cdots \tau_m = 1$, and
- (d2) $n_{\mathcal{C}}(\boldsymbol{\tau}) = 4\nu_{[\mathcal{C}]}$, for each $\mathcal{C}$.

Let $n = \frac{r}{4l}$, and assume $n > \mu_{[\mathcal{C}]}$ for each $[\mathcal{C}]$. In the last step with $\nu_{[\mathcal{C}]} = n - \mu_{[\mathcal{C}]}$,

$$e + m = 4 \sum_{\mathcal{C} \in \mathcal{L}(G)} \mu_{[\mathcal{C}]} + 4 \sum_{\mathcal{C} \in \mathcal{L}(G)} \nu_{[\mathcal{C}]} = 4 \sum_{\mathcal{C} \in \mathcal{L}(G)} n = 4ln = r.$$

Substituting (c1), (c2), (d1), and (d2) in the expressions for (c) shows (c) holds. In fact, $n_{\mathcal{C}}(\boldsymbol{\sigma}, \boldsymbol{\tau}) = 4n = \frac{r}{l}$ for each $\mathcal{C} \in \mathcal{L}(G)$.

B. Construction of $\boldsymbol{\sigma}$. If $I = \{\epsilon\}$, let $e = 0$ and $\boldsymbol{\sigma} = ()$. Otherwise put $\boldsymbol{\sigma} = (\epsilon_1 \epsilon_2, \epsilon_2 \epsilon_3, \ldots, \epsilon_e \epsilon_1)$, where $\epsilon_1 = \epsilon, \epsilon_2, \ldots, \epsilon_e \in I$, not necessarily distinct, but $\epsilon_1 \neq \epsilon_2 \neq \cdots \neq \epsilon_e \neq \epsilon_1$. Then $\boldsymbol{\sigma}$ satisfies (a1) and (c1). Furthermore, if $I = \{\epsilon_1, \ldots, \epsilon_e\}$, then $\boldsymbol{\sigma}$ also satisfies (b). To construct such $\epsilon_1, \ldots, \epsilon_e$, let $n' = \frac{e}{2|I \smallsetminus \{\epsilon\}|}$. Note: n' is an integer divisible by 4, because $|I \smallsetminus \{\epsilon\}| < |G|$. Let

$\epsilon_i = \epsilon$ for odd i, and choose $\epsilon_2, \epsilon_4, \ldots, \epsilon_e$ so that each element of $I \smallsetminus \{\epsilon\}$ occurs in this sequence exactly n' times.

Let $g \in G$, and let $n_g(\boldsymbol{\sigma})$ be the number of indices $1 \le i \le e$ for which $\sigma_i = g$. From the above, $4 | n_g(\boldsymbol{\sigma})$. Moreover, $\sigma_{2i-1}^{\epsilon} = \epsilon(\epsilon\epsilon_{2i})\epsilon = \sigma_{2i-1}^{-1} = \sigma_{2i}$, for each $1 \le i \le e/2$. Hence, $n_g(\boldsymbol{\sigma}) = n_{g^{-1}}(\boldsymbol{\sigma}) = n_{g^\epsilon}(\boldsymbol{\sigma}) = n_{g^{-\epsilon}}(\boldsymbol{\sigma})$. This yields (d1).

C. Construction of τ. For each $[\mathcal{C}]$ let $\nu_{[\mathcal{C}]}$ be a positive integer. Choose k and $\mathbf{g} = (g_1, \ldots, g_k) \in \dot{G}^k$ such that $n_{\mathcal{C}}(\mathbf{g}) = \nu_{[\mathcal{C}]}$, for each $\mathcal{C}$. In particular, $\mathbf{g}$ contains an entry from each $\mathcal{C} \in \mathcal{L}(G)$. A proper subgroup of G misses some conjugacy classes of G [FJ, Lemma 12.4]. Therefore, $G = \langle g_1, \ldots, g_k \rangle$. Furthermore, $k = \sum_{\mathcal{C} \in \mathcal{L}(G)} \nu_{[\mathcal{C}]} = \frac{m}{4}$. Put

$$\boldsymbol{\tau} = (g_1, g_1^{-1}, \ldots, g_k, g_k^{-1}, g_k^{\epsilon}, g_k^{-\epsilon}, \ldots, g_1^{\epsilon}, g_1^{-\epsilon}).$$

This choice satisfies (a2) and (c2), and, for each $\mathcal{C}$,

$$n_{\mathcal{C}}(\boldsymbol{\tau}) = n_{\mathcal{C}}(\mathbf{g}) + n_{\mathcal{C}^{-1}}(\mathbf{g}) + n_{\mathcal{C}^{\epsilon}}(\mathbf{g}) + n_{\mathcal{C}^{-\epsilon}}(\mathbf{g}) = 4\nu_{[\mathcal{C}]}. \qquad \square$$

LEMMA 3.2. *Let $1 \to G \to H \xrightarrow{\pi} C \to 1$ be an exact sequence of finite groups, and let I be a set of involutions in $H \smallsetminus G$. There exists a commutative diagram*

(1)
$$
\begin{array}{ccccccccc}
1 & \longrightarrow & \tilde{G} & \longrightarrow & \tilde{H} & \xrightarrow{\ \tilde{\pi}\ } & C & \longrightarrow & 1 \\
& & \downarrow & & \downarrow & & \| & & \\
1 & \longrightarrow & G & \longrightarrow & H & \xrightarrow{\ \pi\ } & C & \longrightarrow & 1
\end{array}
$$

with exact rows and surjective vertical maps such that the Schur multiplier of $\tilde{G}$ is generated by commutators and $C_{\tilde{H}}(\tilde{G}) = 1$. Finally, every involution in I lifts to at least two involutions in $\tilde{H}$.

PROOF. Choose a presentation $1 \to \mathcal{R} \to \mathcal{F} \to H \to 1$, where $\mathcal{F}$ is the free product of a free group of finite rank with finitely many groups of order 2, say $\langle \delta_1 \rangle, \ldots, \langle \delta_e \rangle$, such that $\{\delta_1, \ldots, \delta_e\}$ maps onto I. The inverse image $\mathcal{F}_1$ of G in $\mathcal{F}$ contains no conjugates of $\delta_1, \ldots, \delta_e$. By the Kurosh Subgroup Theorem [M, Theorem VII.5.1 and Proposition VII.5.3] it is a free of finite rank. Let $\mathcal{N} = [\mathcal{F}_1, \mathcal{R}]$ be the group generated by commutators $[f, r]$ with $f \in \mathcal{F}_1, r \in \mathcal{R}$. Set $F = \mathcal{F}/\mathcal{N}$, $F_1 = \mathcal{F}_1/\mathcal{N}$, and $R = \mathcal{R}/\mathcal{N}$. Then $1 \to R \to F_1 \to G \to 1$ is a central extension.

Schur multiplier theory [Hu, Kap.5, §23] shows R is the direct product of the Schur multiplier $M(G) = R \cap (F_1)'$ and a free abelian group A. Let A_0 be the intersection of all the F-conjugates of A. Then $A_0 \lhd F$. Since $(R : A) = |M(G)| < \infty$, also $(F : A_0) < \infty$. Set $\tilde{H} = F/A_0$, $\tilde{G} = F_1/A_0$, and $S = R/A_0$, to get diagram (1). The image $\tilde{I}$ of $\{\delta_1, \ldots, \delta_e\}$ in $\tilde{H}$ maps onto I. Notice that S is the direct product of $S \cap (\tilde{G})' \cong M(G)$ and A/A_0. As in the proof of [FV3, Lemma 2] the Schur multiplier of $\tilde{G}$ is generated by commutators.

Replace G, H, and I by $\tilde{G}$, $\tilde{H}$, and $\tilde{I}$, to assume that the Schur multiplier of G is generated by commutators. Let T be a non-abelian finite simple group with trivial Schur multiplier. For example, take $T = \mathrm{SL}_2(8)$ [Hu, Satz 25.7].

Form the regular wreath product $\tilde{H}$ of H with T (e.g., [**Hu**, Def. 15.6]). Thus $\tilde{H} = T^j \rtimes H$, with $j = |H|$, and H acts on T^j by permuting the factors in its regular representation. Let $\tilde{G}$ be $T^j \rtimes G \leq \tilde{H}$. Clearly, $C_{\tilde{H}}(T^j) = 1$, and hence $C_{\tilde{H}}(\tilde{G}) = 1$. If $\epsilon \in I$ and $\tau \in T^j$, then $\tau^{-1}\epsilon\tau$ is an involution in $\tilde{H}$ that maps to ϵ. This proves the last assertion of the lemma.

Since $M(T) = 1$, every central extension of T splits. This implies that every representation group of $\tilde{G}$ has a normal subgroup isomorphic to T^j such that the quotient by this subgroup is a representation group of G. Therefore, $M(\tilde{G}) \cong M(G)$ is generated by commutators. □

LEMMA 3.3. *Let $\pi\colon H \to \overline{H}$ be an epimorphism of finite groups, and let $I_1, \ldots, I_m \subseteq H$ and $\overline{I}_1, \ldots, \overline{I}_m \subseteq \overline{H}$ be sets of involutions such that $\pi(I_j) = \overline{I}_j$. Then there exists a finite group $\tilde{H}$, a surjection $\rho\colon \tilde{H} \to H$, and sets of involutions $\tilde{I}_1, \ldots, \tilde{I}_m \subseteq \tilde{H}$ such that $\rho(\tilde{I}_j) = I_j$ for every j, and every automorphism $\bar{\alpha}$ of $\overline{H}$ that satisfies $\bar{\alpha}(\overline{I}_j) = \overline{I}_j$ for all j, lifts to an automorphism $\tilde{\alpha}$ of $\tilde{H}$ (that is, $(\pi \circ \rho) \circ \tilde{\alpha} = \bar{\alpha} \circ (\pi \circ \rho)$) that satisfies $\tilde{\alpha}(\tilde{I}_j) = \tilde{I}_j$ for all j. Moreover, if the I_j are conjugacy domains in H, then the $\tilde{I}_j$ can be taken conjugacy domains in $\tilde{H}$.*

PROOF. Let K be a set of cardinality $\mathrm{Ker}(\pi)$. Consider the free product

$$\tilde{H} = \Big(\underset{\substack{\bar{h}\in\overline{H}\\k\in K}}{\mbox{\Large$*$}} \langle x_{\bar{h},k}\rangle\Big) * \Big(\underset{\substack{\bar{\epsilon}\in\overline{I}\\k\in K}}{\mbox{\Large$*$}} \langle\tilde{\epsilon}_{\bar{\epsilon},k,1}\rangle\Big) * \Big(\underset{\substack{\bar{\epsilon}\in\overline{I}\\k\in K}}{\mbox{\Large$*$}} \langle\tilde{\epsilon}_{\bar{\epsilon},k,2}\rangle\Big) * \cdots * \Big(\underset{\substack{\bar{\epsilon}\in\overline{I}\\k\in K}}{\mbox{\Large$*$}} \langle\tilde{\epsilon}_{\bar{\epsilon},k,m}\rangle\Big),$$

of cyclic groups. Here $\langle x_{\bar{h},k}\rangle \cong \mathbb{Z}$ and $\langle\tilde{\epsilon}_{\bar{\epsilon},k,j}\rangle \cong \mathbb{Z}/2\mathbb{Z}$, and let $\tilde{I}_j = \{\tilde{\epsilon}_{\bar{\epsilon},k,j}|\ \bar{\epsilon} \in \overline{I}_j, k \in K\}$. (Of course, $\tilde{H}$ is not yet finite.) Define a surjection $\rho\colon \tilde{H} \to H$ by mapping $\{x_{\bar{h},k}|\ k \in K\}$ onto $\{h \in H|\ \pi(h) = \bar{h}\}$ and $\{\tilde{\epsilon}_{\bar{\epsilon},k,j}|\ \bar{\epsilon} \in \overline{I}_j\}$ onto $\{\epsilon \in I_j|\ \pi(\epsilon) = \bar{\epsilon}\}$. Then $\rho(\tilde{I}_j) = I_j$. Every automorphism $\bar{\alpha}$ of $\overline{H}$ that satisfies $\bar{\alpha}(\overline{I}_j) = \overline{I}_j$ for all j, lifts to an automorphism $\tilde{\alpha}$ of $\tilde{H}$ defined by $x_{\bar{h},k} \mapsto x_{\bar{\alpha}(\bar{h}),k}$ and $\tilde{\epsilon}_{\bar{\epsilon},k,j} \mapsto \tilde{\epsilon}_{\bar{\alpha}(\bar{\epsilon}),k,j}$. Clearly $\tilde{\alpha}(\tilde{I}_j) = \tilde{I}_j$. If I_j and $\overline{I}_j$ are conjugacy domains, we can replace $\tilde{I}_j$ by the conjugacy domain that it generates in $\tilde{H}$.

Thus $\tilde{H}$ satisfies the requirements of the lemma, except that it is not finite. To make $\tilde{H}$ finite, replace it by its quotient $\tilde{H}/N$, and ρ by the induced quotient map, where N is a characteristic subgroup of finite index in $\tilde{H}$, contained in $\mathrm{Ker}(\rho)$. For example, take N to be the intersection of all normal subgroups M of $\tilde{H}$ with $\tilde{H}/M \cong H$. □

4. Points over ordered fields

Let G be a finite group with a trivial center such that the Schur multiplier of G is generated by commutators. Identify G with the subgroup $\mathrm{Inn}(G)$ of $\mathrm{Aut}(G)$. Fix a sufficiently large integer r that satisfies (2.13) and the assertions of Lemma 3.1. Associate with G and r the moduli spaces $\mathcal{H}^{\mathrm{in}}$ and $\mathcal{H}^{\mathrm{ab}}$ from (2.3).

Our aim is to choose Hurwitz spaces $\mathcal{H}$ and $\bar{\mathcal{H}}$ and some points $\mathbf{q} = [\mathbf{b}, 0, f_0]$ on $\mathcal{H}$ as in (2.4). First, let $e = 8 \cdot |G|!$ and $m = \frac{r-e}{2}$, so $r = e + 2m$. Define the base point $\mathbf{b} = \{b_1, \ldots, b_r\}$ in $\mathcal{U}_r$ by

$$b_1 = 1, \ldots, b_e = e, \text{ and } b_{e+j} = -3 + (2m + 1 - 2j)\sqrt{-1}, \text{ for } j = e, \ldots, 2m.$$

Next, fix generators of $\Pi_1(\mathbb{P}^1 \smallsetminus \mathbf{b}, 0)$. For each $1 \le j \le r$ let D_j be the disc of diameter $\frac{1}{2}$ around b_j (so that $D_1, \ldots, D_r$ are disjoint). Define loops $\gamma_1, \ldots, \gamma_r$ in the complex plane with the initial and final point 0 in the following way:

(1) $\gamma_1 = \beta_1$, $\gamma_2 = \beta_1^{-1}\beta_2$, $\ldots$, $\gamma_e = \beta_{e-1}^{-1}\beta_e$, where β_j is the circle in the counterclockwise direction with diameter $[0, b_j + \frac{1}{2}]$ on the real axis;

(2) for $e < j \le r$ the path γ_j goes on a straight line from 0 towards b_j, then travels on a circle of diameter $\frac{1}{2} < \rho < 1$ in the counterclockwise direction around b_j, and returns on a straight line to 0.

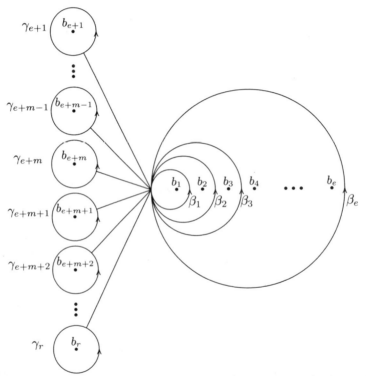

These loops are homotopic to the loops those from (2.12). Therefore they represent generators of the fundamental group $\Pi_1(\mathbb{P}^1 \smallsetminus \bigcap D_j, 0)$, subject only to the relation $\gamma_1 \cdots \gamma_r = 1$. If $\mathbf{a}$ is an r-tuple with $|\mathbf{a} \cap D_j| = 1$ for $j = 1, \ldots, r$, then $\gamma_1, \ldots, \gamma_r$ also represent generators of $\Gamma = \Pi_1(\mathbb{P}^1 \smallsetminus \mathbf{a}, 0)$. Indeed, $\Pi_1(\mathbb{P}^1 \smallsetminus \bigcap D_j, 0) \cong \Gamma$ via the inclusion $\mathbb{P}^1 \smallsetminus \bigcap D_j \to \mathbb{P}^1 \smallsetminus \mathbf{a}$. Furthermore, for such $\mathbf{a}$, we may use $\gamma_1, \ldots, \gamma_r$ also to represent free generators of $\tilde{\Gamma} = \Pi_1(\mathbb{P}^1 \smallsetminus (\mathbf{a} \cup \{\infty\}), 0)$. The canonical epimorphism $\lambda_*: \tilde{\Gamma} \to \Gamma$ induced by the inclusion $\lambda: \mathbb{P}^1 \smallsetminus (\mathbf{a} \cup \{\infty\}) \to \mathbb{P}^1 \smallsetminus \mathbf{a}$ maps the class of γ_j in $\tilde{\Gamma}$ onto the class of γ_j in Γ. Using $\mathbf{b}$ and $\gamma_1, \ldots, \gamma_r$, define the Hurwitz spaces $\mathcal{H}$ and $\tilde{\mathcal{H}}$ and the maps Λ, Ψ, and $\bar{\Psi}$ as in (2.13).

Finally, assume G has a non-inner automorphism ϵ of order 2. Let $G \rtimes \langle \epsilon \rangle$ be the subgroup of $\mathrm{Aut}(G)$ generated by G and ϵ. In particular, the centralizer of G in $G \rtimes \langle \epsilon \rangle$ is trivial. Let $I \subseteq G \rtimes \langle \epsilon \rangle \setminus G$ be a set of involutions, with $\epsilon \in I$ and $|I| \geq 2$. Lemma 3.1 (with m replaced by $2m$) produces an r-tuple $(\sigma_1, \ldots, \sigma_r) \in \mathcal{E}^{(r)}(G)$ (see (2.12)) with the following properties:

(3) $\sigma_j^\epsilon = \sigma_1 \cdots \sigma_{j-1} \sigma_j^{-1} \sigma_{j-1}^{-1} \cdots \sigma_1^{-1}$, for each $1 \leq j \leq e$;

(4) $\sigma_{e+j}^\epsilon = \sigma_{e+(2m+1-j)}^{-1}$ for $j = 1, \ldots, 2m$;

(5) $I = \{\epsilon, \epsilon\sigma_1, \epsilon\sigma_1\sigma_2, \ldots, \epsilon\sigma_1\sigma_2 \cdots \sigma_e\}$.

Fix (for each ϵ and each I) such an r-tuple $(\sigma_1, \ldots, \sigma_r)$. As $\sigma_1 \cdots \sigma_r = 1$, there is a unique epimorphism $f_0 \colon \Pi_1(\mathbb{P}^1 \setminus \mathbf{b}, 0) \to G$ with $f_0(\gamma_j) = \sigma_j$, for $j = 1, \ldots, r$.

DEFINITION 4.1. The point $\mathbf{q} = [\mathbf{b}, 0, f_0] \in \mathcal{H}$ is called **the basic point associated with** G, ϵ, and I. The neighborhood

$$\mathcal{N} = \{\mathbf{p} = [\mathbf{a}, 0, f] \in \mathcal{H}^{in} \mid |\mathbf{a} \cap D_j| = 1, f(\gamma_j) = f_0(\gamma_j) = \sigma_j, \text{ for } j = 1, \ldots, r\}$$

of $\mathbf{q}$ in $\mathcal{H}$ is called **the basic neighborhood of** $\mathbf{q}$.

REMARK 4.2. *Properties of a basic neighborhood.*

(a) A priori, $\mathcal{N}$ is a neighborhood of $\mathbf{q}$ in $\mathcal{H}^{in}$ (see (2.6)). Yet, $\mathcal{N}$ is connected. Hence, $\mathcal{N} \subseteq \mathcal{H}$.

(b) The point $\mathbf{b}$ is $\mathbb{Q}$-rational (2.7). Hence $\mathbf{q}$ is algebraic over $\mathbb{Q}$.

(c) Let $\mathbf{p} \in \mathcal{N}$, and let $\mathbf{a} = \{a_1, \ldots, a_r\} = \Psi(\mathbf{p})$. Then without loss of generality $a_j \in D_j$, for $j = 1, \ldots, r$. If $\mathbf{a}$ is $\mathbb{R}$-rational (i.e., $(X - a_1) \cdots (X - a_r) \in \mathbb{R}[X]$), then $a_1 < \cdots < a_e$ are real, and $a_{e+(2m+1-j)}$ is the complex conjugate of a_{e+j}, for $j = 1, \ldots, m$.

(d) Let c be the complex conjugation. As $\mathcal{H}$ is an affine variety, we may embed it in a fixed affine space $\mathbb{A}^n$. Then the complex topology on it is given by the norm $|| - ||_c$ defined in Definition 1.4. There are only finitely many choices of ϵ and I. Hence there are only finitely many basic points associated with G. Thus there is a positive rational number ν (that depends only on G) such that if $\mathbf{q}$ is a basic point, $\mathbf{p} \in \mathcal{H}$, and $||\mathbf{p} - \mathbf{q}||_c^2 < \nu^2$, then $\mathbf{p}$ is in the basic neighborhood $\mathcal{N}$ of $\mathbf{q}$.

LEMMA 4.3. *Let $\mathbf{p} \in \mathcal{N}$ such that $\Psi(\mathbf{p})$ is $\mathbb{R}$-rational. Then $\delta_\epsilon(\mathbf{p}) = c(\mathbf{p})$, where c is complex conjugation.*

PROOF. Write $\mathbf{p}$ as $[\mathbf{a}, 0, f]$. Then $\mathbf{a} = \Psi(\mathbf{p})$. We have $\delta_\epsilon(\mathbf{p}) = [\mathbf{a}, 0, \epsilon \circ f]$ by (2.14) and $c(\mathbf{p}) = [c(\mathbf{a}), 0, cf] = [\mathbf{a}, 0, cf]$ by (2.11). It remains to show that $cf = \epsilon \circ f$.

Observe that $c\beta_j = \beta_j^{-1}$, for $j = 1, \ldots, e$. Recursively:

(6) $\quad c\gamma_1 = \gamma_1^{-1}, \ c\gamma_2 = \gamma_1 \gamma_2^{-1} \gamma_1^{-1}, \ldots, c\gamma_e = \gamma_1 \cdots \gamma_{e-1} \gamma_e^{-1} \gamma_{e-1}^{-1} \cdots \gamma_1^{-1}.$

Furthermore,

(7) $$c\gamma_{e+j} = \gamma_{e+(2m+1-j)}^{-1}, \text{ for } j = 1, \ldots, 2m.$$

Recall that $(cf)(c\gamma) = f(\gamma)$. Combine (6) and (7) with (3) and (4) to get $cf(\gamma_j) = f(c\gamma_j) = \sigma_j^\epsilon = (\epsilon \circ f)(\gamma_j)$, for each $1 \leq j \leq r$.   □

PROPOSITION 4.4. *Let (K, P) be an ordered field and ι an involution in $G(K)$ inducing P on K. Assume $\widetilde{K} \subseteq \mathbb{C}$. Let $\lambda \in G(\mathbb{Q})$ and $\mathbf{p} \in \mathcal{H}(\widetilde{K})$ with*

$$\|\mathbf{p} - \lambda(\mathbf{q})\|_\iota^2 < \nu^2 \text{ in } \widetilde{K}(\iota), \tag{8}$$

and $\bar{\mathbf{p}} = \Lambda(\mathbf{p})$ is K-rational. Let $h_1: G(F/K(x)) \to \mathrm{Aut}(G)$ be the embedding corresponding to $\mathbf{p}$ over K with $F/K(x)$ Galois as given in (2.17). Put $L = K(\mathbf{p})$ and let H be the image of h_1. The following hold:

(a) $\delta_\epsilon(\mathbf{p}) = \iota(\mathbf{p})$;
(b) *P does not extend to L; in particular, P does not extend to F;*
(c) *$G \rtimes \langle \epsilon \rangle \leq H$, and therefore $I \subseteq H$;*
(d) *$h_1(I_P(F/K(x))) = \mathrm{Con}_H(I)$.*

PROOF. By (2.14), $\delta_\epsilon(\mathbf{p}) \neq \mathbf{p}$. Therefore (a) implies $\iota(\mathbf{p}) \neq \mathbf{p}$. Hence $L \not\subseteq \widetilde{K}(\iota)$, and this implies (b).. Furthermore, the criterion of (2.17) implies that $\epsilon \in H$. Since $G \leq H$, $G \rtimes \langle \epsilon \rangle \leq H$. So it suffices to prove (a) and (d).

Part I. *Reduction to K with archimedian orderings dense in $X(K)$.* Let K_0 be a finitely generated subfield of K, containing the finitely generated subfield $\mathbb{Q}(\bar{\mathbf{p}})$ of K. Let P_0 be the restriction of P to K_0. This ordering is induced from the restriction $\iota_0 \in G(K_0)$ of ι. Let $F_0/K_0(x)$ be the Galois extension and $(h_1)_0: G(F_0/K_0(x)) \to \mathrm{Aut}(G)$ the embedding corresponding to $\mathbf{p}$ over K_0. We may assume that $F = F_0 \cdot K$ from (2.18). If K_0 is sufficiently large, then the restriction map $\mathrm{res}_{F_0}: G(F/K(x)) \to G(F_0/K_0(x))$ is an isomorphism. If we can show that the assertions hold for K_0, P_0, ι_0, $(h_1)_0$, then, by (2.19) and since $\iota_0(\mathbf{p}) = \iota(\mathbf{p})$, they also hold for K, P, ι, h_1. Lemma 1.6(a) shows the set of archimedian orderings on K_0 is dense in $X(K_0)$. So we may assume that K enjoys this property.

Part II. *Reduction to P archimedian.* By Remark 1.8, if P' is an (archimedian) ordering of K sufficiently near to P, then $I_P(F/K(x)) = I_{P'}(F/K(x))$. We may assume an involution $\iota' \in G(K)$, so near to ι, induces P' that

$$\|\mathbf{p} - \lambda(\mathbf{q})\|_\iota^2 = \|\mathbf{p} - \lambda(\mathbf{q})\|_{\iota'}^2 \quad \text{and} \quad \iota(\mathbf{p}) = \iota'(\mathbf{p}).$$

Thus we may replace P by P' and ι by ι'.

Part III. *Reduction to $K = \mathbb{R}$ and $\lambda = 1$.* Assume that P is archimedian. Extend λ^{-1} to an automorphism β of $\mathbb{C}$, and let $\iota' = \beta\iota\beta^{-1}$. Then $\beta(\widetilde{K}(\iota)) = (\beta(\widetilde{K}))(\iota')$ is a real closure of $(\beta(K), \beta(P))$. Hence it is also archimedian. Thus we may assume that $(\beta(\widetilde{K}))(\iota') \subseteq \mathbb{R}$. Hence $\beta\iota\beta^{-1} = \iota' = \mathrm{res}_{\beta(\widetilde{K})}c$, where c is complex conjugation on $\mathbb{C}$.

Since $\mathbf{q}$ is algebraic over $\mathbb{Q}$, $\beta\lambda(\mathbf{q}) = \mathbf{q}$. So, application of β to (8) yields

$$\|\beta(\mathbf{p}) - \mathbf{q}\|_c^2 < \nu^2 \text{ in } \mathbb{R}. \tag{8'}$$

As $\bar{\mathbf{p}}$ is K-rational, $\iota(\bar{\mathbf{p}}) = \bar{\mathbf{p}}$. Thus, $c(\beta(\bar{\mathbf{p}})) = \beta(\bar{\mathbf{p}})$ and $\beta(\bar{\mathbf{p}})$ is $\mathbb{R}$-rational. Also, as δ_ϵ is defined over $\mathbb{Q}$, it commutes with β. Therefore (a) is equivalent to

(a') $$\delta_\epsilon(\beta(\mathbf{p})) = c(\beta(\mathbf{p})).$$

Finally, let $F'/\mathbb{R}(x)$ be the Galois extension and $h'_1 : G(F'/\mathbb{R}(x)) \to \mathrm{Aut}(G)$ the embedding corresponding to $\beta(\mathbf{p})$ over $\mathbb{R}$, and let H' be the image of h'_1. Then by (2.19), condition (d) follows from

(d') $$h'_1(I_{P'}(F'/\mathbb{R}(x))) = \mathrm{Con}_{H'}(I),$$

where P' is the unique ordering of $\mathbb{R}$. Thus, replacing K by $\mathbb{R}$ and $\mathbf{p}$ by $\beta(\mathbf{p})$, we may assume that $K = \mathbb{R}$ and $\lambda = 1$.

Part IV. $K = \mathbb{R}$ *and* $\lambda = 1$. By Remark 4.2(d) we have $\mathbf{p} \in \mathcal{N}$. Write $\mathbf{p}$ as $[\mathbf{a}, 0, f]$. Then $\mathbf{a} = \Psi(\mathbf{p}) = \bar{\Psi}(\bar{\mathbf{p}})$ is $\mathbb{R}$-rational. Lemma 4.3 gives assertion (a).

Part V. *Proof of (d).* By (c)—follows from (a)—we have $G \rtimes \langle \epsilon \rangle \leq H$. Check:

$$|G \rtimes \langle \epsilon \rangle| = 2 \cdot |G| = 2 \cdot [F : \mathbb{C}(x)] = [F : \mathbb{R}(x)] = |H|.$$

So $H = G \rtimes \langle \epsilon \rangle$.

Write $\mathbf{p}$ in the form $\mathbf{p} = [\chi, h]$, with $\chi : X \to \mathbb{P}^1$ (2.1). Fix a point $y \in \chi^{-1}(0)$. Let $Y_0 = \mathbb{P}^1 \setminus (\mathbf{a} \cup \{\infty\})$, and let $\psi : \hat{Y}_0 \to Y_0$ be the universal unramified covering of Y_0. Fix a point $\hat{y} \in \psi^{-1}(0)$. Put $Y = \chi^{-1}(Y_0) \subseteq X$. As $\chi : Y \to Y_0$ is unramified, there exists a unique covering $\phi : \hat{Y}_0 \to Y$ such that $\chi \circ \phi = \psi$ and $\phi(\hat{y}) = y$. Let $\hat{F}$ be the field of algebraic meromorphic functions on $\hat{Y}_0$ (in the sense of [**KN**, p. 199]). Then the field extension $\hat{F}/\mathbb{C}(x)$ induced by ψ is the maximal extension of $\mathbb{C}(x)$ unramified in Y_0.

Let $F = \mathbb{C}(X) = \mathbb{C}(Y)$. We identify $G(F/\mathbb{C}(x))$ with G via h_0, and $G(F/\mathbb{R}(x))$ with H via h_1 (see (2.16) and (2.17)). Then, $h : \mathrm{Aut}(X/\mathbb{P}^1) \to G$ is the canonical isomorphism $\mathrm{Aut}(X/\mathbb{P}^1) \to G(F/\mathbb{C}(x))$ sending $\alpha \in \mathrm{Aut}(X/\mathbb{P}^1)$ to the element $f \mapsto f \circ \alpha^{-1}$ of $G(F/\mathbb{C}(x))$. Similarly, let $\hat{G} = G(\hat{F}/\mathbb{C}(x))$, and let $\hat{h} : \mathrm{Aut}(\hat{Y}_0/Y_0) \to \hat{G}$ be the canonical map sending $\hat{\alpha}$ to the element $\hat{f} \mapsto \hat{f} \circ \hat{\alpha}^{-1}$.

Let $\iota : \Pi_1(\mathbb{P}^1 \setminus \mathbf{a}, 0) \to \mathrm{Aut}(X/\mathbb{P}^1)$ be the epimorphism associated to the point $y \in \chi^{-1}(0)$ (see (2.2)). Similarly define $\hat{\iota} : \Pi_1(Y_0, 0) \to \mathrm{Aut}(\hat{Y}_0/Y_0)$, associated to the point $\hat{y} \in \psi^{-1}(0)$. Then there is a commutative diagram

$$
\begin{array}{ccccc}
\Pi_1(Y_0, 0) & \xrightarrow{\hat{\iota}} & \mathrm{Aut}(\hat{Y}_0/Y_0) & \xrightarrow{\hat{h}} & \hat{G} \\
\downarrow{\scriptstyle\lambda_*} & & \downarrow{\scriptstyle\phi_*} & & \downarrow{\scriptstyle\mathrm{res}_F} \\
\Pi_1(\mathbb{P}^1 \setminus \mathbf{a}, 0) & \xrightarrow{\iota} & \mathrm{Aut}(X/\mathbb{P}^1) & \xrightarrow{h} & G
\end{array}
$$

where λ_* is induced from the inclusion $\lambda : Y_0 \to \mathbb{P}^1 \setminus \mathbf{a}$.

Put $\hat{\sigma}_j = \hat{h} \circ \hat{\iota}(\gamma_j)$, for $j = 1, \dots, r$. Then

(9) $$\mathrm{res}_F \hat{\sigma}_j = h \circ \iota \circ \lambda_*(\gamma_j) = f(\gamma_j) = \sigma_j, \quad \text{for } j = 1, \dots, r.$$

Let $n = (r + e)/2 = e + m$. By Remark 4.2(c) we may assume $a_1 < \cdots < a_e$ are real, and $a_{e+(2m+1-j)}$ is the complex conjugate of a_{e+j}, for $j = 1, \dots, m$.

Observe that $\hat{F}$ is the maximal extension of $\mathbb{R}(x)$ unramified outside the primes of $\mathbb{R}(x)$ induced by $a_1, \ldots, a_r, \infty$. In this situation the proof of [**KN**, Satz 2] shows that there is $\hat{\epsilon} \in \hat{H} = G(\hat{F}/\mathbb{R}(x))$ such that $\hat{\epsilon}, \hat{\sigma}_1, \ldots, \hat{\sigma}_n$ form a system of generators for $\hat{H}$ with the defining relations

$$(10) \qquad \hat{\epsilon}^2 = 1 \quad \text{and} \quad \hat{\sigma}_j^{\hat{\epsilon}} = \hat{\sigma}_1^{-1} \cdots \hat{\sigma}_{j-1}^{-1} \hat{\sigma}_j^{-1} \hat{\sigma}_{j-1} \cdots \hat{\sigma}_1, \quad \text{for } 1 \leq j \leq e.$$

Further, $\hat{\sigma}_{e+j}^{\hat{\epsilon}} = \hat{\sigma}_{e+(2m+1-j)}^{-1}$ for $j = 1, \ldots, 2m$. By (3) and (4) this implies that $\mathrm{res}_F \hat{\epsilon}$ and ϵ act on G in the same way. Since H is a subgroup of $\mathrm{Aut}(G)$, this implies that $\mathrm{res}_F \hat{\epsilon} = \epsilon$.

Each involution of $\hat{H}$ is conjugate to one of $\hat{\epsilon}, \hat{\sigma}_1 \hat{\epsilon}, \hat{\sigma}_2 \hat{\sigma}_1 \hat{\epsilon}, \ldots, \hat{\sigma}_e \cdots \hat{\sigma}_2 \hat{\sigma}_1 \hat{\epsilon}$. Indeed, by [**HJ2**, Lemma 4.2 (Part E)], $\hat{H}$ is the free profinite product of the free profinite group $\langle \hat{\sigma}_{e+1}, \ldots, \hat{\sigma}_n \rangle$ of rank $n - e = m$ with $e + 1$ groups

$$\langle \hat{\epsilon} \rangle, \ \langle \hat{\sigma}_1 \hat{\epsilon} \rangle, \ \langle \hat{\sigma}_2 \hat{\sigma}_1 \hat{\epsilon} \rangle, \ \ldots, \langle \hat{\sigma}_e \cdots \hat{\sigma}_2 \hat{\sigma}_1 \hat{\epsilon} \rangle$$

that are of order two. Thus by [**HR**, Theorem A′] the elements of finite order in $\hat{H}$ are the conjugates of the elements of these $e + 1$ subgroups.

By Lemma 4.5 below, all involutions of $\hat{H}$ are real. Using (9) and (5) conclude

$$I(F/\mathbb{R}(x)) = \mathrm{res}_F I(\hat{F}/\mathbb{R}(x)) = \mathrm{res}_F \mathrm{Con}_{\hat{H}}(\{\hat{\epsilon}, \hat{\sigma}_1 \hat{\epsilon}, \ldots, \hat{\sigma}_e \cdots \hat{\sigma}_1 \hat{\epsilon}\}) =$$

$$= \mathrm{Con}_H(\{\epsilon, \sigma_1 \epsilon, \ldots, \sigma_e \cdots \sigma_1 \epsilon\}) = \mathrm{Con}_H(I). \qquad \square$$

LEMMA 4.5. *Let S be a finite set of finite prime divisors of the field $\mathbb{R}(x)$. Let $\mathbb{R}(x)^S$ be the maximal extension of $\mathbb{R}(x)$ unramified outside $S \cup \{\infty\}$, and set $G^S = G(\mathbb{R}(x)^S/\mathbb{R}(x))$. Then all involutions of G^S are real.*

PROOF. By [**KN**, Satz 3] the absolute Galois group $\mathcal{G}$ of $\mathbb{R}(x)$ has generators $\{\delta, \tau_p \mid p \text{ a finite prime of } \mathbb{R}(x)/\mathbb{R}\}$ with the defining profinite relations

$$\delta^2 = 1 \quad \text{and} \quad \tau_p^{\delta} = \Big(\prod_{p'<p} \tau_{p'}^{-1} \Big) \tau_p^{-1} \Big(\prod_{p'<p} \tau_{p'}^{-1} \Big)^{-1} \quad \text{for all real } p.$$

Here $\prod_{p'<p} \tau_{p'}^{-1}$ is the unique accumulation point of the products $\tau_{p_1}^{-1} \cdots \tau_{p_r}^{-1} \in \mathcal{G}$ for real primes $p_1, \ldots, p_r$ with $p_1 < \cdots < p_r < p$. Furthermore, [**KN**] constructs this system of generators in such a way that for every finite set S of finite primes and every finite prime $p \notin S$ the natural restriction map $\mathcal{G} \to G^S$ maps τ_p onto 1 [**KN**, p. 207].

Let $p_1 < \cdots < p_e$ be the real, and $p_{e+1}, \ldots, p_n$ the complex primes of S. Let $\hat{\sigma}_j$ be the image of τ_{p_j}, for $j = 1, \ldots, n$, and let $\hat{\epsilon}$ be the image of δ in G^S. Then $\hat{\epsilon}, \hat{\sigma}_1, \ldots, \hat{\sigma}_n$ generate G^S and satisfy (10). These are in fact defining relations for G^S by [**KN**, Satz 2]. As in the last part of the proof of Proposition 4.4, each involution of G^S is conjugate to some $\hat{\sigma}_j \cdots \hat{\sigma}_1 \hat{\epsilon}$, where $0 \leq j \leq e$. Thus it suffices to show that each $\hat{\sigma}_j \cdots \hat{\sigma}_1 \hat{\epsilon}$ lifts to an involution of $\mathcal{G}$. To this end put

$$\delta_0 = \delta \quad \text{and} \quad \delta_j = \Big(\prod_{p'\leq p_j} \tau_{p'}^{-1} \Big)^{-1} \delta = \tau_{p_j} \Big(\prod_{p'<p_j} \tau_{p'}^{-1} \Big)^{-1} \delta$$

for $1 \leq j \leq e$. Then δ_j maps onto the involution $\hat{\sigma}_j \cdots \hat{\sigma}_1 \hat{\epsilon}$ in G^S. In fact, given another finite set S' of finite primes of $\mathbb{R}(x)$ that contains S, the same argument shows that δ_j restricts to an involution in $G^{S'}$. As $\mathcal{G} = \varprojlim_{S'} G^{S'}$, we get that δ_j is an involution in $\mathcal{G}$. $\square$

5. The regular real embedding problem over a PRC field

Let K be a PRC field of characteristic 0, and let $\nu > 0$ be a rational number. We proceed as with PAC fields [**FV2**, Section 1] with some extra care.

Let $X_1, \ldots, X_m$ be a partition of $X(K)$ into disjoint clopen subsets. Fix a closed system $\mathcal{X}$ of representatives of the conjugacy classes of involutions in $G(K)$; then $\{\widetilde{K}(\iota) | \iota \in \mathcal{X}\}$ is a closed subset of real closures of K, one for each ordering of K (see Section 1). Put $\mathcal{X}_j = \mathcal{X} \cap I_{X_j}(\widetilde{K}/K)$. Then $\mathcal{X}_1, \ldots, \mathcal{X}_m$ is a partition of $\mathcal{X}$ into disjoint clopen subsets.

LEMMA 5.1. *Let* $\Lambda : \mathcal{H} \to \bar{\mathcal{H}}$ *be an unramified Galois cover of absolutely irreducible, non-singular varieties defined over* K. *Assume that all the automorphisms of* $\mathcal{H}/\bar{\mathcal{H}}$ *are defined over* K. *Let* $\beta : G(K) \to \mathrm{Aut}(\mathcal{H}/\bar{\mathcal{H}})$ *be a homomorphism, and let* L *be the fixed field of* $\ker(\beta)$. *Assume that* L *is not formally real. Let* $\mathbf{q}_1, \ldots, \mathbf{q}_m \in \mathcal{H}(\widetilde{K})$ *satisfy the following.*

(1) $\iota \mathbf{q}_j = \beta(\iota)(\mathbf{q}_j)$, *for each* $\iota \in \mathcal{X}_j$, *for* $j = 1, \ldots, m$.

Then there exists $\mathbf{p} \in \mathcal{H}(\widetilde{K})$ *such that*

(2) $\sigma \mathbf{p} = \beta(\sigma)(\mathbf{p})$ *for each* $\sigma \in G(K)$;

(3) $\|\mathbf{p} - \mathbf{q}_j\|_\iota^2 \leq \nu^2$ *in* $\widetilde{K}(\iota)$, *for each* $\iota \in \mathcal{X}_j$, *for* $j = 1, \ldots, m$;

(4) *the point* $\Lambda(\mathbf{p})$ *of* $\bar{\mathcal{H}}$ *is* K-*rational and* $K(\mathbf{p}) = L$.

PROOF. First notice that (4) follows from (2). Indeed, an automorphism of the unramified cover $\mathcal{H} \to \bar{\mathcal{H}}$ has no fixed points. If (2) holds, then for each $\sigma \in G(K)$ we have $\sigma(\Lambda(\mathbf{p})) = \Lambda(\sigma(\mathbf{p})) = \Lambda(\beta(\sigma)(\mathbf{p})) = \Lambda(\mathbf{p})$ and

$$\sigma(\mathbf{p}) = \mathbf{p} \iff \beta(\sigma)(\mathbf{p}) = \mathbf{p} \iff \beta(\sigma) = 1 \iff \sigma \in G(L).$$

The rest is a straightforward modification of the proof of [**FV2**, Lemma 1]. Apply Weil's descent [**W**, Theorem 3] to the maps $f_{\tau,\rho} = \beta(\tau) \circ \beta(\rho)^{-1}$ to get a variety $\mathcal{H}'$ defined over K, and a linear isomorphism $f : \mathcal{H}' \to \mathcal{H}$ defined over L with these properties. The map $\Lambda \circ f : \mathcal{H}' \to \bar{\mathcal{H}}$ is defined over K and $\sigma f = \beta(\sigma) \circ f$, for each $\sigma \in G(K)$. In particular, suppose that $\mathbf{q}' \in \mathcal{H}'(\widetilde{K})$ and $\mathbf{q} = f(\mathbf{q}') \in \mathcal{H}(\widetilde{K})$. Then, for every $\sigma \in G(K)$

$$\beta(\sigma)(\mathbf{q}) = (\beta(\sigma) \circ f)(\mathbf{q}') = (\sigma f)(\mathbf{q}') = \sigma(f(\sigma^{-1}\mathbf{q}')).$$

Conclude that

(5) $\qquad \sigma \mathbf{q} = \beta(\sigma)(\mathbf{q}) \quad \iff \quad \mathbf{q}' = \sigma \mathbf{q}' \quad \iff \quad \mathbf{q}' \in \mathcal{H}'(\widetilde{K}(\sigma)).$

Use (5) and equation (3) of Section 1 to translate (1)–(3) via f from $\mathcal{H}$ to $\mathcal{H}'$. Let $\mathbf{q}'_j = f^{-1}(\mathbf{q}_j) \in \mathcal{H}'(\widetilde{K})$, for $j = 1, \dots, m$. Then,

(1′) $\mathbf{q}'_j \in \mathcal{H}'(\widetilde{K}(\iota))$, for each $\iota \in \mathcal{X}_j$, for $j = 1, \dots, m$.

We must find $\mathbf{p}' \in \mathcal{H}'(\widetilde{K})$ such that

(2′) $\mathbf{p}' \in \mathcal{H}'(\widetilde{K}(\sigma))$ for each $\sigma \in G(K)$, that is, $\mathbf{p}' \in \mathcal{H}'(K)$;

(3′) $\|\mathbf{p}' - \mathbf{q}'_j\|^2_\iota \leq \nu^2 / \|f\|^2_\iota$ in $\widetilde{K}(\iota)$, for each $\iota \in \mathcal{X}_j$, for $j = 1, \dots, m$.

Suppose $\iota \in \mathcal{X}_j$ for some $1 \leq j \leq m$. As $\|f\|^2_\iota$ is algebraic over K, there is $a_j \in K$ with $a_j^2 > \|f\|^2_\iota$ in $\widetilde{K}(\iota)$. Replace $\mathcal{X}_1, \dots, \mathcal{X}_m$ by a finer partition to assume this is true for each $\iota \in \mathcal{X}_j$. Thus (3′) follows from a stronger statement:

(3″) $\|\mathbf{p}' - \mathbf{q}'_j\|^2_\iota \leq \nu^2 / a_j^2$ in $\widetilde{K}(\iota)$, for each $\iota \in \mathcal{X}_j$, for $j = 1, \dots, m$.

By Proposition 1.2 there is $\mathbf{p}' \in \mathcal{H}'(K)$ such that (3″) holds. □

THEOREM 5.2. *Let L/K be a finite Galois extension with L not formally real and let $\pi: H \to G(L/K)$ be an epimorphism of finite groups. For each $1 \leq j \leq m$ let $I_j \subseteq H$ be a conjugacy domain of involutions such that $\pi(I_j) = I_{X_j}(L/K)$. Then there exists a regular extension F of L, Galois over $K(x)$, and an isomorphism $h_1: G(F/K(x)) \to H$ that maps $I_{X_j}(F/K(x))$ onto I_j. In addition, the following diagram commutes.*

$$
\begin{array}{ccc}
G(F/K(x)) & \xrightarrow{\;\;h_1\;\;} & H \\
& \searrow_{\mathrm{res}_L} \quad \swarrow_{\pi} & \\
& G(L/K) &
\end{array}
$$

In particular, h_1 maps $I(F/K(x))$ onto $\bigcup_j I_j$.

PROOF. By Skolem-Löwenheim Principle [**FJ**, Proposition 6.4] we may assume that $K \subseteq \mathbb{C}$. We divide the proof into five parts.

Part 1. *Weakening of commutativity.* Let $G = \mathrm{Ker}(\pi)$. Instead of commutativity of the diagram it suffices to show that h_1 maps $G(F/L(x))$ onto G. Indeed, apply Lemma 3.3. This gives an epimorphism of finite groups $\rho: \widetilde{H} \to H$ and conjugacy domains of involutions $\widetilde{I}_1, \dots, \widetilde{I}_m \subseteq \widetilde{H}$ with $\rho(\widetilde{I}_j) = I_j$. In addition, every automorphism of $G(L/K)$ that preserves the $I_{X_j}(L/K)$ lifts (under $\rho \circ \pi$) to an automorphism of $\widetilde{H}$ that preserves the $\widetilde{I}_j$. Let $\widehat{G} = \mathrm{Ker}(\rho \circ \pi)$.

Assume that we can find a regular extension $\widehat{F}$ of L, Galois over $K(x)$, and an isomorphism $\hat{h}_1: G(\widehat{F}/K(x)) \to \widehat{H}$ that maps $G(\widehat{F}/L(x))$ onto $\widehat{G}$ and the $I_{X_j}(\widehat{F}/K(x))$ onto the $\widehat{I}_j$. In particular, $\mathrm{Ker}(\pi \circ \rho \circ \hat{h}_1) = G(\widehat{F}/L(x)) = \mathrm{Ker}(\mathrm{res}_L)$. Hence there exists an automorphism α of $G(L/K)$ such that $\alpha \circ \pi \circ \rho \circ \hat{h}_1 = \mathrm{res}_L$ and α preserves the $I_{X_j}(L/K)$. We can lift α to an automorphism $\hat{\alpha}$ of $\widehat{H}$ that preserves the $\widehat{I}_j$. Thus, by composing h_1 with $\hat{\alpha}$ we may assume that $(\pi \circ \rho) \circ \hat{h}_1 = \mathrm{res}_L$.

Now let F be the fixed field of $\mathrm{Ker}(\rho)$ in $\widehat{F}$. Then h_1 induces an isomorphism $h_1: G(F/K(x)) \to H$ with the required properties.

Part 2. *Reduce to commutators generate $M(G)$, $H \subseteq \mathrm{Aut}(G)$, and $|I_j| \geq 2$.* As L is not formally real, $1 \notin I_{X_j}(L/K) = \pi(I_j)$, for each $1 \leq j \leq m$. Thus $I_j \subseteq H \smallsetminus G$. Let $\widetilde{H}$ and $\widetilde{G}$ be as in Lemma 3.2, and let $\widetilde{I}_j$ be the inverse image of I_j in the set of involutions of $\widetilde{H}$. Suppose there is $\hat{F}$ regular over L, with $\hat{F}/K(x)$ is Galois, and an isomorphism $\hat{h} : G(\hat{F}/K(x)) \to \widetilde{H}$ that maps $G(\hat{F}/L(x))$ onto $\widetilde{G}$ and $I_{X_j}(\hat{F}/K(x))$ onto $\widetilde{I}_j$. As in Part 1, the subfield of $\hat{F}$ corresponding to the kernel of the map $\widetilde{H} \to H$ (sending $\widetilde{G}$ to G) is the desired F.

Thus, assume that commutators generate $M(G)$, $C_H(G) = 1$, and $|I_j| \geq 2$ for each j. In particular, the conjugation action of H on G induces a monomorphism $H \to \mathrm{Aut}(G)$. Identify H with its image in $\mathrm{Aut}(G)$ (and G with $\mathrm{Inn}(G)$). Then $G(L/K)$ is a subgroup of $\mathrm{Out}(G) = \mathrm{Aut}(G)/\mathrm{Inn}(G)$, and $\pi : H \to G(L/K)$ is the restriction of the quotient map $\pi : \mathrm{Aut}(G) \to \mathrm{Out}(G)$ to H.

Part 3. *Construction.* Let $\Lambda : \mathcal{H} \to \bar{\mathcal{H}}$ be the cover of Hurwitz spaces, associated with G, defined in Section 4. Let $\beta : G(K) \to \mathrm{Aut}(\mathcal{H}/\bar{\mathcal{H}})$ be the composition of the restriction $G(K) \to G(L/K) \leq \mathrm{Out}(G)$ with the isomorphism $\delta : \mathrm{Out}(G) \to \mathrm{Aut}(\mathcal{H}/\bar{\mathcal{H}})$ (2.14). Furthermore, let ν be as in Remark 4.2(d).

Let M be the field generated over $\mathbb{Q}$ by $\sqrt{-1}$ and the conjugates of basic points associated with G, ϵ, and I as in Definition 4.1, for all possible ϵ and I. This is a finite extension of $\mathbb{Q}$ (Remark 4.2(b)). Refine the partition $X_1, \dots, X_m$ of $X(K)$, and hence also the corresponding partition $\mathcal{X}_1, \dots, \mathcal{X}_m$ of $\mathcal{X}$, so that for each $1 \leq j \leq m$ there are unique $\bar{\epsilon}_j \in G(L/K)$ and $\bar{\iota}_j \in G(M/\mathbb{Q})$ such that $\mathrm{res}_L \mathcal{X}_j = \{\bar{\epsilon}_j\}$ and $\mathrm{res}_M \mathcal{X}_j = \{\bar{\iota}_j\}$.

Fix $1 \leq j \leq m$. Put $I'_j = \{\epsilon \in I_j | \pi(\epsilon) = \bar{\epsilon}_j\}$ and choose $\epsilon_j \in I'_j$. Then $I'_j \subseteq G \rtimes \langle \epsilon_j \rangle$. Let $\mathbf{q}$ be the basic point associated with G, ϵ_j, and I'_j. Then $\mathbb{Q}(\mathbf{q}) \subseteq M$. As the real involutions in $G(M/\mathbb{Q})$ are conjugate, there is $\lambda_j \in G(M/\mathbb{Q})$ with $\lambda_j^{-1} \bar{\iota}_j \lambda_j = \mathrm{res}_M c$: c is complex conjugation. Set $\mathbf{q}_j = \lambda_j(\mathbf{q})$. By Lemma 4.3, $\delta_\epsilon(\mathbf{q}) = c(\mathbf{q})$. Therefore, $\delta_\epsilon(\mathbf{q}_j) = \delta_\epsilon \lambda_j(\mathbf{q}) \lambda_j \delta_\epsilon(\mathbf{q}) = (\lambda_j c \lambda_j^{-1})(\mathbf{q}_j) = \bar{\iota}_j(\mathbf{q}_j)$.

Let $\iota \in \mathcal{X}_j$. Then $\delta_{\bar{\epsilon}_j} = \delta(\pi(\epsilon_j)) = \delta(\bar{\epsilon}_j) = (\delta \circ \mathrm{res}_L)(\iota) = \beta(\iota)$. So $\iota(\mathbf{q}_j) = \bar{\iota}_j(\mathbf{q}_j) = \delta_{\epsilon_j}(\mathbf{q}_j) = \beta(\iota)(\mathbf{q}_j)$. Thus $\mathbf{q}_1, \dots, \mathbf{q}_m$ satisfy (1). Therefore there exists $\mathbf{p} \in \mathcal{H}(\widetilde{K})$ that satisfies (2)-(4). Let $F/K(x)$ be the Galois extension and $h_1 : G(F/K(x)) \to \mathrm{Aut}(G)$ the embedding associated with $\mathbf{p}$ over K (2.17).

Part 4. *The image of h_1.* Let $\tau \in H$. There is $\sigma \in G(K)$ such that $\mathrm{res}_L \sigma = \pi(\tau)$. By (2), $\sigma(\mathbf{p}) = \delta(\mathrm{res}_L \sigma)(\mathbf{p}) = \delta_\tau(\mathbf{p})$. Hence by the criterion of (2.17), τ is in the image of h_1. Thus $H \leq \mathrm{im}(h_1)$. But

$$|H| = |G| \cdot |G(L/K)| = [F : L(x)] \cdot [L : K] = [F : K(x)] = |\mathrm{im}(h_1)|,$$

and hence $H = \mathrm{im}(h_1)$.

Part 5. $h_1(I_{X_j}(F/K(x))) = I_j$. Let $1 \leq j \leq m$ and $P \in X_j$. By Prop. 4.4(d), $h_1(I_P(F/K(x))) = \mathrm{Con}_H(I'_j)$. As $\mathrm{Con}_{G(K)}(\mathcal{X}_j) = I_{X_j}(\widetilde{K}/K)$, $\mathrm{Con}_{G(L/K)}(\bar{\epsilon}_j) = I_{X_j}(L/K)$. Conclude: $\mathrm{Con}_H(I'_j) = I_j$. Thus $h_1(I_P(F/K(x))) = I_j$. $\square$

Theorem 5.2 tells about the structure of the absolute Galois group of $K(x)$. For instance, every finite group is realizable over $K(x)$. This isn't new [**DF2**, Theorem 5.7]. Still, the precise information about real involutions gives more.

THEOREM 5.3. *Let K be a formally real PRC field. Let G be a finite group, and let G_0 be a normal subgroup of G generated by involutions. There is a Galois extension $N/K(x)$ with Galois group G such that the fixed field of G_0 in N is the maximal totally real extension of $K(x)$ in N.*

PROOF. Let $\langle \epsilon \rangle$ be a group of order 2. Put $H = G \times \langle \epsilon \rangle$ and $H_0 = G_0 \times \langle \epsilon \rangle$, and let $\pi \colon H \to G(K(\sqrt{-1})/K)$ be the epimorphism with kernel G. The set I_0 of involutions in G_0 generates G_0. Therefore $I_1 = (I_0 \cup \{1\}) \times \{\epsilon\}$ generates H_0. It is a conjugacy domain in H. Theorem 5.2 (with $m = 1$) gives a Galois extension F of $E = K(x)$ that contains $\sqrt{-1}$ such that $G(F/E) = H$, $G(F/E(\sqrt{-1})) = G$, and I_1 is the set of real involutions in $G(F/E)$. Let N be the fixed field of ϵ and N' the fixed field of $H_0 = \langle I_1 \rangle$. The last condition means that N' is the maximal totally real extension of E in F, and, thererefore, also in N. Clearly $G(N/E) \cong G$ and N' is the fixed field of G_0 in N.   □

6. Totally real Hilbertian fields

As in the preceding sections, all fields are of characteristic 0. Let S/R be a Galois cover of rings [**FJ**, Definition 5.4] with F/E the corresponding Galois extension of quotient fields. Thus R is an integrally closed domain and there is $z \in S$ integral over R such that $S = R[z]$ and the discriminant $d_E(z)$ of z over E is a unit of R. We call such z a **primitive element** for S/R. Assume S/R is **real** [**HL**, Definition 4.2]: R is a regular ring and F is not formally real.

LEMMA 6.1. *The integral closure S' of R in each intermediate extension F' of F/E is also a regular ring.*

PROOF. Observe that S/S' is also a Galois cover. By [**R**, p. 75] it suffices to show that S'/R is étale. i.e., flat and unramified. We have $S = \oplus_{i=0}^{d-1} Rz^i$, where $d = [F : E]$, and so S/R is faithfully flat. Similarly S/S' is faithfully flat. By the descent property [**Ma**, (4.B)], S'/R is (faithfully) flat.

To show that S'/R is unramified, let q be a prime of S, and let $p = q \cap S'$ and $m = q \cap R$. Replace R, S', and S by their localizations at these primes to assume that they are local rings. Then S/S' and S'/R are still faithfully flat. As S/S' is unramified, the field extension $(S/q)/(R/m)$ is separable and finite. Hence so is its subextension $(S'/p)/(R/m)$. As S/R and S/S' are unramified, $mS = q$ and $pS = q$. Thus $(mS')S \cap S' = pS \cap S'$. But S/S' is faithfully flat, hence [**Ma**, (4.C)], $mS' = p$.   □

Let M be a field. Every homomorphism $\phi \colon R \to M$ extends to a homomorphism $\psi \colon S \to \widetilde{M}$, and ψ induces a group homomorphism $\psi^* \colon G(M) \to G(F/E)$:

$$(1) \qquad\qquad \psi\big(\psi^*(\sigma)(x)\big) = \sigma(x) \qquad \text{for } x \in S.$$

[**FJ**, Lemma 5.5]. If ψ' is another extension of ϕ, then ψ'^* and ψ^* differ by an inner automorphism of $G(F/E)$ [**L**, Corollary 1 on p. 247]. In particular, for $\sigma \in G(M)$ and a conjugacy domain $I \subseteq G(F/E)$ we have $\psi'^*(\sigma) \in I$ if and only if $\psi^*(\sigma) \in I$. This allows us to abuse the notation and write $\phi^*(\sigma) \in I$ instead of $\psi^*(\sigma) \in I$. (Cf. also [**HL**, Remark 4.1].)

REMARK 6.2. Let S_0/R_0 be another real Galois cover of rings, with L/K the corresponding extension of the quotient fields. Assume that R is finitely generated over R_0, the field K is algebraically closed in E, and L is the algebraic closure of K in F. Furthermore, $R_0 \subseteq M$ and $\phi: R \to M$ is an R_0-homomorphism.

We may choose the extension ψ of ϕ to be an S_0-homomorphism. From (1):

(a) Composition of ψ^* with $\operatorname{res}_L: G(F/E) \to G(L/K)$ is the restriction map $\operatorname{res}_L: G(M) \to G(L/K)$.

(b) For each $P \in X(K)$, $\phi^*(I_P(M)) \subseteq I_P(F/E)$. Thus, $\phi^*(I(M)) \subseteq I(F/E)$.

Indeed, let $\epsilon \in I_P(M)$, and let $\psi: S \to \widetilde{M}$ be an extension of ϕ. It follows from (1) that ψ maps the integral closure S' of R in $F(\psi^*(\epsilon))$ into $\widetilde{M}(\epsilon)$. The latter field is real closed. Thus, by Knebusch's Theorem [**HL**, Proposition 1.2], P extends to an ordering on $F(\psi^*(\epsilon))$.

Through this section and for an ambient field K consider the following setup: x is transcendental over K, $E = K(x)$ and $R = K[x, h(x)^{-1}]$. Also, S is a real Galois cover of R with F/E the corresponding extension of quotient fields.

(2) Let $X_1, \ldots, X_m$ be a partition of $X(K)$ into disjoint clopen subsets. For each $1 \le j \le m$, let $Q_j \in X(E)$ so that $\operatorname{res}_K Q_j \in X_j$.

Therefore, $a \in K$ with $h(a) \ne 0$ defines $\phi_a: R \to K$ by $x \mapsto a$.

LEMMA 6.3. *In (2) let Q be an ordering on E with P its restriction to K. Denote the real closure of (K, P) by $\overline{K}$. There exist branch points $x_1 < x_2$ in $\overline{K}$ of the extension F/E with no other $\overline{K}$ branch points between them. They have this property. For each $a \in K$ in the interval (x_1, x_2), $\phi_a^*(I_P(K)) \subseteq I_Q(F/E)$.*

PROOF. Let $\epsilon \in I_Q(F/E)$. Choose a primitive element y for $F(\epsilon)/E$, integral over $K[x]$. Let $f_\epsilon = \operatorname{irr}(y, E) \in K[x, Y]$ and S' the integral closure of R in $F(\epsilon)$. The sentence $(\exists X, Y)[f_\epsilon(X, Y) = 0 \wedge \frac{\partial f}{\partial Y}(X, Y) \ne 0]$ holds in $F(\epsilon)$, and therefore also in the real closure of (E, Q). By Tarski's principle it is valid in $\overline{K}$. Thus there is $\overline{a} \in \overline{K}$ such that $f_\epsilon(\overline{a}, Y)$ has a simple root in $\overline{K}$. This certainly remains true if $\overline{a}$ is replaced by a in the neighborhood $\overline{U}$ of $\overline{a}$ in $\overline{K}$ determined by nearest branch points of F/E in $\overline{K}$.

Let ι be the generator of $G(\overline{K})$, and let S' be the integral closure of R in $F(\epsilon)$. Let $a \in \overline{U} \cap K$. Then $\phi_a: R \to K \subseteq \overline{K}$ extends to a homomorphism $\psi: S' \to \overline{K}$. It follows that its extension $\psi: S \to \widetilde{K}$ satisfies $\psi^*(\iota) = \epsilon$. As $I_P(K)$ and $I_Q(F/E)$ are the conjugacy classes of ι and ϵ in the respective groups, we have $\phi_a^*(I_P(K)) \subseteq I_Q(F/E)$. □

DEFINITION 6.4. A formally real field K is **totally real Hilbertian**, if in each setup (2) the following holds. When $G(F/E) = \langle \bigcup_{j=1}^m I_{Q_j}(F/E) \rangle$, then there exists a K-homomorphism $\phi\colon R \to K$ with $\phi^*(G(K)) = G(F/E)$ and $\phi^*(I_{X_j}(K)) = I_{Q_j}(F/E)$, for each $1 \leq j \leq m$. Thus, $\phi^*(I(K)) = I(F/E)$.

COROLLARY 6.5. *If K is a number field, then K is totally real Hilbertian.*

PROOF. Consider (2). For each j put $P_j = \operatorname{res}_K Q_j$. Since $|X(K)| < \infty$, we may refine $X_1, \ldots, X_m$ to assume $X_j = \{P_j\}$. As K is dense in each of its real closures, Lemma 6.3 gives a nonempty open subset U_j of K (with respect to P_j) so that $\phi_a^*(I_{P_j}(K)) \subseteq I_{Q_j}(F/E)$ for each $a \in U_j$. Consider the Hilbert set

$$H_K = \{a \in K \mid h(a) \neq 0 \text{ and } \phi_a^*(G(K)) = G(F/E)\}$$

[**FJ**, Lemma 12.12]. By [**G**, Lemma 3.4], H_K is dense in K in the product topology induced by $P_1, \ldots, P_n$, that is, there is $a \in H_K \cap U_1 \cap \cdots \cap U_m$. Observe that $I_{Q_j}(F/E)$ is a conjugacy class in $G(F/E)$. The surjectivity of ϕ_a^* implies that $\phi_a^*(I_{P_j}(K)) = I_{Q_j}(F/E)$. □

PROPOSITION 6.6. *Let $K = \mathbb{Q}^{\mathrm{tr}}$. Then K is totally real Hilbertian.*

PROOF. Assume (2). Let z be a primitive element for the cover S/R. Let $K' \subseteq K$ be a number field such that $h(x) \in K'[x]$. Put $R' = K'[x, h(x)^{-1}]$ and $S' = R'[z]$, and let E' and F' be their quotient fields. For each j let Q_j' be the restriction of Q_j to E', and let X_j' be the restriction of X_j to K'. Take K' sufficiently large to assume the following.

(i) S'/R' is a real Galois cover.

(ii) $[F' : E'] = [F : E]$, and therefore K and F' are linearly disjoint over K'.

(iii) The sets $X_1', \ldots, X_m'$ are distinct.

Then $X_1', \ldots, X_m'$ is a partition of $X(K')$ and $G(F'/E') \cong G(F/E)$.

By Corollary 6.4 there exists $a \in K'$ such that an extension $\psi'\colon S' \to \widetilde{K}$ of $\phi_a\colon R' \to K'$ satisfies $\psi'^*(I_{X_j'}(K)) = I_{Q_j'}(F'/E')$. Extend ϕ_a to the K-homomorphism $\phi_a\colon R \to K$. As K and S' are linearly disjoint over K', it is possible to extend this ϕ_a and ψ' to the same K-homomorphism $\psi\colon S \to \widetilde{K}$. By (1), the following diagram commutes.

$$
\begin{array}{ccccc}
G(F/E) & \xleftarrow{\;\psi^*\;} & G(K) & \longleftarrow & I(K) \\
\downarrow{\scriptstyle \operatorname{res}_{F'}} & & \downarrow & & \downarrow \\
G(F'/E') & \xleftarrow{\;\psi'^*\;} & G(K') & \longleftarrow & I(K')
\end{array}
$$

From (ii), the left vertical map is an isomorphism. As $K = \mathbb{Q}^{\mathrm{tr}}$, $I(K) = I(\mathbb{Q}) = I(K')$. Thus, the right vertical inclusion is surjective and maps $I_{X_j}(K)$ onto $I_{X_j'}(K')$. Diagram chasing yields $\psi^*(I_{X_j}(K)) = I_{Q_j}(F/E)$. But $G(F/E) = \langle \bigcup_{j=1}^m I_{Q_j}(F/E) \rangle$, and hence $\psi^*(G(K)) = G(F/E)$. □

7. Absolute Galois group of the totally real algebraic numbers

In this section we consider the following category. An **involutory structure** is a pair $(G, I_G) = \mathbf{G}$ for short, where G is a profinite group and I_G is a closed set of involutions in G. A **morphism** of involutory structures $\phi\colon \mathbf{G} \to \mathbf{H}$ is a continuous homomorphism of groups $\phi\colon G \to H$ such that $\phi(I_G) \subseteq I_H$. We say that $\phi\colon \mathbf{G} \to \mathbf{H}$ is an **epimorphism** if $\phi(G) = H$ and $\phi(I_G) = I_H$.

EXAMPLE 7.1(A). Let L/K be a Galois extension with L not formally real. Then $\mathbf{G}(L/K) = (G(L/K), I(L/K))$ is an involutory structure. Let E be an extension of K, and let F/E be a Galois extension such that $L \subseteq F$. Then the restriction $\mathrm{res}_L\colon \mathbf{G}(F/E) \to \mathbf{G}(L/K)$ is a morphism. Moreover, suppose that E/K is regular and totally real: every ordering on K extends to E. Then res_L is an epimorphism (cf. [**HJ1**, Lemma 3.5]).

EXAMPLE 7.1(B). Let S/R be a real Galois cover with F/E the corresponding Galois extension of fields. Let M be a field and let $\phi\colon R \to M$ be a homomorphism. Extend ϕ to a homomorphism $\psi\colon S \to \widetilde{M}$. Then the group homomorphism $\psi^*\colon G(M) \to G(F/E)$ is a morphism of involutory structures $\psi^*\colon \mathbf{G}(M) \to \mathbf{G}(F/E)$ (Remark 6.2(b)).

A **finite image** of $\mathbf{G}$ is a finite involutory structure $\mathbf{H}$ for which there exists an epimorphism $\phi\colon \mathbf{G} \to \mathbf{H}$. Clearly, up to an isomorphism, it is of the form $(G/N, \{\epsilon N/N \mid \epsilon \in I_G\})$, where N is an open normal subgroup of G not meeting I_G. Let $\mathrm{Im}\mathbf{G}$ be the class of all finite images of $\mathbf{G}$.

A **finite embedding problem** for $\mathbf{G}$ consists of an epimorphism $\pi\colon \mathbf{H} \to \mathbf{A}$ of finite involutory structures, together with an epimorphism $\phi\colon \mathbf{G} \to \mathbf{A}$. A **solution** is an epimorphism $\psi\colon \mathbf{G} \to \mathbf{H}$ such that $\pi \circ \psi = \phi$. We say that $\mathbf{G}$ has the **embedding property** if every finite embedding problem $(\pi\colon \mathbf{H} \to \mathbf{A}, \phi\colon \mathbf{G} \to \mathbf{A})$ for $\mathbf{G}$, in which H is a finite image of $\mathbf{G}$, has a solution.

EXAMPLE 7.2. Let D be the free profinite product $\coprod_{x \in X_\omega} \langle \epsilon_x \rangle$ of groups of order 2 over X_ω (Remark 1.5), and let $I_D^0 = \{\epsilon_x \mid x \in X_\omega\}$. A finite involutory structure (A, I_A) is a finite image of (D, I_D^0) if and only if A is generated by I_A. Furthermore, (D, I_D^0) has the embedding property.

With D and I_D^0 as above, put $\mathbf{D} = (D, I_D)$, where I_D is the conjugacy domain $\mathrm{Con}_D(I_D^0)$ of D generated by I_D^0.

LEMMA 7.3. (a) I_D *is all involutions in D, and D is of rank $\aleph_0$.*

(b) *A finite involutory structure $\mathbf{A}$ is in $\mathrm{Im}\mathbf{D}$ if and only if $I_A \neq \emptyset$ is a conjugacy domain in A and $A = \langle I_A \rangle$.*

(c) $\mathbf{D}$ *has the embedding property.*

PROOF OF (a). See [**HJ2**, Corollary 3.2 and Lemma 2.2]

PROOF OF (b). Immediate from Example 7.2.

PROOF OF (c). Let $\pi\colon \mathbf{H} \to \mathbf{A}$, $\phi\colon \mathbf{D} \to \mathbf{A}$ be a finite embedding problem for $\mathbf{D}$. Then $I_A \subseteq A$ and $I_H \subseteq H$ are conjugacy domains. Let $I_A^0 = \phi(I_D^0)$ and let $I_H^0 = \{\epsilon \in I_H \mid \pi(\epsilon) \in I_A^0\}$. As $\mathrm{Con}_D(I_D^0) = I_D$, we have $\mathrm{Con}_A(I_A^0) = I_A$; it follows that $\mathrm{Con}_H(I_H^0) = I_H$.

By Example 7.2, I_A^0 generates A. But $\mathbf{H} \in \mathrm{Im}\mathbf{D}$ implies that I_H generates H. We have $\pi(I_H^0) = I_A^0$ and $\mathrm{Con}_H(I_H^0) = I_H$. By an analogue of Gaschütz' lemma [**HL**, Lemma 3.3 with $n = 0$], I_H^0 generates H.

By Example 7.2 there is an epimorphism $\psi\colon D \to H$ such that $\pi \circ \psi = \phi$ and $\psi(I_D^0) = I_H^0$. Clearly $\psi(I_D) = I_H$.   □

THEOREM 7.4. *Let K be a totally real Hilbertian PRC field. Assume that K has no proper totally real algebraic extensions and $X(K)$ has no isolated points. Put $\mathbf{G} = (G, I_G)$, where G is the absolute Galois group of K and I_G is the conjugacy domain of all involutions in G. Then*

(a) *A finite embedding problem $(\pi\colon \mathbf{H} \to \mathbf{A}, \; \phi\colon \mathbf{G} \to \mathbf{A})$ for $\mathbf{G}$ has a solution, if (*) $I_H \neq \emptyset$ is a conjugacy domain in H and $H = \langle I_H \rangle$.*

(b) *A finite involutory structure $\mathbf{H}$ is in $\mathrm{Im}\mathbf{G}$ if and only if (*) holds.*

(c) *$\mathbf{G}$ has the embedding property.*

PROOF. The fixed field of I_G in G is totally real over K. Thus $G = \langle I_G \rangle$.

PROOF OF (a). As $1 \notin I_A = \phi(I_G)$, $\mathrm{Ker}(\phi) \cap I_G = \emptyset$. Therefore the fixed field L of $\mathrm{Ker}(\phi)$ is not formally real. Without loss of generality $\mathbf{A} = \mathbf{G}(L/K)$ and ϕ is the restriction map.

Theorem 5.2 (with $m = 1$, $X_1 = X(K)$, and $I_1 = I_H$) identifies $\pi\colon \mathbf{H} \to \mathbf{A}$ with the restriction map $\mathrm{res}_L\colon \mathbf{G}(F/E) \to \mathbf{G}(L/K)$, where E is a simple transcendental extension of K, and F is a Galois extension of E that contains L and is regular over L.

Therefore, $G(F/E) = \langle I(F/E) \rangle$. Choose $Q_1, \ldots, Q_m \in X(E)$ with

$$I(F/E) = \bigcup_{j=1}^{m} I_{Q_j}(F/E).$$

We may assume their restrictions $P_1, \ldots, P_m \in X(K)$ to K are distinct. Indeed, each P_j is not isolated in $X(K)$, and hence there is $P \in X(K)$ distinct from $P_1, \ldots, P_m$ and arbitrarily close to P_j. By Remark 1.8(b) we may assume that $I_P(F/E) = I_{P_j}(F/E)$. As $I_P(F/E) = \bigcap_{\substack{Q \in X(E) \\ Q \supseteq P}} I_Q(F/E)$, there is $Q \in X(E)$ above P such that $I_Q(F/E) = I_{Q_j}(F/E)$. We replace Q_j by Q.

Let $X_1, \ldots, X_m$ be a partition of $X(K)$ into disjoint clopen sets such that $P_j \in X_j$. This gives the setup (2) of Section 6. As K is totally real Hilbertian, there is $a \in K$ and an epimorphism $\phi_a^*\colon \mathbf{G}(K) \to \mathbf{G}(F/E)$. By Remark 6.2(a), ϕ_a^* is a solution to our embedding problem.

PROOF OF (b). Condition (*) is necessary, since $I_G \neq \emptyset$ is a conjugacy domain in G and $G = \langle I_G \rangle$. Conversely, assume (*). Let $A = \langle a \rangle = G(K(\sqrt{-1})/K)$ and $\mathbf{A} = (A, \{a\})$, where a is the generator of A, and let $\phi\colon \mathbf{G} \to \mathbf{A}$ be the restriction map. We construct below a finite involutory structure $\hat{\mathbf{H}}$ that satisfies (*), with epimorphisms $\hat{\mathbf{H}} \to \mathbf{H}$ and $\pi\colon \hat{\mathbf{H}} \to \mathbf{A}$. By (a) there is an epimorphism $\psi\colon \mathbf{G} \to \hat{\mathbf{H}}$, and hence $\mathbf{H} \in \mathrm{Im}\mathbf{G}$.

If there is an epimorphism $\pi: \mathbf{H} \to \mathbf{A}$, let $\hat{\mathbf{H}} = \mathbf{H}$. If not, let $\hat{\mathbf{H}} = (H \times A, I_H \times \{a\})$. Both $\mathbf{A}$ and $\mathbf{H}$ are quotients of $\hat{\mathbf{H}}$. Observe that $(*)$ holds for $\hat{\mathbf{H}}$. Otherwise $I_H \times \{a\}$ generates a proper subgroup Γ of $H \times A$ such that the projection $H \times A \to H$ maps Γ onto $H = \langle I_H \rangle$. Thus $\Gamma = \{(h, \pi(h)) \mid h \in H\}$, where $\pi: H \to A$ is an epimorphism. As $I_H \times \{a\} \subseteq \Gamma$, we have $\pi(I_H) = a$. Thus π induces an epimorphism $\mathbf{H} \to \mathbf{A}$, a contradiction.

PROOF OF (c). Clear from (a) and (b). □

If, in addition to the assumptions of the theorem, K is countable, then G is of rank at most $\aleph_0$. Thus the involutory structures $\mathbf{G}$ and $\mathbf{D}$ are very similar, by Lemma 7.3 and Theorem 7.4. In fact, we have the following straightforward modification of [**FJ**, Lemma 24.1].

LEMMA 7.5. *Let $\mathbf{G}$ and $\mathbf{H}$ be involutory structures with embedding property, such that G and H are of rank at most $\aleph_0$. If $\mathrm{Im}\,\mathbf{G} = \mathrm{Im}\,\mathbf{H}$, then $\mathbf{G} \cong \mathbf{H}$.*

THEOREM 7.6. *Let K be a countable totally real Hilbertian PRC field. Assume that K has no proper totally real algebraic extensions and $X(K)$ has no isolated points. Then $\mathbf{G}(K) \cong \mathbf{D}$, and hence $G(K) \cong D$.*

The field $\mathbb{Q}^{\mathrm{tr}}$ of totally real algebraic numbers is PRC by [**P**]. (We remark that although Pop [**P**] states this result, he only gives the proof for an analog. Therefore in all our results about $\mathbb{Q}^{\mathrm{tr}}$ the reference [**P**] should be replaced by a subsequent version, where this omission will be remedied.) It is clearly countable. By Proposition 6.6 it is totally real Hilbertian, and by Remark 1.5, $X(\mathbb{Q}^{\mathrm{tr}})$ has no isolated points. Therefore:

COROLLARY 7.7. *The absolute Galois group of the field $\mathbb{Q}^{\mathrm{tr}}$ of totally real algebraic numbers is the free profinite product D of groups of order 2 over the universal Boolean space $X_\omega = \{0, 1\}^{\aleph_0}$ of weight $\aleph_0$.*

8. Real Frobenius fields

Let S/R be a real Galois ring cover, and let F/E be the corresponding field extension. Let K be a subfield of R and L the algebraic closure of K in F.

The following definitions are valuable [**HL**, Definition 4.2].

(a) S/R is **regular over** K, if the extension E/K is regular. In that case L/K is a finite Galois extension.

(b) S/R is **finitely generated** over K, if R and S are finitely generated rings over K.

(c) F/E is **amply real** over K if E/K is a regular extension, the algebraic closure L of K in F is not formally real, and the extension $F(\epsilon)/L(\epsilon)$ is totally real for every real involution $\epsilon \in G(F/E)$.

DEFINITION 8.1. A field M is said to be **real Frobenius** if it satisfies the following condition: Let S/R be a real Galois ring cover, finitely generated and regular over M, with F/E the corresponding field extension amply real over

M. Let N be the algebraic closure of M in F. Let $\mathbf{H} \leq \mathbf{G}(F/E)$ such that $\mathbf{H} \in \mathrm{Im}\mathbf{G}(M)$ and $\mathrm{res}_N \mathbf{H} = \mathbf{G}(N/M)$. Then there exists an M-homomorphism $\psi \colon S \to \widetilde{M}$ such that $\psi(R) = M$ and $\psi^*(\mathbf{G}(M)) = \mathbf{H}$.

PROPOSITION 8.2. *Let M be a PRC field. If $\mathbf{G}(M)$ has the embedding property, then M is real Frobenius.*

PROOF. (Cf. [**HL**, Proposition 5.6].) Let S/R, F/E, N, and $\mathbf{H}$ be as in Definition 8.1. The embedding property gives an epimorphism of involutory structures $h \colon \mathbf{G}(M) \to \mathbf{H}$ with $\mathrm{res}_N \circ h = \mathrm{res}_N$. Put $L = \widetilde{M}F$. Then

$$G(L/E) = G(\widetilde{M}E/E) \times_{G(NE/E)} G(F/E) = G(M) \times_{G(N/M)} G(F/E).$$

Let D be the fixed field of $\Delta = \{(\delta, h(\delta)) \mid \delta \in G(M)\}$ in L. Then D/M is regular, $DF = D\widetilde{M} = L$, and $D \cap F = E$ [**FJ**, p. 354]. We show that D/M is totally real. Let P be an ordering on M. There is $\epsilon \in I(M)$ such that P is the restriction of P_ϵ from $\widetilde{M}(\epsilon)$. Then $h(\epsilon) \in I_H \subseteq I(F/E)$. Observe that $\widetilde{M}(\epsilon)$ and $F(h(\epsilon))$ are linearly disjoint over $N(\epsilon)$ and $L(\epsilon, h(\epsilon)) = D(\epsilon)F(h(\epsilon))$ contains D. By assumption there is an ordering Q of $F(h(\epsilon))$ such that $\mathrm{res}_{N(\epsilon)}Q = \mathrm{res}_{N(\epsilon)}P_\epsilon$. Therefore P_ϵ and Q extend to an ordering of $L(\epsilon, h(\epsilon))$ [**J**, p. 241]. The restriction of this ordering to D extends P.

The integral closure U of R in D is finitely generated over M [**FJ**, p. 354] and hence U is the coordinate ring of an absolutely irreducible variety V defined over M. Since M is PRC, there exists an M-homomorphism $\psi_0 \colon U \to M$. Extend ψ_0 to an $\widetilde{M}$-epimorphism $\widetilde{M}U \to \widetilde{M}$, and let $\psi \colon S \to \widetilde{M}$ be its restriction to S. Then $\psi(R) = M$, and, by [**FHJ**, Remark on p. 9], we may arrange it so that $\psi^* \colon G(M) \to G(F/E)$ coincides with h. Therefore $\psi^*(\mathbf{G}(M)) = \mathbf{H}$. □

By Corollary 7.7 and Lemma 7.3(c), $\mathbf{G}(\mathbb{Q}^{\mathrm{tr}})$ has the embedding property. By [**P**], $\mathbb{Q}^{\mathrm{tr}}$ is PRC. Therefore:

COROLLARY 8.3. *$\mathbb{Q}^{\mathrm{tr}}$ is real Frobenius.*

9. Real Galois Stratification

This section gives a quantifier elimination procedure for the theory of real Frobenius fields in the language below. The procedure is similar to [**FJ**, Chapter 25] and almost the same as in [**HL**]. So we only comment on the differences.

A **Galois ring/set cover** C/A over a field K [**FJ**, p. 403] is **real** if A is nonsingular, $\mathrm{char}(K) = 0$, and $K(C)$ is not formally real. Put $\mathbf{G}(C/A) = \mathbf{G}(K(C)/K(A))$ (Example 7.1(a)) and let $\mathrm{Sub}[C/A]$ be the involutory substructures of $\mathbf{G}(C/A)$.

Let $K \subseteq M$ be a field. Each $\mathbf{a} \in A(M)$ determines a K-homomorphism $\phi \colon K[A] \to M$, and therefore (see Section 6) a homomorphism $\phi^* \colon G(M) \to G(C/A)$ (unique up to an inner automorphism of $G(C/A)$). Example 7.1(b) says that $\phi^*(\mathbf{G}(M) \leq \mathbf{G}(C/A)$. Omitting the reference to C and M, define the **Artin symbol** $\mathbf{Ar}(A, \mathbf{a})$ as the set $\{\phi^*(\mathbf{G}(M)^\sigma \mid \sigma \in G(C/A)\}$. This is a conjugacy class in $\mathrm{Sub}[C/A]$. For properties of the Artin symbol see [**FJ**, Section 5.3].

For $n \geq 0$ let $\pi \colon \mathbb{A}^{n+1} \to \mathbb{A}^n$ be the projection on the first n coordinates. Let $A \subseteq \mathbb{A}^{n+1}$ and $B \subseteq \mathbb{A}^n$ be two non-singular basic sets [**FJ**, p. 244] such that $\pi(A) = B$. Then $K[B] \subseteq K[A]$. Let $\mathbf{x}$ and $(\mathbf{x}, y)$ be generic points of B and A, respectively. Then $K(A) = K(B)(y)$. Furthermore, let C/A and D/B be real Galois covers such that $K(D)$ contains the algebraic closure of $K(B)$ in $K(C)$.

DEFINITION 9.1 [**HL**, Definition 7.1]. Let M be a field extension of K. An M-**specialization** of the pair $(C/A, D/B)$ is a K-homomorphism ϕ from C into an overfield of M with these properties: $\phi(K[B]) \subseteq M$; and if y is transcendental over $K(B)$, then $\phi(y)$ is transcendental over M.

For such a specialization put $y' = \phi(y)$, $N = M[\phi(D)]$, $R = M[\phi(K[A])]$, $E = M(y')$ (the quotient field of R), $S = M[\phi(C)]$, and $F = E[\phi(C)]$ (the quotient field of S). Then ϕ induces an embedding $\phi^* \colon G(F/E) \to G(C/A)$.

Assume that $\dim A = \dim B + 1$. The pair $(C/A, D/B)$ is **specialization compatible** if the following properties hold for every M and each M-specialization ϕ as above.

   (i) $K(D)$ is the algebraic closure of $K(B)$ in $K(C)$. and for every M and each M-specialization ϕ as above $[K(C) : K(D)(y)] = [F : N(y')]$.
   (ii) The cover $K(C)/K(A)$ is amply real over $K(B)$.
   (iii) For each involution $\epsilon \in G(F/E)$ with $\phi^*(\epsilon)$ real the extension $F(\epsilon)/N(\epsilon)$ is totally real.

Assume that $\dim A = \dim B$. The pair $(C/A, D/B)$ is said to be **specialization compatible** if $K[A]$ is integral over $K[B]$ and $C = D$.

LEMMA 9.2. *Assume that* $\dim A = \dim B + 1$ *and that* $(C/A, D/B)$ *is specialization compatible. Let* $\mathrm{Con}(A)$ *be a conjugacy domain in* $\mathrm{Sub}[C/A]$, *and let* $\mathcal{S}$ *be a a a set of (isomorphism types of) involutory structures. Define*

$$\mathrm{Con}(B) = \begin{cases} \mathrm{res}_{K(D)}(\mathcal{S} \cap \mathrm{Con}(A)) & \text{if } \dim A = \dim B + 1; \\ \{\mathbf{G}^\sigma \mid \mathbf{G} \in \mathrm{Con}(A), \ \sigma \in G(C/B)\} & \text{if } \dim A = \dim B. \end{cases}$$

Let M *be a real Frobenius field that contains* K, *and let* $\mathbf{b} \in B(M)$. *Assume* $\mathrm{Im}\mathbf{G}(M) \cap \mathrm{Sub}[C/A] = \mathcal{S}$. *Then* $\mathbf{Ar}(B, \mathbf{b}) \subseteq \mathrm{Con}(B)$ *if and only if there is* $\mathbf{a} \in A(M)$ *such that* $\pi(\mathbf{a}) = \mathbf{b}$ *and* $\mathbf{Ar}(A, \mathbf{a}) \subseteq \mathrm{Con}(A)$.

PROOF. See [**HL**, Lemma 7.2] in case $\dim A = \dim B + 1$ and [**HL**, Lemma 7.3] in case $\dim A = \dim B$. (Replace everywhere the e-structures of [**HL**] by our involutory structures.) □

LEMMA 9.3 [**HL**, Lemma 7.5]. *Let* K_1 *be a finite extension of* $K(D)$. *There are Zariski open subsets* $A' \subseteq A$, $B' \subseteq B$ *and a specialization compatible pair of real Galois covers* $(C'/A', D'/B')$ *such that* $K(C) \subseteq K(C')$ *and* $K_1 \subseteq K(D')$.

From now on we can proceed exactly as in [**FJ**, Chapter 25]. Replace Galois covers with real Galois covers, and conjugacy classes of subgroups of $G(C_i/A_i)$

with conjugacy classes of involutory substructures of $\mathbf{G}(C_i/A_i)$ (cf. [**HL**, Sections 8 and 9]).

This includes the definition of Galois stratification and Galois formulas [**FJ**, p. 410]. Thus, a **real** Galois stratification

$$\mathcal{A} = \langle \mathbb{A}^n, C_i/A_i, \mathrm{Con}(A_i)| \ i \in I \rangle,$$

is a partition of the affine space $\mathbb{A}^n$ over K as a finite disjoint union $\mathbb{A}^n = \bigcup_{i \in I} A_i$ of nonsingular K-basic sets, each of them equipped with a real Galois cover C_i/A_i and a conjugacy domain $\mathrm{Con}(A_i)$ in $\mathrm{Sub}[C_i/A_i, \mathbf{P}_0]$. The corresponding **real** Galois formula is a formal expression $\mathbf{Ar}(\mathcal{A}, \mathbf{X}) \subseteq \mathrm{Con}(\mathcal{A})$ with the following interpretation. For be an extension M of K and $\mathbf{a} \in M^n$ write $M \models \mathbf{Ar}(\mathcal{A}, \mathbf{a}) \subseteq \mathrm{Con}(\mathcal{A})$ if $\mathbf{Ar}(A_i, \mathbf{a}) \subseteq \mathrm{Con}(A_i)$ for the unique i such that $\mathbf{a} \in A_i(M)$.

If K is a presented field with elimination theory [**FJ**, Definition 17.9], we get an effective elimination of quantifiers for the theory of real Frobenius fields in this language. Moreover, every formula in the language $\mathcal{L}(K)$ of rings with parameters from K is equivalent to a real Galois formula (cf. [**FJ**, Remark 25.8]). The corresponding real Galois stratification may satisfy $C_i = K[A_i][\sqrt{-1}]$, for each $i \in I$. Conclude the following.

PROPOSITION 9.4 [**HL**, Theorem 9.2(a)]. *Let K be a presented field with elimination theory, and let ϑ be a sentence in $\mathcal{L}(K)$. We can effectively find a finite Galois extension L of K with $\sqrt{-1} \in L$, a finite family $\mathcal{H} \supseteq \mathrm{Sub}[L/K]$ of (isomorphism types of) finite involutory structures, and for each $\mathcal{S} \subseteq \mathcal{H}$ a conjugacy domain $\mathrm{Con}(\mathcal{S})$ in $\mathrm{Sub}[L/K]$ contained in $\mathcal{S}$ with the following property. For every real Frobenius field M that contains K and satisfies $\mathrm{Im}\mathbf{G}(M) \cap \mathcal{H} = \mathcal{S}$ we have $M \models \vartheta$ if and only if $\mathrm{res}_L \mathbf{G}(M) \in \mathrm{Con}(\mathcal{S})$.*

In particular, Proposition 9.4 holds for $K = \mathbb{Q}$.

10. Model theoretic results.

Let K' be a given finite extension of $\mathbb{Q}$, say as $K' = \mathbb{Q}[X]/(f)$, where $f \in \mathbb{Z}[X]$ is a given monic irreducible polynomial. Then K' is formally real (resp. $K' \subseteq \mathbb{Q}^{\mathrm{tr}}$) if and only if f has a root in $\mathbb{R}$ (resp. f splits over $\mathbb{R}$). We can effectively decide whether this condition holds [**L**, p. 276].

In particular, given a finite Galois extension L of $\mathbb{Q}$, we can effectively find the field $L \cap \mathbb{Q}^{\mathrm{tr}}$ and the involutory structure $\mathbf{G}(L/L \cap \mathbb{Q}^{\mathrm{tr}})$.

Let $\mathcal{L}(K)$ denote the elementary language of fields with parameters from K.

THEOREM 10.1. *The elementary theory of $\mathbb{Q}^{\mathrm{tr}}$ is effectively decidable.*

PROOF. Apply Proposition 9.4. The field $\mathbb{Q}^{\mathrm{tr}}$ is real Frobenius (Corollary 8.3) and $\mathrm{Im}\mathbf{G}(\mathbb{Q}^{\mathrm{tr}}) = \mathrm{Im}\mathbf{D}$ (Corollary 7.7) is the family of finite involutory structures $\mathbf{H}$ in which $I_H \neq \emptyset$ is a conjugacy domain in H and $H = \langle I_H \rangle$ (Lemma 7.3(b)). Furthermore, if $L/\mathbb{Q}$ is a finite Galois extension and $K' =$

$L \cap \mathbb{Q}^{\mathrm{tr}}$, then $\mathbb{Q}^{\mathrm{tr}}/K'$ is totally real, and hence $\mathrm{res}_{K'} X(\mathbb{Q}^{\mathrm{tr}}) = X(K')$. Therefore $\mathrm{res}_L \mathbf{G}(\mathbb{Q}^{\mathrm{tr}}) = (G(L/K'), I(L/K')) = \mathbf{G}(L/K')$.

Let ϑ be a sentence in $\mathcal{L}(\mathbb{Q})$. Proposition 9.4 effectively gives a finite Galois extension L of $\mathbb{Q}$ with $\sqrt{-1} \in L$, a finite family $\mathcal{H}$ of (isomorphism types of) finite involutory structures, and, for

$$\mathcal{S} = \{\mathbf{H} \in \mathcal{H} \mid I_H \neq \emptyset \text{ is a conjugacy domain in } H \text{ and } H = \langle I_H \rangle\},$$

a conjugacy domain $\mathrm{Con} = \mathrm{Con}(\mathcal{S})$ in $\mathrm{Sub}[L/\mathbb{Q}]$ contained in $\mathcal{S}$. For these, $\mathbb{Q}^{\mathrm{tr}} \models \vartheta$ if and only if $\mathbf{G}(L/L \cap \mathbb{Q}^{\mathrm{tr}}) \in \mathrm{Con}$. This condition is checkable, by the remarks preceding this corollary. $\quad \square$

LEMMA 10.2. *There is a formula* $\theta(X_1, \ldots, X_n) \in \mathcal{L}(\mathbb{Q})$ *with the following property. Let* M *be a PRC field, let* $\mathbf{a} = (a_1, \ldots, a_n) \in M^n$, *and put* $f = Z^n + a_1 Z^{n-1} + \cdots + a_n \in M[Z]$. *Then* $M \models \theta(\mathbf{a})$ *if and only if*

(*) *f has a root α in $\widetilde{M}$ such that $M(\alpha)$ is formally real.*

PROOF. Condition (*) is equivalent to this: There is an ordering P on M such that f has a root in the real closure of (M, P). By Tarski's principle [**HL**, Proposition 1.4] this is equivalent to a finite disjunction of statements of this form: There is an ordering P on M with $\bigwedge_{i=1}^r f_i(\mathbf{a}) = 0 \ \wedge \ \bigwedge_{j=1}^m g_j(\mathbf{a}) \in P$, where $f_1, \ldots, f_r, g_1, \ldots, g_m \in \mathbb{Z}[X_1, \ldots, X_n]$ do not depend on M and $\mathbf{a}$.

Put $g_0 = 1$, and let Δ be the set of finite sums of squares in M. By [**P1**, Corollary 1.6], $\sum_{i,j=0}^m g_i(\mathbf{a})g_j(\mathbf{a})\Delta$ is the intersection of all orderings on M that contain $g_1(\mathbf{a}), \ldots, g_m(\mathbf{a})$. Therefore the last statement is equivalent to: $\bigwedge_{i=1}^r f_i(\mathbf{a}) = 0 \ \wedge \ -1 \notin \sum_{i,j=0}^m g_i(\mathbf{a})g_j(\mathbf{a})\Delta$. As Δ is the set of sums of two squares in the PRC field M [**P2**, Proposition 1.5], a formula in $\mathcal{L}(\mathbb{Q})$ expresses this statement. $\quad \square$

PROPOSITION 10.3. *Every real Galois formula θ over a field K is equivalent to a formula in $\mathcal{L}(K)$, modulo the theory of PRC fields M containing K.*

PROOF. It suffices to express in $\mathcal{L}(K)$ the statement $\mathbf{Ar}(A, \mathbf{X}) \in \mathrm{Con}$, with C/A a real Galois ring/set cover over K and $\mathrm{Con} = \{\mathbf{H}^\sigma \mid \sigma \in G(C/A)\}$, where $\mathbf{H} = (H, I_H)$ is a involutory substructure of $\mathbf{G}(C/A)$. Let $E = K(A)$ and $F = K(C)$ be the quotient fields. For each $G \leq G(C/A) = G(F/E)$ let $F(G)$ be the fixed field of G in F, and let z_G be a primitive element for $F(G)/E$. Replacing A by an open subset A' (that is, replacing the given Galois stratification by its refinement) we may assume that $K[A][z_G]/K[A]$ is a ring cover [**FJ**, Definition 5.4] and z_G is a primitive element for it.

Write $K[A]$ as $K[\mathbf{x}, g(\mathbf{x})^{-1}]$, where $\mathbf{x}$ is a generic point of A over K. Let f_G be a polynomial over K such that $f_G(\mathbf{x}, g(\mathbf{x})^{-1}, Z) = \mathrm{irr}(z_G, E)$. For every $\epsilon \in H$ let h_ϵ be a polynomial over K with $h_\epsilon(\mathbf{x}, g(\mathbf{x})^{-1}, z_H, Z') = \mathrm{irr}(z_{\langle \epsilon \rangle}, F(H))$.

Then $M \models \mathbf{Ar}(A, \mathbf{a}) \in \mathrm{Con}$ means the following conditions hold.

(a) $\mathbf{a} \in A$: there is a specialization $\mathbf{x} \to \mathbf{a}$ such that $g(\mathbf{a}) \neq 0$.

(b) $\mathbf{x} \to \mathbf{a}$ extends to a homomorphism $\psi \colon C \to \widetilde{M}$ such that $\psi^*(G(M)) = H$.

  (c) $\psi^*(I(M)) = I_H$: for every involution $\epsilon \in H$, $\epsilon \in I_H$ if and only if

  (c_ϵ) $\epsilon \in \psi^*(I(M))$.

Assume (a). Then (b) means the conjunction of the following two statements [**FJ**, Remark 25.14]:

  (b1) $f_H(\mathbf{a}, g(\mathbf{a})^{-1}, Z)$ has a root $c \in M$; and

  (b2) if $G < H$, then $f_G(\mathbf{a}, g(\mathbf{a})^{-1}, Z)$ has no root in M.

Assume (a) and (b), and let $\epsilon \in H$ be an involution. Condition (c_ϵ) says

  (c_ϵ') $h_\epsilon(\mathbf{a}, g(\mathbf{a})^{-1}, c, Z')$ has a root $\alpha \in \widetilde{M}$ such that $M(\alpha)$ is formally real.

Therefore the assertion follows by Lemma 10.2.   □

PROPOSITION 10.4. *The following collection of conditions on a field M is equivalent to a primitive recursive set of elementary sentences in $\mathcal{L}(\mathbb{Q})$.*

  (1) *M is PRC.*

  (2) *$M \cap \tilde{\mathbb{Q}} = \mathbb{Q}^{\mathrm{tr}}$.*

  (3) *$\mathrm{Im}\mathbf{G}(M) = \{\mathbf{H}|\ H = \langle I_H \rangle\}$.*

  (4) *$\mathbf{G}(M)$ has the embedding property.*

  (5) *$M/\mathbb{Q}^{\mathrm{tr}}$ is totally real.*

PROOF. For (1) see [**P2**, Theorem 4.1]. Condition (2) says each irreducible polynomial $f \in \mathbb{Q}[X]$ has a root in M if an only if f has a root in $\mathbb{Q}^{\mathrm{tr}}$. This is equivalent to f splits over the real closure of $\mathbb{Q}$. Express the latter in $\mathcal{L}(\mathbb{Q})$ either by Tarski's principle [**HL**, Proposition 1.4] or by Sturm's Theorem [**L**, Chapter XI, §2]. Conditions (3) and (4) easily follow from Lemma 10.2.

    Assume (1) and (2). By Remark 1.8(b), the image X of restriction $X(M) \to X(\mathbb{Q}^{\mathrm{tr}})$ is closed in $X(\mathbb{Q}^{\mathrm{tr}})$. Thus, (5) is equivalent to X dense in $X(\mathbb{Q}^{\mathrm{tr}})$. Now, $X(\mathbb{Q}^{\mathrm{tr}})$ has a basis from the sets $\{P \in X(\mathbb{Q}^{\mathrm{tr}})|\ P$ extends to $\mathbb{Q}^{\mathrm{tr}}(\alpha)\}$, where α runs through the elements of $\tilde{\mathbb{Q}}$. Indeed, by Remark 1.8(b) these sets are clopen. By [**P1**, Corollary 9.2], $\mathbb{Q}^{\mathrm{tr}}$ is SAP. That is, the sets

$$H(c) = \{P \in X(\mathbb{Q}^{\mathrm{tr}})|\ c \in P\} = \{P \in X(\mathbb{Q}^{\mathrm{tr}})|\ P \text{ extends to } \mathbb{Q}^{\mathrm{tr}}(\sqrt{c})\}$$

form a basis for the Harrison topology on $X(\mathbb{Q}^{\mathrm{tr}})$, as c varies on $\mathbb{Q}^{\mathrm{tr}}$.

    It suffices to consider only $\alpha \in \tilde{\mathbb{Q}}$ with $\mathbb{Q}(\alpha)$ formally real; otherwise the corresponding set of orderings is empty. Thus (5) is equivalent to this statement: If $Q(\alpha)$ is a finite formally real extension of $\mathbb{Q}$, then $M(\alpha)$ is formally real. Now use Lemma 10.2.   □

COROLLARY 10.5. *A field M is a model of $\mathrm{Th}(\mathbb{Q}^{\mathrm{tr}})$ if and only if it satisfies conditions (1)–(5).*

PROOF. The conditions hold for $M = \mathbb{Q}^{\mathrm{tr}}$. By Remark 10.4 they thus hold for each model M of $\mathrm{Th}(\mathbb{Q}^{\mathrm{tr}})$. Conversely, assume (1)–(5). Then M is a real Frobenius field (Proposition 8.3), and $\mathrm{Im}\mathbf{G}(M) = \mathrm{Im}\mathbf{G}(\mathbb{Q}^{\mathrm{tr}})$, by (3). So, (2) and (5) imply $\mathrm{res}_{\tilde{\mathbb{Q}}}\mathbf{G}(M) = \mathbf{G}(\mathbb{Q}^{\mathrm{tr}})$. By Proposition 9.4 (with $K = \mathbb{Q}$) the fields M and $\mathbb{Q}^{\mathrm{tr}}$ satisfy the same sentences in $\mathcal{L}(\mathbb{Q})$.   □

REFERENCES

[C]   T.C. Craven, *The topological space of orderings of a rational function field*, Duke Math. J. **41** (1974), 339–347.

[DF1]   P. Debes and M.D. Fried, *Rigidity and real residue class fields*, Acta Arith. **56** (1990), 1–31.

[DF2]   P. Debes and M.D. Fried, *Nonrigid constructions in Galois theory*, Pac. Jour. **163** #1 (1994), 81–122.

[ELW]   R. Elman, T.Y. Lam and A.R. Wadsworth, *Orderings under field extensions*, J. Reine Angew. Math. **306** (1979), 7–27.

[F]   M.D. Fried, *Fields of definition of function fields and Hurwitz families – groups as Galois group*, Comm. Alg. **5** (1977), 17–82.

[FHJ]   M.D. Fried, D. Haran and M. Jarden, *Galois stratification over Frobenius fields*, Advances in Math. **51** (1984), 1–35.

[FHV]   M.D. Fried, D. Haran and H. Völklein, *Absolute Galois group of the totally real numbers*, C.R. Acad. Sci. Paris **317** (1993), 995–999.

[FJ]   M.D. Fried and M. Jarden, *Field Arithmetic*, Ergebnisse der Mathematik III, vol. 11, Springer Verlag, Heidelberg, 1986.

[FV1]   M.D. Fried and H. Völklein, *The inverse Galois problem and rational points on moduli space*, Math. Ann. **290** (1991), 771–800.

[FV2]   M.D. Fried and H. Völklein, *The embedding problem over a Hilbertian PAC-field*, Annals of Math. **135** (1992), 469–481.

[FV3]   M.D. Fried and H. Völklein, *The absolute Galois group of a Hilbertian PRC field*, Israel J. Math. **85** (1994), 85–101.

[G]   W.-D. Geyer, *Galois groups of intersections of local fields*, Israel J. Math. **30** (1978), 382–396.

[HJ1]   D. Haran and M. Jarden, *The absolute Galois group of a pseudo real closed field*, Annali della Scuola Normale Superiore — Pisa, Serie IV **12** (1985), 449–489.

[HJ2]   D. Haran and M. Jarden, *Real free groups and the absolute Galois group of $\mathbb{R}(t)$*, J. of pure and applied algebra **37** (1985), 155–165.

[HL]   D. Haran and L. Lauwers, *Galois stratification over e-fold ordered Frobenius fields*, Israel J. Math. **85** (1994), 169–197.

[H]   R. Hartshorne, *Algebraic Geometry*, Graduate Texts in Mathematics, Springer-Verlag, 1977.

[HR]   W. Herfort and L. Ribes, *Torsion elements and centralizers in free products of profinite groups*, J. für die reine und angewandte Mathematik **358** (1985), 155–161.

[J]   M. Jarden, *The elementary theory of large e-fold ordered fields*, Acta mathematica **149** (1982), 239–260.

[KN]   W. Krull and J. Neukirch, *Die Struktur der absoluten Galoisgruppe über dem Körper $\mathbb{R}(t)$*, Math. Ann. **193** (1971), 197–209.

[L]   S. Lang, *Algebra*, Addison-Wesley, Reading, Mass., 1970.

[M]   W.S. Massey, *Algebraic topology: An introduction*, Graduate Texts in Mathematics, Springer-Verlag, 1977.

[Ma]   H. Matsumura, *Commutative Algebra*, Benjamin/Cummings, Reading, Mass., 1980.

[P]   F. Pop, *Fields of totally Σ-adic numbers*, preprint (1992).

[P1]   A. Prestel, *Lectures on formally real fields*, IMPA Publications, vol. 1093, Springer Verlag, 1984.

[P2]   A. Prestel, *Pseudo real closed fields*, Set Theory and Model Theory, Lecture Notes in Mathematics, vol. 872, Springer Verlag, 1981, pp. 127–156.

[P3]   A. Prestel, *On the axiomatization of PRC-fields*, Methods of Mathematical Logic, Lecture Notes in Mathematics, vol. 1130, Springer Verlag, 1985, pp. 351–359.

[R]   M. Raynaud, *Lecture Notes in Mathematics*, vol. 169, Springer-Verlag, 1970.

[R1]   Julia Robinson, *The undecidability of algebraic rings and fields*, Proc. Amer. Math. Soc. **10** (1959), 950–957.

[R2]   Julia Robinson, *On the decision problem for algebraic rings*, Studies in math. analysis and related topics, Stanford Univ. Press, Stanford, Calif., 1962, pp. 297–304.

[W]   A. Weil, *The field of definition of a variety*, Amer. J. of Math. **78** (1956), 509–524.

UC IRVINE, IRVINE, CA 92717, USA
EMAIL: mfried@math.uci.edu

RAYMOND AND BEVERLY SACKLER FACULTY OF EXACT SCIENCES,
TEL AVIV UNIVERSITY, TEL AVIV 69978, ISRAEL
EMAIL: haran@math.tau.ac.il

UNIVERSITY OF FLORIDA, GAINESVILLE, FL 32611, USA
EMAIL: helmut@math.ufl.edu

Contemporary Mathematics
Volume **174**, 1994

GALOIS GROUPS WITH PRESCRIBED RAMIFICATION

DAVID HARBATER

ABSTRACT. The paper studies Galois groups with a given set of ramified places, in both the function field and number field cases. In the geometric case, it is shown in characteristic p that the fundamental group of a curve of genus g with $r > 0$ points deleted depends upon the choice of curve and points, and not just on g and r (unlike in characteristic 0). In the arithmetic case, certain groups are shown to occur over $\mathbf{Q}$ with given ramification, or are shown not to occur, particularly when the only ramified prime is 2.

INTRODUCTION

This paper concerns the problem of finding which groups occur as Galois groups with prescribed ramification. The problem can appear both in arithmetic and in geometric settings, and can be interpreted as a problem of finding fundamental groups. Specifically, given a Dedekind domain D and a finite set S of primes, we may consider the fundamental group $\pi_1(\operatorname{Spec}(D) - S)$, and the related set $\pi_A(\operatorname{Spec}(D) - S)$ of finite quotients of π_1. Here π_A consists of the finite groups that can occur as Galois groups over (the fraction field of) D with ramification only at S. We then wish to understand π_A and, if possible, π_1, as well as obtaining information about which subgroups of a given group $G \in \pi_A$ can occur as inertia groups. In this paper, we consider these four situations (which are in turn less and less well understood):

(i) complex affine curves;
(ii) affine curves over an algebraically closed field of characteristic p;
(iii) affine curves over a finite field;
(iv) open subsets $U_n = \operatorname{Spec}(\mathbf{Z}[1/n])$ of $\operatorname{Spec}(\mathbf{Z})$.

Section 1 considers (i) - (iii). Situation (i) is the most classical, of course, and in particular π_1 is known, by Riemann's Existence Theorem. But even there, there is no known *explicit* version of Riemann's Existence Theorem, and as a result one can rarely write down algebraically an extension of $\mathbf{C}(x)$ with given group and ramification. Situation (ii) was until recently wide open, but the recent proof [Ra], [Ha3] of Abhyankar's Conjecture [Ab1] has answered the question of what π_A is. In particular, it is now known that π_A of an affine curve U of the form (genus g) $-$ (r points) depends only on the numbers g and r. But the profinite group π_1 is still very much unknown, and in section 1 we show that $\pi_1(U)$ does *not*

1991 *Mathematics Subject Classification*. Primary 11R32, 14H30; Secondary 13B05, 12F10.
Key words and phrases. Galois, ramification, cover, fundamental group.
The author was supported in part by NSA grant # MDA 904-92-H-3024.
This paper is in final form, and no version of it will be submitted for publication elsewhere.

just depend on g and r (Theorem 1.8). This raises the question of to what extent $\pi_1(U)$ determines the curve U. In section 1 we give some results concerning this question, and pose some open problems, as well as deriving other consequences of Abhyankar's Conjecture. We also discuss situation (iii), about which less is known.

Section 2 concerns situation (iv), which is the most wide-open of the four. While much is known about the realization of groups as Galois groups over $\mathbf{Q}$ ("inverse Galois theory"), and about *abelian* groups as Galois groups with prescribed ramification (class field theory), much less is known about which general finite groups occur as Galois groups over $\mathbf{Q}$ with prescribed ramification. While a full solution is of course far off in the future, we present some results of the sort that are currently being sought in connection with situation (iii). Motivated by the analogy with the geometric situation, we make a conjecture on how the tame fundamental group of $U_n = \mathrm{Spec}(\mathbf{Z}[1/n])$ grows with n (Conjecture 2.1), and give some evidence for this (Theorem 2.6). Concerning the opposite part of π_1, viz. the p-part of $\pi_1(U_p)$, after showing that this is cyclic for odd p (Theorem 2.11) we study the more involved case of $p = 2$ (where, e.g., the dihedral group D_4 is in $\pi_A(U_2)$ but not the quaternion group; cf. 2.12(b), 2.14). We also show (cf. Corollary 2.7) that all groups in $\pi_A(U_2)$ are quasi-2 groups (which is analogous to the result for situations (ii) and (iii) in the case of the affine line), but that there are also further restrictions, such as the fact that a non-2 Galois extension ramified only at 2 must have a high index of wild ramification (Theorem 2.23). In fact, we show that all small groups in $\pi_A(U_2)$ are 2-groups, and we find the smallest non-2-group in $\pi_A(U_2)$ (of order 272). In particular, we show that the four smallest non-abelian simple groups do not lie in $\pi_A(U_2)$ (Example 2.21); this relates to Serre's Conjecture. In connection with these results, we also state some open questions and speculations.

I would like to thank Robert Coleman for posing to me the problem of how $\pi_1(U_n)$ grows with n; Hendrik Lenstra for a number of discussions concerning techniques that can be used in studying $\pi_1(U_n)$; and J.-P. Serre for his comments on an earlier version of this paper. I would also like to thank Ram Abhyankar, Michael Larsen, Karl Rubin, Alice Silverberg, John Tate, Jaap Top, and Larry Washington for useful comments and suggestions.

SECTION 1. GEOMETRIC GALOIS GROUPS

Let k be a field, and consider smooth connected affine curves U over k. Each such U is of the form $X - S$, where X is a smooth projective k-curve of some genus g, and $S \neq \emptyset$ consists of finitely many closed points of X. For each U, we wish to understand $\pi_1(U)$ and $\pi_A(U)$. In particular, we may ask how π_1 and π_A vary as U changes – viz. as the choice of S changes, or as the projective curve X varies in moduli.

The most classical case is that of $k = \mathbf{C}$. In this case, we know by Riemann's Existence Theorem (e.g. [Gr, XIII, Cor.2.12]) that $\pi_A(U)$ is the set of finite quotients of the topological fundamental group $\pi_1^{\mathrm{top}}(U)$, and that $\pi_1(U)$ is the profinite completion of $\pi_1^{\mathrm{top}}(U)$. Thus if S consists of n points (so $n > 0$), then $\pi_1^{\mathrm{top}}(U)$ is the group generated by elements $a_1, \ldots, a_g, b_1, \ldots, b_g, c_1, \ldots, c_n$ subject only to the relation that $\prod_{j=1}^g [a_j, b_j] \prod_{i=1}^n c_i = 1$, where $[a, b]$ denotes the commutator $aba^{-1}b^{-1}$. This is isomorphic to the free group on $2g + n - 1$ generators; so $\pi_A(U)$ consists of the finite groups having $2g + n - 1$ generators, and the algebraic fundamental group $\pi_1(U)$ is isomorphic to the free profinite group on $2g + n - 1$ generators.

If X is fixed and the n points of S are allowed to vary, then the above fundamental groups do not change, up to isomorphism. (But there is no canonical isomorphism between the π_1's of the old and new U's, since an isomorphism depends on a choice of homotopy basis.) More generally, consider the moduli space $\mathcal{M}_{g,n}$ of projective curves of genus g and n marked points. For each point $[(X, S)]$ of $\mathcal{M}_{g,n}$, we may consider $\pi_1(U)$ and $\pi_A(U)$, where $U = X - S$. Again, these do not depend on the point of $\mathcal{M}_{g,n}$, up to isomorphism.

Similarly, one may consider the case where k is a general algebraically closed field of characteristic 0. For such k, Grothendieck used the technique of specialization to show [Gr, XIII, Cor.2.12] that the fundamental group of $U = X - S$ is given by the same expression as in the case of ground field $\mathbf{C}$. So again $\pi_1(U)$ depends only on the genus of the curve X and the number of points in S – not on the specific choice of curve X or position of the points S, and not on the field k.

For the rest of this section we consider the situation in characteristic $p > 0$. For now, assume that the field k is algebraically closed. Then π_1 behaves differently than over $\mathbf{C}$. In particular, the affine line is no longer simply connected, since there are Artin-Schreier covers. For example, for each non-zero $c \in k$, there is a $\mathbf{Z}/p$-Galois cover of the affine x-line given by $y^p - y = cx$. Moreover, these covers are non-isomorphic (as $\mathbf{Z}/p$-Galois covers of the x-line) for distinct values of $c \in k$. This points out another difference between characteristics 0 and p: In characteristic p, coverings have "moduli"; and as a result, the group π_1 depends on the choice of algebraically closed ground field k. Indeed, if one enlarges k, then π_1 also becomes enlarged, in characteristic p.

More generally, by taking towers of $\mathbf{Z}/p$-covers, it is possible to realize every finite p-group as a Galois group over any affine k-curve. Many other finite groups also occur in π_A, as the following result states. (For a finite group G, the notation $p(G)$ denotes the (normal) subgroup of G generated by the subgroups of p-power order.)

Theorem 1.1. *Let X be a smooth connected projective curve over k, and let $S \subset X$ consist of n points $(n > 0)$. Then $\pi_A(U) = \{G \mid G/p(G)$ has $2g + n - 1$ generators$\}$.*

This result was originally conjectured in 1957 by Abhyankar [Ab1]. Equivalently, he conjectured that a finite group G occurs over U if and only if every prime-to-p quotient of G occurs over an analogous curve over $\mathbf{C}$ (i.e. a curve of the same genus with the same number of punctures). After partial results by Nori [Ka], Abhyankar [Ab2], and Serre [Se2], the theorem was proven by Raynaud [Ra] in the case that U is the affine line. In this case, the result asserts that π_A consists of all the finite *quasi-p groups*, i.e. the finite groups G such that $G = p(G)$. The more general case of the theorem was proven by the present author in [Ha3]. Moreover that paper proved even more, viz. that for a given G that is predicted to lie in π_A, the cover may be chosen so that its smooth completion is wildly ramified only over one particular point of S. Both [Ra] and [Ha3] rely on patching techniques (rigid or formal analysis) to construct covers with desired Galois groups.

Remark. While a full description of the proof of Theorem 1.1 is beyond the scope of this paper, here is a brief outline. Raynaud's proof of the case $U = \mathbf{A}^1$ [Ra] proceeds by induction, and considers three cases:

(i) G has a non-trivial normal p-subgroup N. Then $G/N \in \pi_A$ by induction, and then $G \in \pi_A$ by Serre's result [Se2].

(ii) A Sylow p-subgroup P of G has the property that G is generated by the set S of proper quasi-p subgroups of G having a Sylow p-subgroup contained in P. Then the groups in S are in π_A by induction, and a patching argument allows the corresponding covers to be pasted together to form a G-Galois cover of $\mathbf{A}^1$.

(iii) Both (i) and (ii) fail. Then an argument involving semi-stable reduction is used to construct a G-cover of $\mathbf{A}^1$.

The proof in the general case [Ha3] proceeds in two steps:

(a) Using [Ra], the result is shown for $\mathbf{P}^1 - \{0, \infty\}$. This is done by patching a $p(G)$-Galois cover of $\mathbf{A}^1$ (which exists by [Ra]) to a cyclic-by-p cover of $\mathbf{P}^1 - \{0, \infty\}$ (which is essentially constructed explicitly).

(b) For a more general affine curve U, an appropriate Galois cover of $\mathbf{P}^1 - \{0, \infty\}$ (given by step (a)) is pasted to a prime-to-p cover of U, to yield a G-Galois cover of U. $\square$

Corollary 1.2. *Let X be a smooth connected projective k-curve of genus g and let S be a non-empty finite subset of X, say having n points. Let $U = X - S$. Then $\pi_A(U)$ depends only on g and n, and not on the choices of X or S. In fact, $\pi_A(U)$ depends only on the value of $2g + n$.*

Proof. Immediate from Theorem 1.1. $\square$

Corollary 1.3. *Fix $g \geq 0$ and $n > 0$. Then π_A of a k-curve of genus g with n points deleted strictly contains π_A of a $\mathbf{C}$-curve of genus g with n points deleted.*

Proof. Denote these two affine curves by U_k and $U_{\mathbf{C}}$. If G is in $\pi_A(U_{\mathbf{C}})$, then G and hence $G/p(G)$ has $2g + n - 1$ generators; so G is in $\pi_A(U_k)$ by 1.1. This shows containment. For strict containment, choose $N > 2g + n - 1$, and let $G = (\mathbf{Z}/p)^N$. Then G is in $\pi_A(U_k)$ by 1.1, since $G/p(G)$ is trivial; but G is not in $\pi_A(U_{\mathbf{C}})$. $\square$

Remarks. (a) Corollary 1.3 is somewhat surprising, for the following reason: A $\mathbf{Z}/3$-Galois cover $E \to \mathbf{P}^1_{\mathbf{C}}$ branched at $\{0, 1, \infty\}$ is an elliptic curve, and its maximal unramified elementary abelian p-cover $F \to E$ satisfies $\mathrm{Gal}(F/E) = (\mathbf{Z}/p)^2$. The composition $F \to \mathbf{P}^1_{\mathbf{C}}$ is Galois, with three-fold ramification over each of its branch points $\{0, 1, \infty\}$, and its Galois group G is a semi-direct product of $(\mathbf{Z}/p)^2$ with $\mathbf{Z}/3$. But there is no such cover of the projective k-line (i.e. no cover with the same group, branch locus, and inertia groups), since any such cover would yield a $(\mathbf{Z}/p)^2$-Galois unramified cover of an elliptic curve in characteristic p. But by Corollary 1.3, there must be some *other* G-Galois cover of $\mathbf{P}^1_k$ branched at $\{0, 1, \infty\}$. Note that such a cover must be wildly ramified somewhere.

(b) Theorem 1.1 and its corollaries do not carry over to the case of *projective* curves (i.e. where we allow $n = 0$). For example, 1.1 and 1.3 fail to hold for genus 1 curves, since $(\mathbf{Z}/p)^2$ is not in π_A of any elliptic curve over k. Similarly, 1.2 fails since $\mathbf{Z}/p$ fails to lie in π_A of a supersingular elliptic curve, although it is in π_A of an ordinary elliptic curve. A further discussion of the projective analog of these questions will appear in the forthcoming Ph.D. thesis of Katherine Stevenson. $\square$

In addition to knowing which groups can occur as Galois groups of unramified covers of an affine curve $U = X - S$ (and hence as Galois groups of branched covers of X unramified away from S), it would be desirable to know what types of ramification can occur over each of the points of S. While the strong form of Theorem 1.1 in [Ha3] provides some information of this sort, the situation is unknown in general. (Cf. also Remark (a) above.) In the simplest situation, that of the affine line, it is easy to see that there is a necessary condition for a group to occur as inertia over infinity:

Proposition 1.4. *Let G be a quasi-p group, let $Y \to \mathbf{P}^1$ be a G-Galois connected branched cover ramified only over infinity, and let $I \subset G$ be an inertia group over infinity. Then I is a semi-direct product $P \rtimes C$, where C is a cyclic group of order prime to p and where P is a p-group whose conjugacy class generates G.*

Proof. Since k is algebraically closed of characteristic p, every inertia group is cyclic-by-p. Now for $I = P \rtimes C$ an inertia group over infinity, let $N \subset G$ be the subgroup generated by the conjugacy class of P. Then N is a normal subgroup, and $Y/N \to \mathbf{P}^1$ is a G/N-Galois connected branched cover, unramified away from infinity. But since N contains P and its conjugates, this cover is at most tamely ramified over infinity. But over any algebraically closed field, the projective line has no non-trivial connected covers that are unramified away from infinity and tamely ramified over infinity. So G/N is trivial, and thus $N = G$. $\square$

Abhyankar has recently suggested that the converse of Proposition 1.4 may be true; i.e. that the conditions on I in the conclusion of 1.4 may imply that I is an inertia group over infinity of some G-Galois unramified cover of $\mathbf{A}^1$ (i.e. of a branched cover of $\mathbf{P}^1$ ramified over infinity). Some evidence for this is the following:

(i) For many quasi-p matrix groups, Abhyankar has shown that the cyclic group of order p occurs as an inertia group over infinity of some unramified cover of $\mathbf{A}^1$.

(ii) By [Ha2, Theorem 2], if $I \subset G$ is a p-subgroup occuring as an inertia group over infinity of a G-Galois cover of $\mathbf{A}^1$, and if $I' \subset G$ is a p-subgroup containing I, then I' also occurs (for some other cover).

(iii) By (ii) and [Ha2, Lemma to Theorem 4], or by [Ra, Cor. 2.2.6], every Sylow p-subgroup of a quasi-p group G is an inertia group over infinity of some G-Galois cover of $\mathbf{A}^1$.

A related problem is to describe the set $\pi_A^{p'\mathrm{ram}}(U)$ of Galois groups of unramified Galois covers of $U = X - S$ whose completion has the property that all of its inertia groups (over the points of S) are of order prime to p. (In characteristic p, $\pi_A^{p'\mathrm{ram}}(U) = \pi_A^{t}(U)$, the set of Galois groups of tamely ramified covers; these groups may have order divisible by p.) With $\pi_A^{p'\mathrm{ram}}(U)$ replacing $\pi_A(U)$, the statement of Theorem 1.1 becomes false because every group in $\pi_A^{p'\mathrm{ram}}(U)$ has $2g + n - 1$ generators [Gr, XIII, Cor.2.12], unlike $\pi_A(U)$; the correct replacement for the assertion is unknown. Remark (a) after Corollary 1.3 shows that the analog of 1.3 for $\pi_A^{p'\mathrm{ram}}$ fails. And the analog of Corollary 1.2 for $\pi_A^{p'\mathrm{ram}}$ also fails, as the following example shows:

Example 1.5. Assume $p \neq 2$ and let $\lambda \in \mathbf{P}^1$. Let $E \to \mathbf{P}^1_k$ be the branched cover having degree 2 and branched precisely at $\{0, 1, \infty, \lambda\}$. Thus E is an elliptic curve, and if λ is chosen so that E is ordinary then there is a unique unramified Galois covering morphism $E^* \to E$ of degree p, where E^* is connected. Thus $E^* \to \mathbf{P}^1$ is Galois, and its Galois group G is a semi-direct product of $\mathbf{Z}/p$ with $\mathbf{Z}/2$. Thus $G \in \pi^t_A(U)$, where $U = \mathbf{P}^1 - \{0, 1, \infty, \lambda\}$. But if now another value of λ is chosen (say λ') yielding a supersingular elliptic curve $E' \to \mathbf{P}^1_k$, then G cannot lie in $\pi^t_A(U')$, where $U' = \mathbf{P}^1 - \{0, 1, \infty, \lambda'\}$. For otherwise, there is a Galois branched cover $E'^* \to \mathbf{P}^1$ with group G, having only tame ramification. By the structure of G, this cover dominates a degree two cover $E'' \to \mathbf{P}^1$ that is unbranched away from $\{0, 1, \infty, \lambda'\}$. Since the characteristic of k is not 2, such a cover has genus at most 1, with equality if and only if $E'' \cong E'$ (in which case E'' is supersingular). Thus E'' has no connected unramified covers of degree p, and so the $\mathbf{Z}/p$-Galois cover $E'^* \to E''$ is totally ramified somewhere. But $E'^* \to E''$ is at most tamely ramified, since $E'^* \to \mathbf{P}^1$ is. This is a contradiction, showing that $G \notin \pi^t_A(U')$, although $G \in \pi^t_A(U)$. $\square$

The above results concerned π_A, the set of finite quotients of the algebraic fundamental group π_1. The profinite group π_1 contains more information than the set π_A, and it would be desirable to have analogs for π_1. But while negative results are known – e.g. that π_1 is not free – there is no reasonable conjecture describing what π_1 is isomorphic to, even in the case of the affine line. And the analog of the last part of Corollary 1.2 is false for π_1; i.e. the fundamental group of a curve of genus g with n points deleted does not just depend on $2g + n$ in characteristic p:

Proposition 1.6. *Let $U = \mathbf{P}^1 - S$, where S is a set of $n + 2$ points $(n \geq 0)$, and let $U' = E - S'$, where E is an ordinary elliptic curve and S' is a set of n points. Then $\pi_1(U)$ is not isomorphic to $\pi_1(U')$.*

Proof. Let $f' : E^* \to E$ be an unramified connected $\mathbf{Z}/p$-Galois cover of elliptic curves, which exists since E is ordinary, and let $U'^* = f'^{-1}(U') \subset E^*$. Let $N' \subset \pi_1(U')$ be the normal subgroup of index p corresponding to the unramified Galois cover $U'^* \to U'$. Thus N' may be identified with $\pi_1(U'^*)$. Note that $E^* - U'^* = f'^{-1}(S')$ consists of np points, since there are p points of E^* over each point of E. Since E^* has genus 1, it follows that the prime-to-p part of N' is free on $2 \cdot 1 + np - 1 = np + 1$ generators.

Now suppose that there is an isomorphism $\phi : \pi_1(U) \to \pi_1(U')$. Let $N = \phi^{-1}(N')$. Thus N is a normal subgroup of index p in $\pi_1(U)$, corresponding to a connected unramified $\mathbf{Z}/p$-Galois cover $U^* \to U$. So N may be identified with $\pi_1(U^*)$. Let P^* be the smooth completion of the affine curve U^*. Thus there is a $\mathbf{Z}/p$-Galois branched cover $f : P^* \to \mathbf{P}^1$, whose branch locus in contained in S. Let i be the number of points of S that are actually ramified; each of these i points is then totally ramified, since p is prime. Here $i > 0$, since there are no connected unramified covers of $\mathbf{P}^1$ of degree greater than 1. Now $f^{-1}(S) = P^* - U^*$ consists of exactly $i + (n + 2 - i)p$ points. Also, if g is the genus of P^*, then by the Riemann-Hurwitz formula in the wild case we get $2g - 2 \geq -2p + ip$, i.e. $2g \geq 2 + (i - 2)p$. So the prime-to-p part of N is free on at least $(2 + (i - 2)p) + (i + (n + 2 - i)p) - 1 = np + i + 1$

generators. But since $N = \phi^{-1}(N')$ and ϕ is an isomorphism, and since $i > 0$, this contradicts the fact that the prime-to-p part of N' is free on only $2 \cdot 1 + np - 1 = np + 1$ generators. $\square$

Moreover, the analog for π_1 of the first part of 1.2 also fails. That is, for a fixed choice of $g \geq 0$ and a positive integer n, two distinct affine curves of the form (genus g) $-$ (n points) can have non-isomorphic fundamental groups, as the following result shows:

Proposition 1.7. *Let E be an ordinary elliptic curve, let E' be a supersingular elliptic curve, and let $n > 0$. Let U and U' be affine curves obtained by deleting n points from E and E' respectively. Then $\pi_1(U)$ and $\pi_1(U')$ are non-isomorphic.*

Proof. Since E is ordinary there is a connected $\mathbf{Z}/p$-Galois unramified cover $f : E^* \to E$. Let $U^* = f^{-1}(U)$, and let N be the normal subgroup of index p in $\pi_1(U)$ corresponding to f. Thus N may be identified with $\pi_1(U^*)$. Since $U^* = E^* - f^{-1}(S)$ and $f^{-1}(S)$ consists of np points, the prime-to-p part of N is free on $2 \cdot 1 + np - 1 = np + 1$ generators.

Now assume that there is an isomorphism $\phi : \pi_1(U) \to \pi_1(U')$, and let $N' = \phi(N)$. Also, let $U'^* \to U'$ be the $\mathbf{Z}/p$-Galois connected unramified cover corresponding to N'; thus N' may be indentified with $\pi_1(U'^*)$. Let E'^* be the smooth completion of U'^*, and let $f' : E'^* \to E'$ be the corresponding branched cover. Let i be the number of branch points of f'. Thus $i \leq n$, and also $i > 0$ since the supersingular elliptic curve E' has no degree p connected unramified covers. Now $E'^* - U'^* = f'^{-1}(E' - U')$ consists of exactly $i + (n - i)p$ points. And by the Riemann-Hurwitz formula in the wild case, we find that $g = \text{genus}(E'^*)$ satisfies $2g - 2 \geq ip$, i.e. $2g \geq ip + 2$. So the prime-to-p part of N' is free on at least $(ip + 2) + (i + (n - i)p) - 1 = np + i - 1$ generators. Since $i > 0$ and N' is isomorphic to N, this is a contradiction. $\square$

Finally, even two affine open subsets of the same projective curve, with the same number of points deleted, can fail to have isomorphic fundamental groups. In particular, this can occur for the projective line with four points deleted:

Theorem 1.8. *Assume that $p \neq 2$. Let $\lambda, \lambda' \in \mathbf{P}^1 - \{0, 1, \infty\}$ be points which respectively have ordinary and supersingular j-invariants. Let $U = \mathbf{P}^1 - \{0, 1, \infty, \lambda\}$ and $U' = \mathbf{P}^1 - \{0, 1, \infty, \lambda'\}$. Then $\pi_1(U)$ and $\pi_1(U')$ are non-isomorphic.*

Proof. Let $f : P^* \to \mathbf{P}^1$ be the two-fold cover that is totally ramified over the points $0, 1, \infty, \lambda$, let $U^* = P^* - f^{-1}(\{0, 1, \infty, \lambda\})$, and let N be the normal subgroup of index 2 in $\pi_1(U)$ corresponding to the unramified cover $U^* \to U$. Thus P^* is an ordinary elliptic curve, $P^* - U^*$ consists of the four ramification points of f, and N may be identified with $\pi_1(U^*)$. Assume that there is an isomorphism $\phi : \pi_1(U) \to \pi_1(U')$, and let $N' = \phi(N)$. Thus N' is a normal subgroup of index 2 in $\pi_1(U')$, corresponding to a two-fold unramified cover $U'^* \to U'$, and isomorphic to $\pi_1(U'^*)$. Let P'^* be the smooth completion of U'^*, and let $f' : P'^* \to \mathbf{P}^1$ be the corresponding branched cover. Since f' is a two-fold cover of the projective line having at most four branch points, and since $p \neq 2$, the Hurwitz formula implies that the number of branch points of f' is either 2 or 4. In the former case P'^* has genus 0, and U'^* is isomorphic to $\mathbf{P}^1 - $ (6 points). And in the latter case, P'^* is a supersingular elliptic curve, and U'^* is isomorphic to $P'^* - (4 \text{ points})$. But ϕ restricts to an isomorphism $\pi_1(U^*) \to \pi_1(U'^*)$. So in the first case this contradicts Proposition 1.6, and in the second case this contradicts Proposition 1.7. $\square$

Remark. The proofs of 1.6-1.8 actually show more. Namely, for any group G, let G' denote the commutator subgroup $[G, G]$. Then under the hypotheses of 1.6 and 1.7, the quotients π_1/π_1'' are non-isomorphic; and under the hypotheses of 1.8, the quotients π_1/π_1''' are non-isomorphic. $\square$

The above theorem suggests the following question:

Question 1.9. For U an affine k-curve, to what extent does the profinite group $\pi_1(U)$ determine U?

A very weak form of Question 1.9, which is nevertheless not known, is this: Consider two affine curves $U_i = X_i - S_i$ ($i = 1, 2$), where X_i is a smooth projective k-curve of genus g_i and S_i has n_i elements. If $\pi_1(U_1) \cong \pi_1(U_2)$, then must $g_1 = g_2$ and $n_1 = n_2$? In light of Proposition 1.6, the first case to consider is that of $X_1 = \mathbf{P}^1$, $n_1 = 3$, $X_2 = $ a supersingular elliptic curve, $n_2 = 1$; we would then wish to show that the π_1's are not isomorphic.

Alternatively, if we instead fix values for g and n, then the first case to consider is that of $U = \mathbf{P}^1 - S$, where S consists of four points. By the triple transitivity of $\mathrm{Aut}(\mathbf{P}^1)$, we may assume that S is of the form $\{0, 1, \infty, \lambda\}$, and the question is then to what extent $\pi_1(U)$ determines $j(\lambda)$. The following result shows two cases in which different values of j can correspond to isomorphic π_1's:

Proposition 1.10. *Let* $\lambda, \lambda' \in \mathbf{P}^1 - \{0, 1, \infty\}$, *and let* $j, j' \in k$ *be the corresponding j-invariants. Also, let* $U = \mathbf{P}^1 - \{0, 1, \infty, \lambda\}$ *and* $U' = \mathbf{P}^1 - \{0, 1, \infty, \lambda'\}$. *Then* $\pi_1(U) \cong \pi_1(U')$ *provided that either:*

(a) j, j' *are algebraic over* $\mathbf{F}_p$ *and lie in the same orbit under Frobenius; or*
(b) j, j' *are both transendental over* $\mathbf{F}_p$.

Proof. In each case, it suffices to show that there is an $\mathbf{F}_p$-isomorphism $U \to U'$, since this would pull back the tower of covers of U' to the tower of covers of U. Now the k-isomorphism classes of U and U' are determined by j and j', and any field automorphism of k induces an $\mathbf{F}_p$-automorphism of $\mathbf{P}^1$. So it suffices to show that there is a field automorphism of k taking j to j'.

In (a), we have that $j, j' \in \overline{\mathbf{F}}_p$. We are supposing that $F^i(j) = j'$, for some i, where $F : \overline{\mathbf{F}}_p \to \overline{\mathbf{F}}_p$ is the Frobenius automorphism. Since k is algebraically closed, there is an extension of F^i to an automorphism of k, also taking j to j'.

In (b) there is a field isomorphism $\overline{\mathbf{F}}_p(j) \to \overline{\mathbf{F}}_p(j')$ taking j to j'. Again this extends to an automorphism of k, taking j to j'. $\square$

This proposition suggests the following more precise form of Question 1.9:

Question 1.11. (a) Does the converse of 1.10 hold? That is, if U and U' are as in 1.10 and if $\pi_1(U) \cong \pi_1(U')$ holds, must (a) or (b) hold?

(b) If $k = \overline{\mathbf{F}}_p$, then more generally for any two affine k-curves U and U', does $\pi_1(U) \cong \pi_1(U')$ imply that U and U' are $\mathbf{F}_p$-isomorphic?

Next, we turn to the case in which k is taken to be a *finite* field of characteristic p, and we consider fundamental groups of geometrically connected affine curves U over k. That is, we consider $\pi_1(U)$ and $\pi_A(U)$, arising from connected unramified Galois covers of U. Since U is geometrically connected, the $\overline{k}$-scheme $\overline{U} = U \times_k \overline{k}$ is connected, and there is a split exact sequence

$$1 \to \pi_1(\overline{U}) \to \pi_1(U) \to \mathrm{Gal}(\overline{k}/k) \to 1.$$

Since $\mathrm{Gal}(\overline{k}/k)$ is a cyclic profinite group generated by the Frobenius automorphism F, to give a splitting is to give the image of F. And once this lifting of F to $\pi_1(U)$ is given, there is an action of F on $\pi_1(\overline{U})$. This action thus determines the group $\pi_1(U)$ as an extension of $\mathrm{Gal}(\overline{k}/k)$ by $\pi_1(\overline{U})$, and hence it determines $\pi_A(U)$. But this action of Frobenius is not currently understood. The previous discussion in the algebraically closed case, however, suggests the following

Question 1.12. Let U, U' be geometrically connected affine k-curves. Let $\phi : \pi_1(U) \to \mathrm{Gal}(\overline{k}/k)$ and $\phi' : \pi_1(U') \to \mathrm{Gal}(\overline{k}/k)$ be as in the exact sequence above. If there is an isomorphism $\alpha : \pi_1(U) \to \pi_1(U')$ such that $\alpha \circ \phi' = \phi$, must $U \cong U'$ as k-schemes?

The analog of this for affine open subsets of the projective line over a *number field* was proven by Nakamura, in [Na].

Another issue in this situation is that connected covers of U need not be geometrically connected. Indeed, a Galois cover $U^* \to U$ will be geometrically connected if and only if it is *regular*; i.e. if and only if k is algebraically closed in the function field of U^*. Thus there is a subset $\pi_A^{\mathrm{reg}}(U) \subset \pi_A(U)$ corresponding to Galois groups of regular covers. We would like to understand $\pi_A^{\mathrm{reg}}(U)$.

If U is a geometrically connected affine k-curve and $U^* \to U$ is a G-Galois unramified cover, then we may extend constants to the algebraic closure $\overline{k}$ of k, and obtain a G-Galois cover $\overline{U}^* \to \overline{U}$ of connected k-curves. Thus $\pi_A^{\mathrm{reg}}(U) \subset \pi_A(\overline{U})$, and the latter set is understood by the algebraically closed case. Now given a connected G-Galois cover $\overline{U}^* \to \overline{U}$, Frobenius acts on the cover, in the sense of inducing a G-Galois cover $\overline{U}^{*F} \to \overline{U}$ by letting F act on the coefficients of the equations defining $\overline{U}^*$ as a cover and also on the Galois automorphisms. By [De], the G-Galois cover $\overline{U}^* \to \overline{U}$ is induced by a G-Galois cover $U^* \to U$ if and only if $\overline{U}^{*F} \to \overline{U}$ is isomorphic, as a G-Galois cover, to $\overline{U}^* \to \overline{U}$. (The corresponding fact is false in the case of covers over number fields, because the field of moduli of a G-Galois cover need not in general be a field of definition; cf. [CH, Example 2.6].) And so a finite group G lies in $\pi_A^{\mathrm{reg}}(U)$ if and only if some G-Galois cover $\overline{U}^* \to \overline{U}$ satisfies the above property. But again, the difficulty in using this criterion to find $\pi_A^{\mathrm{reg}}(U)$ explicitly is that the action of Frobenius on $\pi_1(\overline{U})$ is not understood.

Recently, Abhyankar has obtained many quasi-p groups as Galois groups of explicit unramified covers of the affine line over $\mathbf{F}_p$. Motivated by this, he has asked the following

Question 1.13. If p is a prime and G is a finite quasi-p group, must G lie in $\pi_A(\mathbf{A}^1_{\mathbf{F}_p})$?

But the patching methods used in [Ra] and [Ha3] to prove Abhyankar's conjecture over algebraically closed fields do not seem applicable to this situation, due to difficulties involved with specialization.

SECTION 2. ARITHMETIC GALOIS GROUPS

This section considers the problem of finding π_1 and π_A in the unequal characteristic case – specifically, for a "curve" of the form $U_n = \mathrm{Spec}(\mathbf{Z}[\frac{1}{n}])$, where n is a square-free positive integer. It was suggested by I.R. Shafarevich [Sh, sect. 3] that $\pi_1(U_n)$ is topologically finitely generated for each n, and R. Coleman has asked how π_A and π_1 grow with n. Presumably every finite group lies in $\pi_A(U_n)$ for some sufficiently large n.

By the analogy between number fields and function fields, we might expect $\pi_A(U_n)$ to behave similarly to $\pi_A(U)$, where U is an open subset of the affine line over $\mathbf{F}_p$. In that situation, a square-free polynomial f of degree d has norm p^d and defines a set of d geometric points. For such an f, if $U \subset \mathbf{A}^1_{\mathbf{F}_p}$ is the set where $f \neq 0$, then the tame part of $\pi_A^{\mathrm{reg}}(U)$ is contained in $\pi_A^{\mathrm{t}}(\overline{U})$, whose elements are each generated by d elements. In the arithmetic situation, $U_n \subset \mathrm{Spec}(\mathbf{Z})$ is the non-vanishing set of a square-free positive integer n, of norm n. So we make the following conjecture:

Conjecture 2.1. There is a constant C such that for every positive square-free integer n, every group in $\pi_A^{\mathrm{t}}(U_n)$ has a generating set with at most $\log n + C$ elements.

That is, if K is a Galois extension of $\mathbf{Q}$ having Galois group G and only tame ramification, and if n is the product of the distinct primes dividing the discriminant, then the number of generators of G is conjectured to be at most $\log n + \mathcal{O}(1)$. More groups would thus be allowed as n increases. This conjecture is consistent with the expectation, suggested at this conference by B. Birch, that every finite group is the Galois group of a tamely ramified extension of $\mathbf{Q}$.

In this section we particularly consider the case where $n = p$ is prime. If G is the Galois group of an extension K of $\mathbf{Q}$ ramified only over p, and if $N \subset G$ is the (normal) subgroup generated by the Sylow p-subgroups of the inertia groups, then N is a quasi-p group, and $L = K^N$ is a tamely ramified Galois extension of $\mathbf{Q}$ with group G/N ramified only over p. Since $N \subset p(G)$ (where, as in section 1, $p(G)$ denotes the quasi-p part of G), we obtain

Proposition 2.2. If Conjecture 2.1 holds, and if $G \in \pi_A(U_p)$ for some prime number p, then $G/p(G)$ is generated by at most $\log p + C$ elements (where C is as in 2.1).

We begin with two examples of Galois extensions K of $\mathbf{Q}$ that are ramified only at a single prime:

Examples 2.3. (a) If a prime number p is chosen so that the class number $h(\mathbf{Q}(\zeta_p)) > 1$ then the Hilbert class field K of $\mathbf{Q}(\zeta_p)$ is ramified only at p. For example we may take p to be 23, with class number 3, showing that the non-trivial semi-direct product $\mathbf{Z}/3 \rtimes (\mathbf{Z}/23)^*$ is in $\pi_A(U_{23})$.

(b) (following ideas of Ken Ribet in his talk at this conference) Let N be a prime such that the genus of $J = J_0(N)$ is at least 1, and take p to be a prime not dividing the numerator of $\frac{N-1}{12}$. The action of $G_\mathbf{Q} = \mathrm{Gal}(\overline{\mathbf{Q}}/\mathbf{Q})$ on the torsion points of J yields a representation $\rho : G_\mathbf{Q} \to GL(2,k)$ for some finite field k of characteristic p, and the image A of ρ is $\{g \in GL(2,k) \mid \det(g) \in \mathbf{F}_p^*\}$. So A is a quotient of $G_\mathbf{Q}$, and taking $p = N$ we get a corresponding Galois extension of $\mathbf{Q}$ with Galois group A. This extension is unramified away from p, since ρ is (because J has good reduction outside p). $\square$

Note that in these examples the prime p must be sufficiently large, and this may reflect the fact that according to 2.1 and 2.2 there should be "more" extensions ramified only over a fixed large prime than there are ramified only over a fixed small prime.

For any group G, let G' be the commutator subgroup of G; let $G^{\mathrm{ab}} = G/G'$ be the abelianization of G; and let G^{solv} be the maximal solvable quotient of G.

By class field theory, the abelianized fundamental group of U_n is given by $\pi_1(U_n)^{\mathrm{ab}} \cong \prod_{p|n} \mathbf{Z}_p^* \cong \prod_{p|n}[(\mathbf{Z}/p)^* \times \mathbf{Z}_p]$. Similarly, the abelianized tame fundamental group is given by $\pi_1^{\mathrm{t}}(U_n)^{\mathrm{ab}} = \prod_{p|n}(\mathbf{Z}/p)^*$, and hence the number of generators of this group is equal to the number of distinct odd primes dividing n.

Remark. In the geometric situation, Theorem 1.8 produced two affine curves of the form $\mathbf{P}^1 - (4 \text{ points})$ whose π_1's were non-isomorphic, even modulo π_1''' (cf. the remark after that result). In the arithmetic situation, much more is true: by the comments above, a square-free positive integer n is determined by $\pi_1(U_n)^{\mathrm{ab}} = \pi_1(U_n)/\pi_1(U_n)'$. But this is less surprising, since no two n's can have the same norm, and so intuitively different U_n's have different numbers of "missing points."

A key difference between the arithmetic and geometric cases concerns the relationship between the discriminant Δ and the index of ramification e. In the geometric case, fixing the degree N and bounding the values of e does not bound $|\Delta|$, when there is wild ramification. Thus, in characteristic p, there are branched covers of $\mathbf{P}_k^1$ of degree p, unramified except at a single point, having arbitrarily large discriminant and hence arbitrarily large genus (e.g. the covers $y^p - y = x^{-n}$, with n prime to p). But in the arithmetic case, for a given degree N, knowing the values of e bounds the discriminant. By combining this upper bound on $|\Delta|$ with a lower bound (e.g. Odlyzko's bounds), it is often possible to show that G is not in $\pi_A(U_n)$.

Specifically, if $K \subset L$ is a totally ramified extension of p-adic fields having ramification index e, and if $\mathbf{p}$ is the prime of $\mathcal{O}_K$ over p, then

$$v_{\mathbf{p}}(\Delta_{L/K}) \le e - 1 + e v_{\mathbf{p}}(e).$$

(This can be seen by writing $\mathcal{O}_L = \mathcal{O}_K[x]/(f(x))$ and then evaluating $v_{\mathbf{q}}(f'(\pi_L))$, where $\mathbf{q}$ is the prime of $\mathcal{O}_L$ over p and π_L is the uniformizer of $\mathcal{O}_L$. See [Se1, p. 568,

Proposition 3].) So for a global finite extension $\mathbf{Q} \subset L$, with ramification indices e_i and residue degrees f_i over p, we have

$$(*) \qquad\qquad v_p(\Delta_{L/\mathbf{Q}}) \le \sum_i f_i\big(e_i - 1 + e_i v_p(e_i)\big).$$

In particular, if L is ramified only over a single prime p and $[L : \mathbf{Q}] = n$, then $|\Delta_{L/\mathbf{Q}}|^{\frac{1}{n}} \le p^{1+\frac{1}{n}\sum(e_i f_i v_p(e_i)-f_i)}$; so if L is Galois over $\mathbf{Q}$ with ramification indices equal to e then

$$(**) \qquad\qquad |\Delta_{L/\mathbf{Q}}|^{\frac{1}{n}} \le p^{1+v_p(e)-\frac{1}{e}}.$$

(Compare [Se1, p. 570, Proposition 6].)

Using this, we obtain evidence for Conjecture 2.1 (Theorem 2.6 below). First we need two lemmas:

Lemma 2.4. *Let p be prime.*

(a) *If the class number $h_p = h(\mathbf{Q}(\zeta_p))$ is equal to 1, then $\pi_1^{\mathrm{t}}(U_p)^{\mathrm{solv}}$ is cyclic of order $p - 1$.*

(b) *If $h_p > 1$, then $\pi_1^{\mathrm{t}}(U_p)^{\mathrm{solv}}$ is not cyclic.*

Proof. (a) Let $G \in \pi_A^{\mathrm{t}}(U_p)^{\mathrm{solv}}$ and let K be a G-Galois extension of $\mathbf{Q}$ that is un-ramified except at p, where it is tamely ramified. We wish to show that $K \subset \mathbf{Q}(\zeta_p)$. The fixed field $K^{G'}$ under the commutator subgroup G' is an abelian extension of $\mathbf{Q}$ ramified only at p, where it is tamely ramified; so $K^{G'} \subset \mathbf{Q}(\zeta_p)$.

Now any non-trivial unramified abelian extension L of $K^{G'}$ is linearly disjoint from $\mathbf{Q}(\zeta_p)$ over $K^{G'}$, since the latter extension is totally ramified. Thus $L(\zeta_p)$ is a non-trivial abelian unramified extension of $\mathbf{Q}(\zeta_p)$; a contradiction, showing that $h(K^{G'}) = 1$. So the abelian extension $K^{G'} \subset K^{G''}$ has no unramified subextensions, and hence is totally ramified. Thus the G/G''-Galois extension $\mathbf{Q} \subset K^{G''}$ is totally, and tamely, ramified over p. So its inertia group, viz. G/G'', is cyclic, and in particular abelian. Thus G'' contains G', and hence $G'' = G'$. But G' is solvable; so G' is trivial and $K = K^{G'} \subset \mathbf{Q}(\zeta_p)$.

(b) Let K be the Hilbert class field of $\mathbf{Q}(\zeta_p)$. Thus K is Galois over $\mathbf{Q}$; let G be the Galois group. If G is cyclic, then $\mathbf{Q} \subset K$ is an abelian extension ramified only at p, where it is tamely ramified. Thus $K \subset \mathbf{Q}(\zeta_p)$, and hence $K = \mathbf{Q}(\zeta_p)$. This contradicts the assumption that $h_p = 1$, showing that G is non-cyclic. Hence the group $\pi_1^{\mathrm{t}}(U_p)^{\mathrm{solv}}$ is also non-cyclic. $\square$

Lemma 2.5. *Let G be a non-solvable group of order ≤ 500, such that every proper quotient of G is abelian. Let $g \in G$ and let e be the order of g. Then one of the following holds:*

   (i) $G \cong A_5$, $e \le 5$;

   (ii) $G \cong S_5$, $e \le 6$;

   (iii) $G \cong \mathrm{PSL}(2,7)$, $e \le 7$;

   (iv) $1 \to \mathrm{PSL}(2,7) \to G \to \mathbf{Z}/2 \to 1$ is exact, $e \le 14$;

   (v) $G \cong A_6$, $e \le 5$.

Proof. Let N be a minimal non-trivial normal subgroup of G. Thus N is of the form H^ν, for some simple group H and some integer $\nu \geq 1$. Let $\overline{G} = G/N$. So $\overline{G}$ is abelian. Since G is not solvable and $\overline{G}$ is solvable, it follows that N and hence H is non-abelian. Thus $|H| \geq 60$. Since $|G| \leq 500$, we have that $\nu = 1$ and so $N = H$.

Since $n \leq 500$, we have that $|H| \leq 500$, and therefore H is isomorphic either to the alternating group A_5 of order 60; to the simple group $\mathrm{SL}(3,2) \cong \mathrm{PSL}(2,7)$ of order 168; or to the alternating group A_6 of order 360. The maximal order of an element in A_5 and A_6 is 5 (corresponding to a five-cycle), and in $\mathrm{PSL}(2,7)$ it is 7 (corresponding to the upper triangular matrix with entries on and above the diagonal equal to 1). Since $|G| \leq 500$, we have that $|\overline{G}| = 1$ if $H \cong A_6$, and $|\overline{G}| \leq 2$ if $H \cong \mathrm{PSL}(2,7)$. If $G \cong S_5$ then the maximal order of an element is 6 (the product of a two-cycle and a three-cycle). And if $H \cong \mathrm{PSL}(2,7)$ and $\overline{G} \cong \mathbf{Z}/2$, then any $g \in G$ satisfies $g^2 \in H$ and so $g^{14} = 1$. So it remains to show that if $H \cong A_5$ and $\overline{G}$ is non-trivial, then $G \cong S_5$.

So take $H = A_5$ and assume $\overline{G} = G/H$ is non-trivial, and let $\rho : \overline{G} \to \mathrm{Out}(A_5) \cong \mathbf{Z}/2$ be the homomorphism induced by the given extension. We claim that ρ is an isomorphism. If not, then since $\overline{G}$ is non-trivial, we have that $\ker(\rho)$ is non-trivial. Let $\overline{g} \neq 1$ be in $\ker(\rho)$, and let $g \in G$ lie over $\overline{g}$. Also, let $\overline{E} \subset \overline{G}$ be the subgroup generated by $\overline{g}$, and let $E \subset G$ be the inverse image of $\overline{E}$. Thus $g \in E$, and E is generated by g and H. Since $\overline{G}$ is abelian, $\overline{E}$ is normal in $\overline{G}$, and hence E is normal in G. By the choice of $\overline{g}$, conjugation of H by g is an inner automorphism of H, say by $h \in H$. Thus $gh^{-1} \in G$ acts trivially on H by conjugation, i.e. gh^{-1} commutes with the elements of H, and in particular with h. Thus g commutes with h, and so gh^{-1} commutes with g. Since E is generated by g and H, we have that gh^{-1} lies in the center Z of E. But $gh^{-1} \neq 1$ (because its image in $\overline{G}$ is $\overline{g} \neq 1$), and so $Z \neq 1$. Since Z is a characteristic subgroup of N, and since N is normal in G, it follows that Z is a non-trivial normal subgroup of G. But Z is abelian, and so G/Z is (like G) non-solvable and hence non-abelian. This contradicts the hypothesis on G, proving the claim that ρ is an isomorphism.

Thus $\rho : \overline{G} \to \mathrm{Out}(A_5) \cong \mathbf{Z}/2$ is an isomorphism. Let $\overline{g}$ be the involution in $\overline{G}$ and let $g \in G$ lie over $\overline{g}$. Thus the conjugation action of g on $H = A_5$ is not an inner automorphism. But $\mathrm{Aut}(A_5) = S_5$, and so this conjugation action agrees with conjugation on A_5 by some odd permutation τ in S_5. Here $\tau = \sigma h$, where $\sigma \in S_5$ is a transposition and $h \in H = A_5$. Replacing the lift g of $\overline{g}$ by gh^{-1}, we are reduced to the case that g has order 2 and it acts on A_5 the same way as the transposition σ. So there is an isomorphism $G \to S_5$ which is the identity on A_5 and takes g to σ. $\square$

Theorem 2.6. (a) *If $p < 23$ is prime, then $\pi_1^{\mathrm{t}}(U_p)$ is cyclic of order $p - 1$.*

(b) *The group $\pi_1^{\mathrm{t}}(U_{23})$ is not cyclic.*

Proof. (a) Since $p \leq 19$, the class number $h(\mathbf{Q}(\zeta_p)) = 1$. So if 2.6 fails, then 2.4(a) implies that there is a non-solvable finite quotient of $\pi_1^{\mathrm{t}}(U_p)$. Let G be such a quotient of smallest possible order n, and let K be a corresponding G-Galois extension of $\mathbf{Q}$ ramified only over p, where it is tamely ramified of index e. For any non-trivial normal subgroup of $N \subset G$, the minimality of G implies that G/N

is solvable. So by 2.4(a), G/N is cyclic of order r dividing $p-1$. This shows that every proper quotient of G is abelian.

In particular, let $N_0 \subset G$ be minimal among the non-trivial normal subgroups of G. The minimality of N_0 implies that N_0 is of the form H^ν, for some simple group H and some integer $\nu \geq 1$. Here H is non-abelian since G is not solvable. Write $\Delta = \Delta_{K/\mathbf{Q}}$.

We claim that $n = |G| < 500$. Namely, since K is tamely ramified over p with ramification index $e < n$, and unramified elsewhere, by the inequality (**) above we have that $|\Delta|^{\frac{1}{n}} < 19^{1-\frac{1}{n}}$. But by [Od, p. 380, 1.11], $|\Delta|^{\frac{1}{n}} \geq 21.8 \cdot e^{-70/n}$. So $21.8 \cdot e^{-70/n} < 19^{1-\frac{1}{n}}$, and so

$$n < \frac{70 - \log(19)}{\log(21.8/19)} < 500.$$

Since the ramification is tame, the ramification index e is equal to the order of some element in G. Since $n = |G| < 500$, and since every proper quotient of G is abelian, by Lemma 2.5 we are in one of the following five cases:

  (i) $n = 60$, $e \leq 5$;
  (ii) $n = 120$, $e \leq 10$;
  (iii) $n = 168$, $e \leq 7$;
  (iv) $n = 2 \cdot 168 = 336$, $e \leq 14$;
  (v) $n = 360$, $e \leq 5$.

In each case, inequality (**) shows that $|\Delta|^{\frac{1}{n}} \leq 19^{1-\frac{1}{e}}$, and [Od, Table 1, pp. 400-401] provides a lower bound for $|\Delta|^{\frac{1}{n}}$. These upper and lower bounds (rounding up, in the case of the upper bounds) are respectively 10.55 and 12.23; 14.16 and 14.38; 12.48 and 15.12; 15.40 and 17.51; 10.55 and 17.94. In each case this is a contradiction.

(b) Since the class number of $\mathbf{Q}(\zeta_{23})$ is 3, by 2.4(b) we have that $\pi_1^{\mathrm{t}}(U_{23})^{\mathrm{solv}}$ is not cyclic. Hence $\pi_1^{\mathrm{t}}(U_{23})$ is also not cyclic. $\square$

As in section 1, for any finite group G and any prime p, we denote by $p(G)$ the subgroup of G generated by the p-subgroups of G, and we say that G is a *quasi-p group* if $p(G) = G$.

Corollary 2.7. *If $p < 23$ is prime, and G is in $\pi_A(U_p)$, then $G/p(G)$ is cyclic of order dividing $p - 1$.*

Proof. Let K be a G-Galois extension of $\mathbf{Q}$ ramified only at p. Then the subfield $K^{p(G)}$ is a $G/p(G)$-Galois extension of $\mathbf{Q}$ ramified only at p, and it is tamely ramified over p since $|G/p(G)|$ is prime to p. So we are done by 2.6(a). $\square$

In particular, every group in $\pi_A(U_2)$ is a quasi-2 group.

Remark. J.-P. Serre has observed to the author that $\pi_1(U_{11})$ is not solvable. In particular, it has a quotient isomorphic to $\mathrm{GL}(2, 11)$, provided by the 11-division points of the elliptic curve of conductor 11 (or equivalently, by the Ramanujan function representation modulo 11). Moreover, the same argument using the Ramanujan function shows that $\pi_1(U_p)$ is not solvable for larger primes p (using modular forms other than delta for $p = 23$ and $p = 691$).

Proposition 2.8. *Let p and q be (possibly equal) prime numbers, let G be a p-group, and let $\mathbf{Q} \subset K$ be a G-Galois extension ramified only at q.*

   (a) *The extension $\mathbf{Q} \subset K$ is totally ramified over q.*

   (b) *The class number of K is prime to p.*

Proof. (a) If not, let I be an inertia group over q. Then I is a proper subgroup of G. But any proper subgroup of a p-group is contained in a proper normal subgroup. So I lies in a proper normal subgroup $N \subset G$. Thus the fixed field K^N is unramifed over $\mathbf{Q}$ with group $G/N \neq 1$, and this is impossible.

   (b) Let C be the class group of K, let $C' = C/N$ be the maximal quotient of C of p-power order, let L be the Hilbert class field of K, and let $L' = L^N$. Then the extension $\mathbf{Q} \subset L'$ is a Galois extension of p-power order, and thus by (a) it is totally ramified over q. Hence so is $K \subset L'$. But $K \subset L'$ is a subextension of $K \subset L$, and so is unramified. Hence the extension $K \subset L'$ is trivial, and so its Galois group C' is also trivial. Thus $N = C$. So the class number of K, which is equal to $|C|$, is prime to p. $\square$

Proposition 2.9. *Under the hypotheses of 2.8, suppose additionally that every proper subfield K' of K that is Galois over $\mathbf{Q}$ has class number 1. Assume that the class group C of K is non-trivial.*

   (a) *There is an injective group homomorphism $\rho : G \to \mathrm{Aut}(C)$, arising from the action of G on $C = \mathrm{Gal}(L/K)$, where L is the Hilbert class field of K.*

   (b) *If G is non-abelian, then C is not cyclic.*

Proof. (a) Let $\Gamma = \mathrm{Gal}(L/\mathbf{Q})$. Thus we have the exact sequence $1 \to C \to \Gamma \to G \to 1$, and by 2.8(b) the kernel and cokernel have relatively prime order. Thus this sequence splits and we may regard $G \subset \Gamma$. Let $\rho : G \to \mathrm{Aut}(C)$ correspond to the induced conjugation action of G on C, and let $N = \ker(\rho) \subset G \subset \Gamma$. We wish to show that N is trivial.

   So assume not. Then N is a non-trivial normal subgroup of G, and the elements of N commute with all the elements of C. So N is normal in Γ. Let $G' = G/N$, $\Gamma' = \Gamma/N$, $K' = K^N$ and $L' = L^N$. Then the extension $\mathbf{Q} \subset K'$ is G'-Galois, and $\mathbf{Q} \subset L'$ is Γ'-Galois. Hence the extension $K' \subset L'$ is C-Galois. Since $N \neq 1$, the field K' is strictly contained in K; so by hypothesis, the class number of K' is 1. Thus the abelian extension $K' \subset L'$ is totally ramified. Also, $K' \subset K$ is totally ramified over q since $\mathbf{Q} \subset K$ is, by 2.8(a). Now the degree of the Galois extension $K' \subset K$ is a power of p, whereas $[L' : K'] = |C|$ is prime to p by 2.8(b). So the fields K and L' are linearly disjoint over K', and thus the compositum $KL' = L$ is totally ramified over K. But L is unramified over K. So $L = K$ and C is trivial, a contradiction.

   (b) If C is cyclic, then $\mathrm{Aut}(C)$ is abelian. Since G is non-abelian, this contradicts the conclusion of (a). $\square$

Corollary 2.10. *Under the hypotheses of 2.9, if $p = 2$ and G is non-abelian, then the class number of K is at least 9.*

Proof. Let C be the class group of K. Then $|C|$ is odd by 2.8(b), and C is non-cyclic by 2.9(b). Thus the abelian group C is a product of at least two cyclic groups, each of order at least 3. So $|C| \geq 9$. $\square$

If p is an odd prime, and $\mathbf{Q} \subset K$ is a G-Galois extension ramified only at p, with G an abelian p-group, then $G = \mathbf{Z}/p^{n-1}$ for some n. For the prime 2, the corresponding assertion holds provided one restricts attention to totally real fields. In fact, these assertions remain true even without assuming that G is abelian. Namely, if we write $\pi_1^{\mathrm{tr}}(U_2)$ for the quotient of $\pi_1(U_2)$ corresponding to totally real extensions, and write $\pi_1^p(U_p)$ [resp. $(\pi_1^{\mathrm{tr}})^p(U_2)$] for the maximal p-quotient of $\pi_1(U_p)$ [resp. of $\pi_1^{\mathrm{tr}}(U_2)$], we have the following easy result:

Theorem 2.11. (a) If p is an odd prime, then $\pi_1^p(U_p) \cong \mathbf{Z}_p$. Equivalently, a finite p-group G is in $\pi_A(U_p)$ if and only if G is cyclic.
(b) $(\pi_1^{\mathrm{tr}})^2(U_2) \cong \mathbf{Z}_2$. Equivalently, a finite 2-group G is in $\pi_A^{\mathrm{tr}}(U_2)$ if and only if G is cyclic.

Proof. Let $\Pi = \pi_A(U_p)$ if p is odd, and let $\Pi = \pi_A^{\mathrm{tr}}(U_2)$ if $p = 2$. It suffices to show that if $G \in \Pi$ then G is cyclic. Now for $G \in \Pi$, the abelianization G^{ab} is also in Π. So by the comment before the theorem, G^{ab} is cyclic of p-power order. But $G^{\mathrm{ab}} = G/G'$, where G' is the commutator, and G' is contained in the Frattini subgroup F of G. So G/F is cyclic. By the Burnside Basis Theorem, so is G. $\square$

From now on we restrict attention to the case of U_2, and study the set $\pi_A(U_2)$. While 2.11(b) does not determine precisely which 2-groups are in $\pi_A(U_2)$, it does provide a necessary condition for a 2-group to lie in $\pi_A(U_2)$:

Proposition 2.12. Let G be a non-trivial 2-group in $\pi_A(U_2)$.
(a) Then G contains an involution i such that the normal subgroup $N \subset G$ generated by i has the property that G/N is cyclic.
(b) The normal subgroup H of G generated by the involutions of G has the property that G/H is cyclic.

Proof. Part (b) is immediate from (a). For (a), assume that $G \in \pi_A(U_2)$, and let $\mathbf{Q} \subset K$ be a corresponding G-Galois extension ramified only over 2. Let $g \in G$ be the image of complex conjugation, under $\mathrm{Gal}(\overline{\mathbf{Q}}/\mathbf{Q}) \to \mathrm{Gal}(K/\mathbf{Q}) = G$. If g is trivial, then K is a real Galois extension of $\mathbf{Q}$, and hence is totally real; so G is cyclic by Proposition 2.11(b), and we may choose i to be the unique involution in G. So now assume that g is non-trivial, take $i = g$, and let N be the normal subgroup generated by i (i.e. the subgroup of G generated by i and its conjugates). Thus K^N is totally real, and so is cyclic over $\mathbf{Q}$ with group G/N. $\square$

Question 2.13. Is condition (a) of Proposition 2.12 necessary and sufficient for a finite 2-group G to lie in $\pi_A(U_2)$?

Using 2.12, various 2-groups can be shown not to lie in $\pi_A(U_2)$. For example, the quaternion group Q of order 8 is not in $\pi_A(U_2)$, since the only involution is -1, and so it does not satisfy (b) of Proposition 2.12. On the other hand, we have:

Example 2.14. The dihedral group $D_4 = \langle \sigma, \tau \mid \sigma^4 = 1, \tau^2 = 1, \tau\sigma = \sigma^{-1}\tau \rangle$ is in $\pi_A(U_2)$. Namely, writing $u = 1 - \sqrt{2}$ for the fundamental unit in $\mathbf{Q}(\zeta_8)$, we have that $D_4 = \mathrm{Gal}(K_\nu/\mathbf{Q})$, $\nu = 1, 2$, where $K_1 = \mathbf{Q}(\zeta_8, \sqrt{u})$ and $K_2 = \mathbf{Q}(\zeta_8, \sqrt{\zeta_8 u}) = \mathbf{Q}(\zeta_8, \sqrt[4]{2})$. The discriminants of these fields satisfy $|\Delta|^{1/8} \leq 8$. Moreover these are the only two D_4-Galois extensions of $\mathbf{Q}$ ramified only at 2, and each has class number 1.

To see this, one notes that any D_4-Galois extension of $\mathbf{Q}$ ramified only at 2 must be a degree 2 Kummer extension of $\mathbf{Q}(\zeta_8)$, ramified only at the prime over 2. Hence such an extension is of the form $\mathbf{Q}(\zeta_8, \sqrt{\alpha})$, where $\alpha = \zeta_8^i u^j \pi^k$, where $\pi = 1 - \zeta_8$ generates the prime of $\mathbf{Q}(\zeta_8)$ over 2, and where i, j, k each equal 0 or 1. Examining the possibilities and ruling out the non-Galois extensions and the abelian extensions of $\mathbf{Q}$, one obtains precisely the above two fields, each of which has Galois group D_4.

To see that the class number h_ν of K_ν is 1, note that otherwise Corollary 2.10 implies that $h_\nu \geq 9$. So K_ν has an unramified abelian extension L of degree ≥ 9, which is Galois over $\mathbf{Q}$. Now since K_ν is obtained by adjoining the square root of a unit to $\mathbf{Q}(\zeta_8)$, we have that the discriminant $\Delta_{K_\nu} = \Delta_{K_\nu/\mathbf{Q}}$ satisfies $|\Delta|^{1/8} \leq 8$. Hence the discriminant Δ_L of L over $\mathbf{Q}$ satisfies $|\Delta_L|^{1/[L:\mathbf{Q}]} \leq 8$. But $[L : \mathbf{Q}] \geq 8 \cdot 9 = 72$, and for extensions of degree $N \geq 72$, Odlyzko's lower bound on $|\Delta|^{1/N}$ [Od, p. 401, Table 1] is 12.84. This is a contradiction. $\square$

By the above, we obtain

Proposition 2.15. *The 2-groups of order ≤ 8 in $\pi_A(U_2)$ are precisely the groups* 1, $\mathbf{Z}/2$, $\mathbf{Z}/2 \times \mathbf{Z}/2$, $\mathbf{Z}/4$, $\mathbf{Z}/8$, $\mathbf{Z}/4 \times \mathbf{Z}/2$, *and the dihedral group* D_4. *Moreover, all of the corresponding field extensions of $\mathbf{Q}$ have class number 1.*

Proof. By Example 2.14, the dihedral group D_4 is in $\pi_A(U_2)$, and the two corresponding field extensions K_1 and K_2 each have class number 1. Also, as noted after Proposition 2.12, the quaternion group Q does not occur in $\pi_A(U_2)$. Since the abelian groups in $\pi_A(U_2)$ are precisely the groups of the form $\mathbf{Z}/2^n$ or $\mathbf{Z}/2^n \times \mathbf{Z}/2$ for $n \geq 0$, corresponding the the quotients of $\mathrm{Gal}\big(\mathbf{Q}(\zeta_{2^{n+1}})/\mathbf{Q}\big)$, the abelian groups that occur are those listed. Each of the corresponding fields is thus a subfield of $\mathbf{Q}(\zeta_{32})$, which has class number 1. If a subfield $K \subset \mathbf{Q}(\zeta_{32})$ has a class number bigger than 1, then by 2.8(b) there is a non-trivial unramified Galois extension $K \subset L$ having odd degree. Thus L and $\mathbf{Q}(\zeta_{32})$ are linearly disjoint over K, and so the compositum $L(\zeta_{32})$ is a non-trivial unramified extension of $\mathbf{Q}(\zeta_{32})$. This is a contradiction. $\square$

In degree 16, a more delicate use of Odlyzko's bounds is required, as in the following example (which is needed in Theorem 2.25 below):

Example 2.16. There are precisely two groups of order 16 lying in $\pi_A(U_2)$ that have D_4 as a quotient, viz. D_8 and the group $E = \langle s, t \mid s^8 = 1, t^2 = 1, tst^{-1} = s^3 \rangle$. Moreover each of these groups corresponds to exactly two non-isomorphic extensions of $\mathbf{Q}$ ramified only at 2, one of which dominates K_1 and the other of which dominates K_2 (where the fields K_ν are as in Example 2.14). These four fields L satisfy $|\Delta_L|^{1/16} \leq 16$, and they have class number 1.

To see this, we begin as in Example 2.14. Namely, any such field L is a degree 4 Kummer extension of $\mathbf{Q}(\zeta_8)$, ramified only at the prime over 2. Hence it is of the form $\mathbf{Q}(\zeta_8, \sqrt[4]{\alpha})$, with $\alpha = \zeta_8^i u^j \pi^k$, where $u = 1 - \sqrt{2}$, $\pi = 1 - \zeta_8$ and $0 \leq i, j, k \leq 3$. Ruling out the non-Galois extensions, one is left with the fields $L_1 = \mathbf{Q}(\zeta_8, \sqrt[4]{u}) = K_1(\sqrt{u_1})$, where $u_1 = \sqrt{u} \in K_1$, and having Galois group E over $\mathbf{Q}$; $L_2 = \mathbf{Q}(\zeta_8, \sqrt[4]{iu}) = K_1(\sqrt{\zeta_8 u_1})$, with group D_8 over $\mathbf{Q}$; $L_3 = \mathbf{Q}(\zeta_8, \sqrt[4]{\zeta_8 u \pi^2}) = \mathbf{Q}(\zeta_8, \sqrt[8]{2})$ containing K_2, with group E over $\mathbf{Q}$; and $L_4 = \mathbf{Q}(\zeta_8, \sqrt[4]{\zeta_8^{-1} u \pi^2})$ containing K_2, with group D_8 over $\mathbf{Q}$.

Now L_1 and L_2 are each obtained from K_1 by adjoining the square root of a unit. Also, letting $v = \pi/\omega^2 \in K_2$, where $\omega = 1 - \sqrt{\zeta_8 u} \in K_2$ generates the prime of K_2 over 2, we have that v is a unit. Writing $u_2 = \sqrt{\zeta_8 u} \in K_2$, we have that $L_3 = K_2(\sqrt{u_2 v})$ and $L_4 = K_2(\sqrt{\zeta_8^{-1} u_2 v})$, each of which is obtained from K_2 by adjoining the square root of a unit. So for $1 \leq \nu \leq 4$, the discriminant Δ_{L_ν} over $\mathbf{Q}$ satisfies $|\Delta_{L_\nu}|^{1/16} \leq 16$, and any unramified extension H of L_ν with $n = [II : \mathbf{Q}]$ satisfies $|\Delta_H|^{1/n} \leq 16$.

Now if the class number $h(L_\nu)$ is greater than 1, then the class group C of L_ν is non-cyclic, by Proposition 2.9 (using Proposition 2.15 to verify the hypotheses of 2.9) and odd (by 2.8). So either $C \cong (\mathbf{Z}/3)^2$ or else $|C| \geq 25$. In the latter case, L_ν has an abelian unramified extension that is Galois over $\mathbf{Q}$ with degree at least $16 \cdot 25 = 400$. But Odlyzko's lower bound for extensions $\mathbf{Q} \subset H$ of degree $n \geq 240$ [Od, Table 1, p.401] is $|\Delta_H|^{1/n} \geq 16.28$. This is a contradiction. Thus $C \cong (\mathbf{Z}/3)^2$.

Let H be the Hilbert class field of L_ν. Thus H is Galois over $\mathbf{Q}$, say with group Γ. Writing $G = \mathrm{Gal}(K_\nu/\mathbf{Q})$, there is an exact sequence $1 \to C \to \Gamma \to G \to 1$, which splits since C is odd and G is a 2-group. By 2.9(a), the induced map $\rho : G \to \mathrm{Aut}(C)$ is injective. But $\mathrm{Aut}(C) \cong \mathrm{GL}(2,3)$, and the highest power of 2 dividing $|\mathrm{GL}(2,3)|$ is 16, so ρ must be an isomorphism from G to a Sylow 2-subgroup of $\mathrm{GL}(2,3)$. Now there is an injection $\rho : E \to \mathrm{GL}(2,3)$, given by

$$s \mapsto \begin{pmatrix} 1 & -1 \\ 1 & 1 \end{pmatrix}, \quad t \mapsto \begin{pmatrix} 0 & 1 \\ 1 & 0 \end{pmatrix},$$

and this is an isomorphism onto a Sylow 2-subgroup S of $\mathrm{GL}(2,3)$. So there is no embedding of D_8 into $\mathrm{GL}(2,3)$, and thus L_2 and L_4 have class number 1.

It remains to show that L_ν has class number 1 for $\nu = 1, 3$. So assume otherwise. Then by the above, Γ must be isomorphic to the semidirect product of $N = (\mathbf{Z}/3)^2$ with E, where E acts on $(\mathbf{Z}/3)^2$ via the above injection. Thus $n = |\Gamma| = 144$.

Let $\wp$ be the unique prime of L_ν over 2. We claim that $\wp$ splits completely in H, into nine primes of norm 2. To see this, let $\mathcal{P}$ be a prime of H over $\wp$, let D be the decomposition group at $\mathcal{P}$, and let I be the inertia group at $\mathcal{P}$. Then I is a Sylow 2-subgroup of Γ, and after altering the choice of $\mathcal{P}$ we may assume that $I = E$ (which we regard as a subgroup of Γ, with conjugation action given via the isomorphism $E \tilde{\to} S$). Now I is normal in D, and to prove the claim it suffices to show that $D = I$, or equivalently that $D \cap N = 1$. So let $d \in D \cap N$. Since $s \in E$, the normality of $I = E$ in D implies that $dsd^{-1} \in E$; so the commutator $[d, s] \in E$. But $[d, s] = d(sd^{-1}s^{-1}) \in N$ since $d \in N$ and N is normal in Γ. Since $E \cap N = 1$ in Γ, we have that $[d, s] = 1$. But conjugation by s (i.e. multiplication by the matrix $\rho(s)$) cyclically permutes the non-identity elements of N. Thus $d = 1$. This proves the claim.

According to Odlyzko's lower bound [Wa, p.221], if real numbers $\sigma, \tilde{\sigma} > 1$ are chosen subject to two inequalities, and $H/\mathbf{Q}$ is any totally complex Galois extension of degree n, then

$$\log(|\Delta_H|) \geq n(\log(2\pi) - \psi(\sigma)) + \frac{n}{2}(2\sigma - 1)\psi'(\tilde{\sigma}) + 2Z(\sigma) + (2\sigma - 1)Z_1(\tilde{\sigma})$$
$$- \frac{2}{\sigma} - \frac{2}{\sigma - 1} - \frac{2\sigma - 1}{\tilde{\sigma}^2} - \frac{2\sigma - 1}{(\tilde{\sigma} - 1)^2}.$$

Here $\psi(s) = \Gamma'(s)/\Gamma(s)$, $Z(s) = -\zeta_H'(s)/\zeta_H(s)$, and $Z_1(s) = -Z'(s)$. Thus for real $s > 1$,

$$Z(s) = \sum_{\mathcal{P}} \frac{\log N\mathcal{P}}{N\mathcal{P}^s - 1}, \quad Z_1(s) = \sum_{\mathcal{P}} \frac{(\log N\mathcal{P})^2 N\mathcal{P}^s}{(N\mathcal{P}^s - 1)^2}.$$

Since all of the terms in the two above summations are positive, the inequality remains true if only those primes over 2 are included. In our situation, there are nine such primes, each of norm 2. So we have that $Z(\sigma) \geq 9(\log 2)/(2^\sigma - 1)$ and $Z_1(\tilde{\sigma}) \geq 9(\log 2)^2 \cdot 2^{\tilde{\sigma}}/(2^{\tilde{\sigma}} - 1)^2$. Taking $\sigma = 1.145$ and $\tilde{\sigma} = \frac{5}{6} + \frac{1}{6}\sqrt{12\sigma^2 - 5}$ (which satisfy the two required inequalities), we thus obtain that $\log|\Delta_H| \geq 404.53$ and so $|\Delta_H|^{1/144} \geq 16.59$. But $|\Delta_H|^{1/144} \leq 16$. This is a contradiction, proving that indeed the class number of L_ν is 1. $\square$

Remarks. (a) If one is willing to assume the Generalized Riemann Hypothesis, then the computations of the above example can be significantly shortened. Namely, by Corollary 2.10 and Proposition 2.15, if the class number of L_ν is greater than 1, then it is at least 9. In that case, L_ν has an abelian unramified extension H that is Galois over $\mathbf{Q}$ with degree $n \geq 16 \cdot 9 = 144$. As before, $|\Delta_H|^{1/n} \leq 16$. But under GRH, Odlyzko's lower bound for extensions $\mathbf{Q} \subset H$ of degree $n \geq 140$ (even without any information about splittings of primes) is $|\Delta_H|^{1/n} \geq 16.67$. This is a contradiction.

(b) J.-P. Serre observed that Odlyzko's method can be systematized and improved by use of Weil's "explicit formulas." This yields better lower bounds for the discriminant, either with or without assuming GRH. For example, the bound 16.28 in the above example can be replaced by 18.81. See [Se1, pp. 240-243 and p. 710] for further details and references.

By Corollary 2.7, every group in $\pi_A(U_2)$ is a quasi-2 group. This can also be seen more directly. Namely, if $G \in \pi_A(U_2)$, then $G/p(G)$ is of odd order, and so is solvable. Thus if $G/p(G)$ is non-trivial, then it has a non-trivial abelian odd quotient, corresponding to a non-trivial abelian odd extension of $\mathbf{Q}$ ramified only at 2. This is impossible, since such an extension would have to lie in some $\mathbf{Q}(\zeta_{2^n})$. So actually $G/p(G)$ is trivial, and thus G is a quasi-2-group.

Strengthening this argument, we obtain the following result (where $K_0 = \mathbf{Q}$ is the case just considered):

Proposition 2.17. *Let K be a Galois extension of $\mathbf{Q}$ ramified only over 2, and let $\mathbf{Q} \subset K_0$ be an intermediate Galois extension whose degree is a power of 2. Then either*

(i) *$\mathrm{Gal}(K/K_0)$ is a quasi-2-group; or*

(ii) *there is a non-trivial abelian unramified extension $K_0 \subset L$ of odd degree such that $L \subset K$ and L is Galois over $\mathbf{Q}$.*

Proof. Let $G = \mathrm{Gal}(K/\mathbf{Q})$ and let $N = \mathrm{Gal}(K/K_0)$. The quasi-2 part $p(N)$ of N is normal in G, since it is characteristic in the normal subgroup $N \subset G$. Replacing G and N by $G/p(N)$ and $N/p(N)$ respectively, we may assume that N has odd order, and hence is solvable. In this situation, it suffices to prove that if $K_0 \neq K$ (i.e. if (i) fails) then the extension $K_0 \subset K$ is not totally ramified. For then (ii) holds, with L taken to be the maximal abelian unramified subextension of $K_0 \subset K$.

So assume otherwise, i.e. that $K_0 \subset K$ is non-trivial, and is totally ramified. Now the extension $\mathbf{Q} \subset K_0$ is unramified away from 2 and is Galois of 2-power degree; hence it is totally ramified over 2. Thus $\mathbf{Q} \subset K$ is totally ramified over 2, with inertia group G. But then $G = P \rtimes C$, where the normal subgroup P is a 2-group and C is cyclic of order prime to 2. Here C is non-trivial, since G is not a 2-group (because $|N| = [K : K_0]$ is odd and greater than 1). So the field of invariants K^P is a non-trivial C-Galois extension of $\mathbf{Q}$ ramified only over 2, which is impossible since C is abelian and odd. $\square$

As a consequence of 2.17, various quasi-2-groups can be shown not to lie in $\pi_A(U_2)$. In particular, we have:

Example 2.18. For any positive integer n, let G be the dihedral group D_n of order $2n$. Then G is a quasi-2 group, and $G^{\mathrm{ab}} \cong \mathbf{Z}/2$ or $(\mathbf{Z}/2)^2$, but G is not in $\pi_A(U_2)$ unless n is a power of 2. Namely, if D_n is the Galois group of a Galois extension $\mathbf{Q} \subset K$ ramified only at the prime 2, then let $N \subset G$ be a cyclic normal subgroup of order n, and let L be the fixed field of N. Then L is either $\mathbf{Q}(i)$, $\mathbf{Q}(\sqrt{2})$, or $\mathbf{Q}(\sqrt{-2})$, since these are the only 2-cyclic extensions of $\mathbf{Q}$ ramified only at 2. Each of these has class number 1, so 2.17 implies that the cyclic group $N = \mathrm{Gal}(K/L)$ is a quasi-2-group. Thus n is a 2-power. $\square$

In the solvable case, 2.17 yields:

Lemma 2.19. *Under the hypotheses of Lemma 2.17, if $G = \mathrm{Gal}(K/\mathbf{Q})$ is solvable, and K_0 is the maximal 2-power subextension of K, then we may replace (i) of 2.17 by the condition that $K = K_0$ (i.e. G is a 2-group).*

Proof. Let $N = \mathrm{Gal}(K/K_0)$. Then N is the minimal normal subgroup of G whose index is a power of 2. If $K \neq K_0$, then $N \neq 1$. Since G is solvable, G has a normal subgroup $N_1 \subset N$ such that N/N_1 is of the form $(\mathbf{Z}/p)^n$ for some prime p and some $n \geq 1$. By the minimality of N, the index $(G : N_1)$ is not a power of 2, and so p is odd. Thus every 2-subgroup of N is contained in the proper subgroup N_1, and so $N = \mathrm{Gal}(K/K_0)$ is not a quasi-2-group. So (i) of 2.17 fails, and thus (ii) holds. $\square$

As a result, we obtain:

Theorem 2.20. *Let G be a solvable group in $\pi_A(U_2)$. Then either G is a 2-group of order < 16, or G has a quotient of order 16.*

Proof. The result is clear if G is a 2-group, so assume not. Let K be a G-Galois extension of $\mathbf{Q}$ ramified only at 2, let K_0 be as in Lemma 2.19, and let $N = \mathrm{Gal}(K/K_0)$. Since G is not a 2-group, $N \neq 1$. So by 2.19, K_0 has class number > 1. Since K_0 is Galois over $\mathbf{Q}$ of 2-power degree, it follows by 2.15 that $|G/N| = [K_0 : \mathbf{Q}] \geq 16$. $\square$

In the non-solvable case, there are the following examples of groups that are not in $\pi_A(U_2)$:

Example 2.21. (a) J. Top [To, Lemma 4.8.2] has shown that the alternating groups A_5 and A_6, and the symmetric groups S_5 and S_6, do not lie in $\pi_A(U_2)$. This follows from showing that there is no extension of $\mathbf{Q}$ of degree 5 or 6 that is ramified only at 2, even without a Galois hypothesis. This was shown by observing that the upper bound (*) given above shows that $v_2(\Delta) \le 11$ for degree 5, and $v_2(\Delta) \le 14$ for degree 6; and by inspecting lists of all fields of degree 5 and 6 with small discriminant. (In the case of A_5, Lenstra and Brumer have each observed that one can instead apply the upper bound (**) to the Galois extension itself, and this contradicts Odlyzko's lower bound.)

(b) The simple group $\mathrm{SL}(2,8)$ is not in $\pi_A(U_2)$. To see this, assume not and apply (**) to such an extension. Since the largest power of 2 dividing $|\mathrm{SL}(2,8)| = 504$ is 8, we have that $v_2(e) \le 3$ (dropping the last term in the exponent), and so $|\Delta|^{\frac{1}{n}} \le 2^4 = 16$. But according to [Od, p. 380, 1.11], for any extension of $\mathbf{Q}$ of degree n we have the inequality $|\Delta| \ge (21.8)^n e^{-70}$ (where in *this* formula $e = 2.718\ldots$). So in our situation, $|\Delta|^{\frac{1}{n}} \ge 21.8 \cdot e^{-70/504} = 18.973\ldots > 16$. This is a contradiction.

(c) The simple group $G = \mathrm{SL}(3,2)$ of order 168 does not lie in $\pi_A(U_2)$. For this, we again proceed by contradiction and apply (**). But now in order to get a contradiction we must proceed more carefully, and need to analyze the possible ramification indices. Since the largest power of 2 dividing $|G| = 168$ is 8, we have that $v_2(e) \le 3$. If $v_2(e) < 3$, then (**) gives $|\Delta|^{\frac{1}{n}} \le 2^3 = 8$. On the other hand if $v_2 = 3$, then each of the Sylow 2-subgroups of G lies in some inertia group as a normal subgroup. But the upper triangular matrices in G form a Sylow 2-subgroup P of G, and the normalizer $N_G(P)$ is equal to P (this being true for the subgroup of upper triangular matrices in $\mathrm{SL}(m,k)$ for any m and any field k). Thus the inertia group containing P is P itself, showing that $e = 8$. Hence in this case (i.e. when $v_2 = 3$), (**) gives that $|\Delta|^{\frac{1}{n}} \le 2^{1+3-\frac{1}{8}} = 14.672\ldots$. So in both cases, $|\Delta|^{\frac{1}{n}} \le 15$. But by [Od, Table 1, p.401], for any Galois extension of degree ≥ 160, $|\Delta|^{\frac{1}{n}} \ge 15.12$. This is a contradiction. $\square$

Remarks. (a) Example 2.21 shows that the four smallest simple groups (viz. A_5, $\mathrm{SL}(3,2)$, A_6, and $\mathrm{SL}(2,8)$) do not lie in $\pi_A(U_2)$. This raises the question of whether there are any simple groups in $\pi_A(U_2)$. It seems probable (to the author) that such groups do exist, but it may be difficult to find them.

(b) Example 2.21 provides instances of Serre's Conjecture. Namely, that conjecture implies that no group G of the form $\mathrm{SL}(2,2^m)$ can lie in $\pi_A(U_2)$. For $m = 1$ we get $G = S_3$, which by Theorem 2.20 (or by Example 2.18) does not lie in $\pi_A(U_2)$. For $m = 2$ we get $G = A_5$, and for $m = 3$ we get $G = \mathrm{SL}(2,8)$; by Example 2.21 these also do not lie in $\pi_A(U_2)$. But in fact, in a 1973 letter to Serre, J. Tate gave a proof of Serre's conjecture in the case of the prime 2. Specifically, he showed that if $G \subset \mathrm{SL}(2,2^m)$ is the Galois group of an extension $K/\mathbf{Q}$ that is ramified only at 2, then $K \subset \mathbf{Q}(\sqrt{-1}, \sqrt{2})$ and G is contained in the matrices of trace 0. His argument used the fact that a Sylow 2-subgroup of G is elementary abelian, along with class field theory and the Minkowski bound on the discriminant. Serre has observed that a similar argument shows that no $\mathrm{SL}(2, 3^m)$ lies in $\pi_A(U_3)$. See [Se1, p.710, note 2 to p.229] for further comments on Tate's argument.

By 2.6(a), there is no non-trivial Galois extension of $\mathbf{Q}$ that is tamely ramified over 2 and unramified elsewhere. In fact more is true, as Theorem 2.23 below shows. First we prove

Lemma 2.22. *If $G \in \pi_A(U_2)$ and $|G| \leq 300$ then G is solvable.*

Proof. Proceeding by induction on $|G|$, we assume the result holds for all strictly smaller groups. If N is any non-trivial normal subgroup in G, then G/N is also in $\pi_A(U_2)$, but it is smaller than G. So G/N is a solvable.

If G is not solvable, then any N as above is also non-solvable (since G/N is solvable), and so N has order ≥ 60. Since $|G| \leq 300$, we have that G/N has order ≤ 5 and so is abelian. Thus G satisfies the hypotheses of Lemma 2.5. Since $|G| \leq 300$, we have that G is isomorphic either to A_5, to S_5, or to $\mathrm{PSL}(2,7) \cong \mathrm{SL}(3,2)$. But these groups do not lie in $\pi_A(U_2)$, by Example 2.21, and this is a contradiction. $\square$

Theorem 2.23. *Let $\mathbf{Q} \subset K$ be a Galois extension ramified only over 2, with Galois group G and ramification index e. Then 16 divides e unless G is a 2-group of order < 16 (in which case the extension is totally ramified).*

Proof. We have already observed that if G is a 2-group then the extension is totally ramified. And by Theorem 2.20, if G is not a 2-group of order < 16 and G is solvable, then G has a quotient H of order 16. In this case the corresponding H-Galois subextension of K is totally ramified, and so 16 divides e. So it remains to consider the case that G is not solvable. Now by Lemma 2.22, $|G| > 300$. But for a Galois field extension $\mathbf{Q} \subset K$ of degree $n \geq 240$, [Od, Table 1, p.401] says that $|\Delta|^{\frac{1}{n}} \geq 16.28$. So by (**), we have that $16.28 < 2^{1+v_2(e)}$ and so $v_2(e) > 3$. Since $v_2(e)$ is an integer, it is at least 4, and so 16 divides e. $\square$

The number 16 in Theorem 2.23 cannot be replaced by a higher power of 2, as the following example shows:

Example 2.24. Let Γ be the semi-direct product $\mathbf{Z}/17 \rtimes (\mathbf{Z}/17)^*$, where the conjugation action of $(\mathbf{Z}/17)^*$ on $\mathbf{Z}/17$ is given by multiplication. Then $\Gamma \in \pi_A(U_2)$, corresponding to the Hilbert class field H of $\mathbf{Q}(i(\zeta_{64} + \zeta_{64}^{-1}))$. Namely, the class number of $F = \mathbf{Q}(\zeta_{64})$ is 17, and so the maximal abelian unramified extension L of F is Galois over F with group $\mathbf{Z}/17$. Now $\mathrm{Gal}(F/\mathbf{Q}) \cong \mathbf{Z}/16 \times \mathbf{Z}/2$, which has a unique subgroup V of the form $(\mathbf{Z}/2)^2$. The three subgroups of order 2 correspond to the intermediate fields $F_1 = \mathbf{Q}(\zeta_{64})^+$, $F_2 = \mathbf{Q}(\zeta_{32})$, and $F_0 = \mathbf{Q}(i(\zeta_{64} + \zeta_{64}^{-1}))$. For $i = 0, 1, 2$ let a_i be the involution in $V_i = \mathrm{Gal}(F/F_i)$. Thus $a_1 a_2 = a_0$. For $i = 1, 2$, F_i has class number 1, and so the unramified $\mathbf{Z}/17$-Galois extension $F \subset L$ does not descend to a $\mathbf{Z}/17$-Galois extension of F_i (which, if it existed, would have to be unramified; a contradiction). Now for $i = 1, 2$, $\mathrm{Gal}(L/F_i)$ is a semidirect product of the form $\mathbf{Z}/17 \rtimes \mathbf{Z}/2$, and since $F \subset L$ does not descend to F_i, this cannot be a direct product. Thus in the semi-direct product, the involution a_i acts on $\mathrm{Gal}(L/F) \cong \mathbf{Z}/17$ non-trivially, and hence as multiplication by -1 (for $i = 1, 2$).

Thus $a_0 = a_1 a_2$ acts trivially on $\mathrm{Gal}(L/F)$. So $\mathrm{Gal}(L/F_0) \cong \mathbf{Z}/17 \times \mathbf{Z}/2$, and the subfield L_0 of L corresponding to the subgroup $\mathbf{Z}/2$ is then a $\mathbf{Z}/17$-Galois unramified extension of F_0. In fact, it is the maximal abelian unramified extension of F_0, and so it is Galois over $\mathbf{Q}$, with group $\mathbf{Z}/17 \rtimes \mathrm{Gal}(F_0/\mathbf{Q})$. The conjugation action of this semi-direct product is given by a homomorphism $\alpha : \mathrm{Gal}(F_0/\mathbf{Q}) \to \mathrm{Aut}(\mathbf{Z}/17)$, and the kernel of α is trivial since the $\mathbf{Z}/17$-Galois extension $F_0 \subset L_0$ does not descend further (by Theorem 2.23). Thus α is an isomorphism, showing that the extension is as asserted. $\square$

In fact, the Γ-Galois extension $\mathbf{Q} \subset H$ of Example 2.24 is the smallest non-2-group extension of $\mathbf{Q}$ with ramification only at 2:

Theorem 2.25. *The group* $\Gamma = \mathbf{Z}/17 \rtimes (\mathbf{Z}/17)^*$*, of order 272, is the smallest non-2-group in* $\pi_A(U_2)$*, and there is a unique extension of* $\mathbf{Q}$ *with this degree that is ramified only at 2.*

Proof. By 2.24, the group Γ is in $\pi_A(U_2)$, corresponding to the Hilbert class field H of $\mathbf{Q}(i(\zeta_{64} + \zeta_{64}^{-1}))$. Let G be the smallest non-2-group in $\pi_A(U_2)$, corresponding to a field extension $\mathbf{Q} \subset L$. We wish to show that $G \cong \Gamma$, and that $L = H$.

Since G is smallest, we have that $|G| \leq 272$, and so G is solvable by Lemma 2.22. Let N be the minimal normal subgroup of 2-power index in G. Thus $N \neq 1$, since $|G|$ is not a power of 2. By minimality of G, the subgroup N is minimal among all the non-trivial normal subgroups of G; hence N is an elementary abelian p-group $(\mathbf{Z}/p)^\nu$, for some odd prime p. By Theorem 2.20, $\overline{G} = G/N$ has order ≥ 16. Let $K = L^N$. Thus $\mathbf{Q} \subset K$ is $\overline{G}$-Galois and is ramified only over 2. Since $|G| \leq 16 \cdot 17$, we have that $|N| \leq 17$.

We claim that the extension $K \subset L$ is unramified. For if not, let $K \subset L_0$ be the maximal unramified intermediate extension. This extension is Galois, say with group G/N_0, where $N_0 \neq 1$. If N_0 is strictly contained in N, then the index $(G : N_0)$ is not a power of 2 (by definition of N), and so G/N_0 is a non-2-group in $\pi_A(U_2)$ which is strictly smaller than G (since $N_0 \neq 1$). This is a contradiction, proving that $N_0 = N$. Thus the extension $K \subset L$ is totally ramified, with inertia group N. Since N has no non-trivial 2-power quotients (by definition of N), this inertia group must be tame, and hence cyclic of odd order. This contradicts Proposition 2.17, proving the claim.

Next, we show that the homomorphism $\rho : \overline{G} \to \mathrm{Out}(N) = \mathrm{Aut}(N)$, induced by the exact sequence $1 \to N \to G \to \overline{G} \to 1$, is injective. Let I be an inertia group of $L/\mathbf{Q}$ over 2. Since $K \subset L$ is unramified, and since $\mathbf{Q} \subset K$ is totally ramified over 2 by 2.8(a), it follows that the homomorphism $G \to \overline{G}$ maps I isomorphically onto $\overline{G}$. This induces a splitting of the exact sequence, and allows us to regard $\overline{G} \subset G$. Let $N_1 = \ker(\rho) \subset \overline{G} \subset G$. Then the elements of N_1 commute with those of N, and N_1 is normal in $\overline{G}$; so N_1 is normal in G, of 2-power order. Thus G/N_1 is a non-2-group in $\pi_A(U_2)$. The minimality of G implies that $N_1 = 1$, as desired.

In order to prove that $G \cong \Gamma$, it suffices to show that $\nu = 1$. For if this is shown, then N is cyclic of order p, so that $\mathrm{Out}(N) = \mathrm{Aut}(N) \cong (\mathbf{Z}/p)^*$ has $p-1$ elements. But $\rho : \overline{G} \to \mathrm{Out}(N)$ is injective, and so $p - 1 \geq 16$; i.e. $|N| = p \geq 17$. But $|N| \leq 17$, and so $p = 17$ and ρ is an isomorphism. Thus $G \cong \Gamma$, as desired.

As shown above, the Galois extension $K \subset L$ is unramified, and has group N. If $\nu > 1$, then $17 \geq |N| \geq p^2$, and so $p = 3$ and $\nu = 2$. Thus $N \cong (\mathbf{Z}/3)^2$, and so

$\text{Out}(N) = \text{Aut}(N) \cong \text{GL}(2,3)$, a group of order $48 = 16 \cdot 3$. Since $\rho : \overline{G} \to \text{Out}(N)$ is injective, and since $\overline{G}$ is a 2-group of order ≥ 16, it follows that $|\overline{G}| = 16$ and ρ defines an isomorphism between $\overline{G}$ and a Sylow 2-subgroup of $\text{GL}(2,3)$. That is, $\overline{G}$ is isomorphic to the group $E = \langle s, t \mid s^8 = 1, t^2 = 1, tst^{-1} = s^3 \rangle$ (cf. Example 2.16). Since K is E-Galois over $\mathbf{Q}$ with ramification only over 2, Example 2.16 says that K has class number 1. But L is an unramified Galois extension of K with group $(\mathbf{Z}/3)^2$. This is a contradiction, and so indeed $\nu = 1$.

Thus $G \cong \Gamma$. As above, L is a 17-cyclic unramified extension of K, which is a 16-cyclic extension of $\mathbf{Q}$ ramified only at 2. But the only such degree 16 extensions of $\mathbf{Q}$ are $\mathbf{Q}(i(\zeta_{64} + \zeta_{64}^{-1}))$ and $\mathbf{Q}(\zeta_{64})^+$. The former has class number 17 and the latter has class number 1. So L must be the Hilbert class field of the former. This proves uniqueness. $\square$

We conclude by returning to the problem of studying $\pi_1^{\text{tr}}(U_2)$ (cf. Theorem 2.11(b)).

Remark. In terms of the parallel between the arithmetic and geometric situations, $\pi_1^{\text{tr}}(U_2)$ may be regarded as an analog of $\pi_1(\mathbf{P}^1 - \{\xi\})$, where ξ is a single point, since 2 is the rational prime of minimal possible degree, and since number theorists traditionally regard totally real number fields as being the ones that are unramified over the prime at infinity. On the other hand, this standard interpretation of totally real fields is a bit arbitrary, since the extension of local fields at infinity could instead be viewed as arising from "an extension of residue fields," rather than from ramification. Indeed, there are reasons for this alternative interpretation [Ha1, section 2, esp. remark after Proposition 2.5]. Under this view, totally real fields ramified only over 2 correspond to Galois covers of $\mathbf{P}_k^1$ that are ramified only over $(x = 0)$, and have a k-rational point over the base point $(x = \infty)$.

Proposition 2.26. *Let G be a non-cyclic solvable group in $\pi_A^{\text{tr}}(U_2)$, corresponding to a totally real field K. Let K_0 be the maximal subfield of K of the form $\mathbf{Q}(\zeta_{2^n})^+$. Then K contains a subfield L which is unramified over K_0 of odd degree > 1.*

Proof. By 2.11(b), the fields of the form $\mathbf{Q}(\zeta_{2^m})^+$ are the only totally real 2-power Galois extensions of $\mathbf{Q}$ ramified only at 2. So K_0 is the maximal subfield of K that is Galois over $\mathbf{Q}$ and whose Galois group over $\mathbf{Q}$ is a 2-group. Also, $K \neq K_0$ since G is not cyclic. So the result follows from 2.19. $\square$

Corollary 2.27. *Let G be a non-cyclic solvable group in $\pi_A^{\text{tr}}(U_2)$. Then G has a cyclic quotient of order 32 (or 64, assuming the Generalized Riemann Hypothesis).*

Proof. Let K and K_0 be as in Proposition 2.26. Then the class number of $\mathbf{Q}(\zeta_{2^n})^+$ is greater than 1. But the class number of $\mathbf{Q}(\zeta_{2^n})^+$ is known to be 1 for $n \leq 6$ (or even $n \leq 7$, assuming GRH). So in K_0, $n > 6$ (resp. $n > 7$), and the conclusion follows. $\square$

Actually, there is no value of n for which $\mathbf{Q}(\zeta_{2^n})^+$ is known to have class number greater than 1. And as H.W. Lenstra and L. Washington have speculated to the author, the ideas of [CL] suggest the possibility that all of these fields might have class number 1. Also, no non-abelian simple groups are known to lie in $\pi_A^{\text{tr}}(U_2)$ (or, for that matter, in $\pi_A(U_2)$; cf. the comment after Example 2.21). Indeed, the only groups known to lie in $\pi_A^{\text{tr}}(U_2)$ are the cyclic 2-groups. So we ask the following

Question 2.28. (a) Is every $h(\mathbf{Q}(\zeta_{2^n})^+) = 1$? Equivalently (by Prop. 2.26), is $(\pi_1^{\mathrm{tr}})^{\mathrm{solv}}(U_2) = \mathbf{Z}_2$?

(b) Is $\pi_1^{\mathrm{tr}}(U_2) = \mathbf{Z}_2$?

Addendum (added April 26, 1994). G. Malle has pointed out to me that an affirmative answer to Question 2.13 follows from a result of H. Markscheitis [Ma]. In fact, [Ma] showed that $\pi_1^2(U_2)$ is the pro-2-group on two generators a, b subject only to the relation $b^2 = 1$. So by the Burnside Basis Theorem, a finite 2-group G is in $\pi_A(U_2)$ if and only if G/F has two generators, one of them of order ≤ 2. (Here F is the Frattini subgroup of G.) Since G/F is abelian, this condition will be satisfied if G is generated by an element g together with the conjugates of an involution i. Hence the condition in 2.12(a) indeed implies that a 2-group G is in $\pi_A(U_2)$. Cf. also [Ko].

REFERENCES

[Ab1] S. Abhyankar, *Coverings of algebraic curves*, Amer. J. Math. **79** (1957), 825–856.

[Ab2] S. Abhyankar, *Galois theory on the line in nonzero characteristic*, Bull. Amer. Math. Soc. (New Ser.) **27** (1992), 68–133.

[CL] H. Cohen and H.W. Lenstra, *Heuristics on class groups of number fields*, Number Theory: Noordwijkerhout 1983 (H. Jager, eds.), Lecture Notes in Math., vol. 1068, Springer-Verlag, Berlin-Heidelberg-New York, pp. 33–62.

[CH] K. Coombes and D. Harbater, *Hurwitz families and arithmetic Galois groups*, Duke Math. J. **52** (1985), 821–839.

[De] E. Dew, *Fields of definition of arithmetic Galois covers*, Ph.D. thesis, University of Pennsylvania, 1991.

[Gr] A. Grothendieck, *Revêtements étales et groupe fondamental*, SGA 1, Lecture Notes in Math., vol. 224, Springer-Verlag, Berlin-Heidelberg-New York, 1971.

[Ha1] D. Harbater, *Arithmetic discriminants and horizontal intersections*, Math. Annalen **291** (1991), 705–724.

[Ha2] D. Harbater, *Formal patching and adding branch points*, Amer. J. Math. **115** (1993), 487–508.

[Ha3] D. Harbater, *Abhyankar's conjecture on Galois groups over curves*, 1993 preprint, Inventiones Math. (to appear).

[Ka] T. Kambayashi, *Nori's construction of Galois coverings in positive characteristics*, Algebraic and Topological Theories, Tokyo, 1985, pp. 640–647.

[Ko] H. Koch, *l-Erweiterungen mit vorgegebenen Verzweigungsstellen*, J. reine angew. Math. **219** (1965), 30–61.

[Ma] H. Markscheitis, *On p-extensions with one critical prime*, Izvestija Akad. Nauk SSSR, Ser. Mat. **27** (1963), 463–466 (Russian).

[Na] H. Nakamura, *Galois rigidity of the étale fundamental groups of punctured projective lines*, J. reine angew. Math. **411** (1990), 205–216.

[Od] A.M. Odlyzko, *On conductors and discriminants*, Algebraic Number Fields (A. Fröhlich, eds.), Durham Symposium, 1975, Academic Press, London, 1977, pp. 377–407.

[Ra] M. Raynaud, *Revêtements de la droite affine en caractéristique p > 0 et conjecture d' Abhyankar*, Inventiones Math. **116** (1994), 425–462.

[Se1] J.-P. Serre, *Oeuvres: Collected Papers, Volume III*, Springer-Verlag, Berlin-Heidelberg-New York-Tokyo, 1986.

[Se2] J.-P. Serre, *Construction de revêtements étales de la droite affine en caractéristique p*, Comptes Rendus **311** (1990), 341–346.

[Sh] I.R. Shafarevich, *Algebraic number fields*, Proc. Intl. Congr. Math., Stockholm, 1962, Inst. Mittag-Leffler, Djursholm, 1963, pp. 163–176 (also in I.R. Shafarevich, Collected Mathematical Papers, Springer-Verlag, Berlin-Heidelberg-New York, 1989, pp. 283–294).

[To]   J. Top, *Hecke L-series related with algebraic cycles or with Siegel modular forms*, Ph.D. thesis, Utrecht, 1989.

[Wa]   L. Washington, *Introduction to Cyclotomic Fields*, GTM, vol. 83, Springer-Verlag, New York, 1982.

DEPARTMENT OF MATHEMATICS, UNIVERSITY OF PENNSYLVANIA, PHILADELPHIA, PA 19104-6395

E-mail address: harbater@rademacher.math.upenn.edu

Contemporary Mathematics
Volume **174**, 1994

Note on the Zeros of p-Adic L-Functions

TAUNO METSÄNKYLÄ

ABSTRACT. Let χ be a Dirichlet character involving a nonprincipal "second factor". A criterion for the p-adic L-function $L_p(s, \chi)$ to have a zero is presented.

This note is a supplement to R. Ernvall's and the author's computational study [2], [3] of the zeros of the Kubota-Leopoldt p-adic L-functions $L_p(s, \chi)$. Let the prime p be odd and assume that the character χ has a nonprincipal "second factor" ψ_n, i.e., a character of order p^n and conductor p^{n+1}, where $n \geq 1$. Write $\chi = \theta\psi_n$, where θ is an even character whose conductor is not divisible by p^2. For $n = 1$ there is a method enabling one to find characters $\chi = \theta\psi_1$ such that $L_p(s, \chi)$ has a zero (see §3 of [2] and §2 of [3]). The present paper shows that this method cannot be extended to $n > 1$. In fact, we will give some evidence suggesting that the zeros of $L_p(s, \theta\psi_n)$ may be very rare when n is large.

Let v_p denote the exponential valuation (with $v_p(p) = 1$) on $\mathbb{C}_p$, the completion of an algebraic closure of the p-adic field $\mathbb{Q}_p$. Denote by $\mathcal{O}_\theta$ the ring of integers in the extension of $\mathbb{Q}_p$ within $\mathbb{C}_p$ generated by the values of θ. We assume that θ is nonprincipal, for $L_p(s, \chi)$ has no zeros otherwise. The conductor of θ is of the form d or dp, where d is prime to p. Let

$$f_\theta(T) \in \mathcal{O}_\theta[[T]]$$

be the Iwasawa power series for $L_p(s, \chi)$, satisfying the equation

$$L_p(s, \chi) = f_\theta(\rho_n(1 + dp)^s - 1),$$

1991 *Mathematics Subject Classification.* Primary 11S40, 11R20, 11R23; Secondary 11Y40.
This paper is in final form and no version of it will be submitted for publication elsewhere

where ρ_n is a primitive p^nth root of 1, $\rho_n = \psi_n(1+dp)^{-1}$. The function $L_p(s, \chi)$ is analytic on the disc $D_s = \{s \in \mathbb{C}_p : v_p(s) > -1 + \frac{1}{p-1}\}$. Similarly, $f_\theta(T)$ is an analytic function of T on $D_T = \{T \in \mathbb{C}_p : v_p(T) > 0\}$. This function has λ_θ zeros in D_T (counting multiplicities), where λ_θ is the λ-invariant of the power series $f_\theta(T)$.

The mapping

$$\kappa_\chi : D_s \rightarrow D_T, \qquad \kappa_\chi(s) = \rho_n(1+dp)^s - 1$$

sends any zero s_0 of $L_p(s, \chi)$ to a zero T_0 of $f_\theta(T)$. Conversely, given a zero T_0 of $f_\theta(T)$, one may ask whether or not the following statement holds true:

Statement (Z). There exists, for some character $\chi = \theta\psi_n$, a zero $s_0 \in D_s$ of $L_p(s, \chi)$ such that $\kappa_\chi(s_0) = T_0$.

We point out that κ_χ maps D_s into the circle

$$C_n = \{T \in \mathbb{C}_p : v_p(T) = 1/\phi(p^n)\},$$

where ϕ denotes Euler's totient function (see [2]). Hence a necessary condition for Statement (Z) to be true is that $T_0 \in C_n$. The following theorem collects results proved in [1] (Theorems 4 and 5) and [2] (Propositions 3 and 4). Observe, however, that the inequality (1) now appears in a slightly revised role.

Let exp and log denote the p-adic exponential and logarithm function, respectively. As above, let n be fixed, $n \geq 1$.

THEOREM 1. *Let $T_0 \in C_n$ be a zero of $f_\theta(T)$.*

(i) *Statement* (Z) *is equivalent to each of the following two conditions:*

(Z1) $v_p(\log(1 + T_0)) > \frac{1}{p-1}$;

(Z2) *there exists a primitive p^nth root of 1, say ζ_{p^n}, such that*

(1)
$$v_p(1 + T_0 - \zeta_{p^n}) > \frac{1}{p-1}.$$

(ii) *If* (Z) *is true, then*

$$s_0 = \frac{\log(1 + T_0)}{\log(1 + dp)}$$

and

(2)
$$\psi_n(1 + dp) = \zeta_{p^n}^{-1},$$

where ζ_{p^n} is defined by (Z2). *Moreover, ζ_{p^n} and ψ_n, determined by* (1) *and* (2), *respectively, are unique, and $\zeta_{p^n} \in \mathbb{Q}_p(T_0)$.*

The next result enables us to modify (Z2) to obtain more precise information.

THEOREM 2. *Let ζ_{p^n} denote a fixed primitive p^nth root of 1.*

(i) *If T_0 is any point in C_n satisfying (Z1), then there is $c \in \mathbb{Z}$ such that*

(3) $$0 \leq c < p^{n-1}, \qquad v_p(1 + T_0 - \zeta_{p^n}^c) \geq \frac{1}{p-1}.$$

Moreover, c is prime to p if $n > 1$.

(ii) *If T_0 is any element with $v_p(T_0) > 0$ of the cyclotomic field $\mathbb{Q}_p(\zeta_{p^n})$ and if there is $c \in \mathbb{Z}$ satisfying (3), then (Z1) holds.*

PROOF. (i) Put $s_0 = \frac{\log(1+T_0)}{\log(1+p)}$. Then $s_0 \in D_s$, by (Z1), and

(4) $$1 + T_0 = \zeta(1 + p)^{s_0},$$

where ζ is a root of 1. Since

(5) $$v_p(1 + T_0 - \zeta) = v_p(\exp(s_0 \log(1 + p)) - 1) = v_p(s_0 \log(1 + p)) > \frac{1}{p-1},$$

we have $v_p(1 - \zeta) = v_p(T_0) = 1/\phi(p^n)$. Hence, ζ is a primitive p^nth root of 1, $\zeta = \zeta_{p^n}^k$ with $0 < k < p^n$, k prime to p.

Write $k = bp^{n-1} + c$ with $0 \leq b < p$, $0 \leq c < p^{n-1}$ (if $n > 1$, then $(c, p) = 1$). Then $\zeta_{p^n}^k = \mu_p^b \zeta_{p^n}^c$, where μ_p is a primitive pth root of 1. Letting $T_0' = (1+p)^{s_0} - 1$ we compute, in view of (4), that

$$1 + T_0 - \zeta_{p^n}^c = \zeta_{p^n}^c(\mu_p^b(1+p)^{s_0} - 1) = \zeta_{p^n}^c(\mu_p^b - 1 + \mu_p^b T_0').$$

Since $v_p(T_0') > \frac{1}{p-1}$ (cf. (5)), we see that $v_p(1 + T_0 - \zeta_{p^n}^c) \geq \frac{1}{p-1}$.

(ii) Consider the prime element $\pi = 1 - \zeta_{p^n}$ in the ring of integers, say $\mathcal{O}$, of the field $\mathbb{Q}_p(\zeta_{p^n})$. Our assumption implies that

$$1 + T_0 - \zeta_{p^n}^c = \beta \pi^{p^{n-1}},$$

where $\beta \in \mathcal{O}$. Consequently,

$$\log(1 + T_0) = \log(1 + \zeta_{p^n}^{-c}\beta\pi^{p^{n-1}}) = \log(1 + \alpha\pi^{p^{n-1}})$$

with $\alpha \in \mathcal{O}$. We may choose $a \in \mathbb{Z}$ so that $\alpha \equiv a \pmod{\pi}$.

It is known that $p = \epsilon\pi^{\phi(p^n)}$, where ϵ is a unit of $\mathcal{O}$, $\epsilon \equiv -1 \pmod{\pi}$ (cf. [2], proof of Proposition 5(ii)). Thus we obtain

$$\log(1 + T_0) \equiv \alpha\pi^{p^{n-1}} + \frac{\alpha^p \pi^{p^n}}{p} \equiv (\alpha + \frac{\alpha^p}{\epsilon})\pi^{p^{n-1}}$$

$$\equiv (a - a^p)\pi^{p^{n-1}} \equiv 0 \pmod{\pi^{p^{n-1}+1}}.$$

This gives us (Z1).

Theorem 2 answers negatively the question, posed in [2] at the end of §3, whether the conditions

(6) $$f_\theta(T_0) = 0, \qquad T_0 \in \mathbb{Q}_p(\zeta_{p^n}), \qquad v_p(T_0) = 1/\phi(p^n),$$

for $n > 1$, are sufficient for Statement (Z) to be true. Indeed, under these conditions it follows from Theorem 2 that (Z1) (and, thus, (Z)) holds if and only if

$$1 + T_0 \equiv (1 - \pi)^c \pmod{\pi^{p^{n-1}}}$$

for some $c \in \mathbb{Z}$, $0 < c < p^{n-1}$, $(c, p) = 1$.

Suppose that the numbers T_0 satisfying (6) are randomly distributed, as regards the coefficients in their π-adic expansions (this naïve hypothesis is supported by some computational evidence). Then the result above shows that the probability for the existence of the corresponding zeros s_0 quickly becomes small as n increases.

In [3] we found several cases of such zeros s_0 for $n = 1$ but none for $n > 1$. There is even a possibility that the functions $L_p(s, \theta\psi_n)$ possess no zeros for $n > 1$.

REFERENCES

1. N. Childress and R. Gold, *Zeros of p-adic L-functions*, Acta Arith. **48** (1987), 63–71.
2. R. Ernvall and T. Metsänkylä, *Computation of the zeros of p-adic L-functions*, Math. Comp. **58** (1992), 815–830; *Supplement*, S37–S53.
3. ———, *Computation of the zeros of p-adic L-functions, II*, Math. Comp. (to appear).

DEPARTMENT OF MATHEMATICS, UNIVERSITY OF TURKU, FIN-20500 TURKU, FINLAND
E-mail address: taumets@sara.cc.utu.fi

Contemporary Mathematics
Volume **174**, 1994

La fonction L p-adique de Kubota-Leopoldt

BERNADETTE PERRIN-RIOU

RÉSUMÉ. Depuis sa construction jusqu'à la démonstration de la "conjecture principale" par Mazur-Wiles puis par Kolyvagin et Rubin, la fonction de Kubota-Leopoldt a longuement été étudiée. Plus généralement, parallèlement aux fonctions L complexes des motifs et aux conjectures de Bloch-Kato sur leurs valeurs, on peut essayer de comprendre ce que sont les fonctions L p-adiques des motifs (ou des représentations p-adiques géométriques), comment exprimer les conjectures sur leurs valeurs spéciales, les conjectures principales ([**Pa**], [**Pb**]). Il est alors naturel de voir ce que cette étude générale donne sur la fonction de Kubota-Leopoldt, c'est ce que j'ai voulu faire dans la note qui suit. J'y reprends ce qui est connu sur la fonction L p-adique attachée à la représentation de Tate $\mathbb{Q}_p(1)$ et à ses "twists" en utilisant les résultats de [**Pa**].

PLAN

1. - Rappels.
2. - Quelques formules sur $\mathcal{L}_\omega$.
    2.1. - Valeur de $\mathcal{L}_\omega$ en $\chi^{-j}\eta$ pour $j > 1$
    2.2. - Valeur de $\mathcal{L}_\omega$ en $\chi^{-j}\eta$ pour $j < 0$
    2.3. - Valeur de $\mathcal{L}_\omega$ en $\chi^{-1}\eta$ et en η
3. - Fonction de Kubota-Leopoldt.
    3.1. - Définition.
    3.2. - Valeurs de $\mathcal{L}_{K-L}$ en $\chi^{-j}\eta$
    3.3. - "Conjecture" principale.
4. - Cohomologie galoisienne.
    4.1. - Notations.
    4.2. - Cohomologie galoisienne et dimensions
    4.3. - Démonstrations par la méthode de Kolyvagin
5. - Conjectures sur les valeurs spéciales.

1991 *Mathematics Subject Classification.* 11E95, 11G40, 11R23, 11R42.
Key words and phrases. théorie d'Iwasawa, cyclotomique, représentations p-adiques.
This paper is in final form and no version of it will be submitted for publication elsewhere.

1. Rappels

1.1. On fixe une clôture algébrique $\overline{\mathbb{Q}}$ de $\mathbb{Q}$ et on note $G_{\mathbb{Q}} = \mathrm{Gal}(\overline{\mathbb{Q}}/\mathbb{Q})$. Soit p un nombre premier impair et m un entier premier à p. Dans la suite, F désigne une extension finie abélienne de $\mathbb{Q}$ de conducteur m et d'ordre premier à p et on pose $G = \mathrm{Gal}(F/\mathbb{Q})$. On note $F_{\infty} = F(\mu_{p^{\infty}})$, $F_n = F(\mu_{p^{n+1}})$ où $\mu_{p^{n+1}}$ est le groupe des racines p^{n+1}-ièmes de l'unité. Ainsi, $\mathbb{Q}_{\infty}$ est la $\mathbb{Z}_p^{\times}$-extension cyclotomique de $\mathbb{Q}$, $F_{\infty} = F\mathbb{Q}_{\infty}$, $G_{\infty} = \mathrm{Gal}(F_{\infty}/F) \cong \mathrm{Gal}(\mathbb{Q}_{\infty}/\mathbb{Q})$ et $\mathrm{Gal}(F_{\infty}/\mathbb{Q}) = G_{\infty} \times G$. On pose $\Lambda = \mathbb{Z}_p[[G_{\infty}]]$, $\Lambda[G] = \mathbb{Z}_p[[\mathrm{Gal}(F_{\infty}/\mathbb{Q})]] = \mathbb{Z}_p[G] \otimes \Lambda$. Soit $\mathbb{Q}_p(1) = \mathbb{Q}_p \otimes_{\mathbb{Z}_p} \varprojlim_n \mu_{p^{n+1}}$ et χ le caractère cyclotomique de $G_{\mathbb{Q}}$: si ζ est une racine de l'unité d'ordre une puissance de p et si $\tau \in G_{\mathbb{Q}}$, $\tau\zeta = \zeta^{\chi(\tau)}$. Pour tout entier j, on note $\mathbb{Q}_p(j)$ la représentation abélienne de $G_{\mathbb{Q}}$ de caractère χ^j. On choisit un système compatible de racines p^{n+1}-ièmes de l'unité $\epsilon = (\zeta_n)$ qui soit une base de $\mathbb{Z}_p(1)$, ainsi ζ_0 est d'ordre p, et une racine de l'unité α d'ordre m.

Si ξ est un caractère de G, on note $\mathbb{Q}(\xi)$ le corps des valeurs de ξ et $\mathbb{Z}[\xi]$ son anneau d'entiers. Si M est un $\mathbb{Z}_p[G]$-module, on note

$$M^{(\xi)} = \{x \in M \otimes \mathbb{Z}_p[\xi] \quad \text{tel} \quad \text{que} \quad \tau x = \xi(\tau)x\}$$

Si ρ est un caractère (abélien) de $G_{\mathbb{Q}}$, on note $\epsilon(\rho) \in \{\pm\}$ le signe de $\rho(c)$ où c est une conjugaison complexe de $G_{\mathbb{Q}}$. On note $\mathbf{1}$ le caractère trivial. Enfin, si ρ est un caractère d'ordre fini de $G_{\mathbb{Q}}$ de conducteur M et si β est une racine de l'unité d'ordre M, on pose

$$G(\rho, \beta) = \sum_{\tau \in \mathrm{Gal}(\mathbb{Q}(\mu_M)/\mathbb{Q})} \rho(\tau)\tau(\beta).$$

1.2. On pose $F_{n,p} = F_n \otimes_{\mathbb{Q}} \mathbb{Q}_p$ et

$$H^i(F_{n,p}, \mathbb{Z}_p(j)) = \oplus_{v|p} H^i(F_{n,v}, \mathbb{Z}_p(j)), \quad Z_{\infty,p}^i(F, \mathbb{Z}_p(j)) = \varprojlim_n H^i(F_{n,p}, \mathbb{Z}_p(j)).$$

Par la théorie de Kummer, $Z_{\infty,p}^1(F, \mathbb{Z}_p(1))$ est isomorphe à la limite projective pour les applications normes du complété p-adique des $F_{n,p}^{\times}$. Il contient $\tilde{Z}_{\infty,p}^1(F, \mathbb{Z}_p(1)) = \varprojlim_n U_{F_{n,p}}$ où $U_{F_{n,p}}$ est le complété p-adique du groupe des unités de $F_{n,p}$. Pour j quelconque, on a

$$Z_{\infty,p}^1(F, \mathbb{Z}_p(j)) = Z_{\infty,p}^1(F, \mathbb{Z}_p(1)) \otimes \mathbb{Z}_p(j-1)$$

et on pose

$$\tilde{Z}_{\infty,p}^1(F, \mathbb{Z}_p(j)) = \tilde{Z}_{\infty,p}^1(F, \mathbb{Z}_p(1)) \otimes \mathbb{Z}_p(j-1).$$

En utilisant la dualité locale, on montre facilement que

$$Z_{\infty,p}^2(F, \mathbb{Z}_p(j)) \sim \oplus_{v|p} \mathbb{Z}_p(j-1)$$

où $M_1 \sim M_2$ signifie qu'il existe un homomorphisme $M_1 \to M_2$ à noyau et conoyau finis. On a la suite exacte de Λ-modules due à Iwasawa

$$0 \to \tilde{Z}^1_{\infty,p}(F, \mathbb{Z}_p(1)) \to Z^1_{\infty,p}(F, \mathbb{Z}_p(1)) \to \oplus_{v|p}\mathbb{Z}_p \to 0$$

ou encore, en tordant par $\mathbb{Z}_p(j-1)$

$$0 \to \tilde{Z}^1_{\infty,p}(F, \mathbb{Z}_p(j)) \to Z^1_{\infty,p}(F, \mathbb{Z}_p(j)) \to \oplus_{v|p}\mathbb{Z}_p(j-1) \to 0\,.$$

Rappelons que $Z^1_{\infty,p}(F, \mathbb{Z}_p(j))$ et $\tilde{Z}^1_{\infty,p}(F, \mathbb{Z}_p(j))$ sont des Λ-modules de rang $[F:\mathbb{Q}]$ dont le sous-module de torsion est isomorphe à $\oplus_{v|p}\mathbb{Z}_p(j)$. Ils sont de plus munis d'une action de G. Le $\Lambda[G]$-module $Z^1_{\infty,p}(F, \mathbb{Z}_p(j))$ est alors de rang 1 (composante par composante). On note $Tw : \Lambda \to \Lambda$ l'opérateur induit par $\tau \mapsto \chi(\tau)\tau$ et $Tw_{1,\mathbb{Q}_p(j)}$ l'opérateur de twist :

$$Z^1_{\infty,p}(F, \mathbb{Z}_p(j)) \cong Z^1_{\infty,p}(F, \mathbb{Z}_p(j)) \otimes \mathbb{Z}_p(1) \cong Z^1_{\infty,p}(F, \mathbb{Z}_p(j+1))$$

induit par $(x_n \mod p^{n+1}) \mapsto (x_n \otimes \zeta_n \mod p^{n+1})$ et noté aussi Tw ou $x \mapsto x \otimes \epsilon$.

1.3. Le φ-module filtré $D_p(\mathbb{Q}_p(j))$ associé à $\mathbb{Q}_p(j)$ sur $\mathbb{Q}_p$ est canoniquement $\mathbb{Q}_p$ muni d'un endomorphisme bijectif $\varphi = p^{-j}$ et de la filtration

$$\mathrm{Fil}^k D_p(\mathbb{Q}_p(j)) = \begin{cases} D_p(\mathbb{Q}_p(j)) & \text{si } k \leq -j \\ 0 & \text{si } k > -j \end{cases}.$$

On note $D_p(\mathbb{Z}_p(j))$ le réseau canonique $\mathbb{Z}_p$ de $\mathbb{Q}_p$. On note e_{-j} la base canonique de $D_p(\mathbb{Q}_p(j))$ (et de $D_p(\mathbb{Z}_p(j))$) : on a $\varphi e_{-j} = p^{-j}e_{-j}$. On a alors un isomorphisme canonique $D_p(\mathbb{Q}_p(j)) \to D_p(\mathbb{Q}_p(j+1))$ donné par $e \mapsto e \otimes e_{-1}$. On pose $D_p(\mathbb{Q}_p(j))_F = F \otimes_{\mathbb{Q}_p} D_p(\mathbb{Q}_p(j)) \cong \oplus_{v|p}F_v \otimes D_p(\mathbb{Q}_p(j))$. Si $\mathcal{O}_F$ est l'anneau des entiers de F, on pose $D_p(\mathbb{Z}_p(j))_{\mathcal{O}_F} = \mathcal{O}_F \otimes_{\mathbb{Z}_p} D_p(\mathbb{Z}_p(j))$. Soit σ l'endomorphisme de Frobenius absolu : il agit sur la plus grande extension abélienne non ramifiée en p par $\sigma(\beta) = \beta^p$ pour toute racine de l'unité β d'ordre premier à p. Il agit en particulier sur F ; on prolonge l'endomorphisme φ en un endomorphisme σ-semi-linéaire de $D_p(\mathbb{Q}_p(j))_F$.

Si K est une extension finie de $\mathbb{Q}_p$, soit

$$\exp_{\mathbb{Q}_p(j)} = \exp_{K,\mathbb{Q}_p(j)} : K \otimes D_p(\mathbb{Q}_p(j)) \to H^1(K, \mathbb{Q}_p(j))$$

l'exponentielle de Bloch-Kato ([**B-K90**], [**F-P91**, I]). L'image de $\exp_{K,\mathbb{Q}_p(j)}$ est par définition le $\mathbb{Q}_p$-espace vectoriel $H^1_f(K, \mathbb{Q}_p(j))$: on a

$$H^1_f(K, \mathbb{Q}_p(1)) = \mathbb{Q}_p \otimes_{\mathbb{Z}_p} U_K$$

et $\exp_{K,\mathbb{Q}_p(1)}$ est l'application exponentielle usuelle ;

$$H^1_f(K, \mathbb{Q}_p) = \mathrm{Hom}_{\mathbb{Z}_p}(\mathrm{Gal}(K^{nr}/K), \mathbb{Q}_p)$$

où K^{nr} est la plus grande extension non ramifiée de K, $H^1_f(K, \mathbb{Q}_p(j))$ est nul pour $j < 0$ et $H^1_f(K, \mathbb{Q}_p(j)) = H^1(K, \mathbb{Q}_p(j))$ pour $j > 1$. On note $H^1_f(K, \mathbb{Z}_p(j))$ l'image réciproque de $H^1_f(K, \mathbb{Q}_p(j))$ dans $H^1(K, \mathbb{Z}_p(j))$.

On pose $\tau.(1+T) = (1+T)^{\chi(\tau)}$ pour $\tau \in G_\infty$ où T est une indéterminée. On fait agir Λ sur $1+T$ par prolongement par linéarité et continuité. Si $f \in \Lambda$, soit $G(T)$ la solution dans $\mathbb{Q}_p[[T]] \otimes_{\mathbb{Q}_p} D_p(\mathbb{Q}_p(j))_F$ de l'équation $(1-\varphi)G = f.(1+T)$ où φ est l'endomorphisme σ-semi-linéaire tel que

$$\varphi(h(T) \otimes d) = h((1+T)^p - 1) \otimes \varphi(d)$$

pour $h \in \mathbb{Q}_p[[T]]$ et $d \in D_p(\mathbb{Q}_p(j))_F$. Rappelons que $\epsilon = (\zeta_n)$ est un système compatible de racines de l'unité. On pose

$$\Xi^\epsilon_{n,\mathbb{Q}_p(j)}(f) = (1 \otimes p\varphi)^{-(n+1)}(G)(\zeta_n - 1) \in \mathbb{Q}_{n,p} \otimes_{\mathbb{Q}_p} D_p(\mathbb{Q}_p(j))_F.$$

Si $\omega \in Z^1_{\infty,p}(\mathbb{Z}_p(j))$, on note ω_n son image dans $H^1(F_{n,p}, \mathbb{Z}_p(j))$; on désigne par $(\Lambda \otimes_{\mathbb{Z}_p} D_p(\mathbb{Z}_p(j))_{\mathcal{O}_F})^{\tilde\Delta_j=0}$ le noyau de $g \mapsto \chi^j(g)$ modulo l'image de $(1 - p^j\sigma)$ avec $g \in \Lambda \otimes_{\mathbb{Z}_p} D_p(\mathbb{Z}_p(j))_{\mathcal{O}_F}$.

1.4.

THÉORÈME. (A) *Pour tout $j \geq 1$, il existe un unique isomorphisme*

$$\Omega^\epsilon_{\mathbb{Q}_p(j),j,F} = \Omega^\epsilon_{\mathbb{Q}_p(j),j} :$$
$$(\Lambda \otimes_{\mathbb{Z}_p} D_p(\mathbb{Z}_p(j))_{\mathcal{O}_F})^{\tilde\Delta_j=0} \to \tilde{Z}^1_{\infty,p}(F, \mathbb{Z}_p(j))/ \oplus_{v|p} \mathbb{Z}_p(j)$$

tel que le diagramme suivant soit commutatif

$$
\begin{array}{ccc}
(\Lambda \otimes_{\mathbb{Z}_p} D_p(\mathbb{Z}_p(j))_{\mathcal{O}_F})^{\tilde\Delta_j=0} & \xrightarrow{\Omega^\epsilon_{\mathbb{Q}_p(j),j}} & \tilde{Z}^1_{\infty,p}(F, \mathbb{Z}_p(j))/ \oplus_{v|p} \mathbb{Z}_p(j) \\
\Xi^\epsilon_{n,\mathbb{Q}_p(j)} \downarrow & & \downarrow \\
F_{n,p} \otimes_{\mathbb{Z}_p} D_p(\mathbb{Z}_p(j)) & \xrightarrow{(j-1)! \exp_{F_{n,p},\mathbb{Q}_p(j)}} & H^1(F_{n,p}, \mathbb{Z}_p(j))'
\end{array}
$$

avec $H^1(F_{n,p}, \mathbb{Z}_p(j))' = H^1(F_{n,p}, \mathbb{Z}_p(j))/H^1(F_{\infty,p}/F_{n,p}, \mathbb{Z}_p(j))$ *et où l'application verticale de droite se déduit de la projection naturelle ;*

(B) *On a*

$$Tw_{1,\mathbb{Q}_p(j)} \circ \Omega^\epsilon_{\mathbb{Q}_p(j),j} \circ Tw \otimes e_1 = -\Omega^\epsilon_{\mathbb{Q}_p(j+1),j+1}.$$

La partie (A) pour $j = 1$ est due à Coleman. Le théorème complet est démontré dans [Pa]. [1]

La relation (B) est fondamentale dans la suite et permet de passer de $\mathbb{Q}_p(1)$ à $\mathbb{Q}_p(j)$.

On déduit du théorème que pour tout $j \in \mathbb{Z}$, il existe un unique isomorphisme

$$\Omega^\epsilon_{\mathbb{Q}_p(j),j} : (\Lambda \otimes_{\mathbb{Z}_p} D_p(\mathbb{Z}_p(j))_{\mathcal{O}_F})^{\tilde\Delta_j=0} \to \tilde{Z}^1_{\infty,p}(F, \mathbb{Z}_p(j))/ \oplus_{v|p} \mathbb{Z}_p(j)$$

---

[1]La formulation un peu différente de celle utilisée pour les courbes elliptiques dans [Pc] vient de ce que la représentation $V = \mathbb{Q}_p(j)$ suffit à obtenir l'existence et l'unicité de $\Omega^\epsilon_{\mathbb{Q}_p(j),j} = \Omega^\epsilon_{V,j}$ et que l'on n'a pas besoin d'agrandir Λ

telle que

$$Tw_{1,\mathbb{Q}_p(j)} \circ \Omega^{\epsilon}_{\mathbb{Q}_p(j),j} \circ Tw \otimes e_1 = -\Omega^{\epsilon}_{\mathbb{Q}_p(j+1),j+1} \,.$$

1.5. Posons $t_{\mathbb{Q}_p(j)} = D_p(\mathbb{Q}_p(j))$, $t_{\mathbb{Z}_p(j)} = D_p(\mathbb{Z}_p(j))$ si $j \geq 0$ (resp. $= 0$ si $j \leq 0$). Pour toute place $v|p$, on a une suite exacte [**B-K90**]

$$0 \to H^0(F_v, \mathbb{Q}_p(j)) \to D_p(\mathbb{Q}_p(j))_{F_v} \to D_p(\mathbb{Q}_p(j))_{F_v} \oplus F_v \otimes t_{\mathbb{Q}_p(j)}$$
$$\to H^1_f(F_v, \mathbb{Q}_p(j)) \to 0$$

où la seconde application est donnée par $x \mapsto (1 - \varphi)x \oplus x \mod \mathrm{Fil}^0$. On pose

$$\mathrm{Tam}_v(\mathbb{Z}_p(j)) = |\det(1 - p^{-j}\sigma|D_p(\mathbb{Q}_p(j))_{F_v})|_p \,[H^1_f(F_v, \mathbb{Z}_p(j)) : t_{\mathbb{Z}_p(j)}]$$

où $|\ |_p$ est la valeur absolue normalisée par $|p|_p = p^{-1}$ et où $[:]$ désigne l'indice généralisé : si M_1 et M_2 sont des $\mathbb{Z}_p$-modules de type fini tels que $\mathbb{Q}_p \otimes M_1 = \mathbb{Q}_p \otimes M_2$, on a

$$[M_1 : M_2] = \sharp(\mathrm{tors}(M_1)(\sharp(\mathrm{tors}(M_2))^{-1}[M_1' : M_1' \cap M_2']/[M_2' : M_1' \cap M_2']$$

avec $\mathrm{tors}(M_i)$ le sous-module de torsion de M_i et $M_i' = M_i/\mathrm{tors}(M_i)$ pour $i = 1, 2$. Une conséquence facile du théorème précédent (dans sa version locale) est que pour $j \geq 1$, on a

$$\mathrm{Tam}_v(\mathbb{Z}_p(j))/\mathrm{Tam}_v(\mathbb{Z}_p(1-j))\mathbb{Z}_p = (j-1)!^{-[F_v:\mathbb{Q}_p]}\mathbb{Z}_p$$

(théorème de Bloch et Kato, [**B-K90**]). L'idée de la démonstration est la suivante : posons $X = (\Lambda \otimes_{\mathbb{Z}_p} D_p(\mathbb{Z}_p(j))_{F_v})^{\tilde{\Delta}_j=0}$, $Y = \tilde{Z}^1_{\infty,v}(F, \mathbb{Z}_p(j))/\mathbb{Z}_p(j)$. On a un isomorphisme $X_{G_\infty} \cong Y_{G_\infty}$. Le $\mathbb{Z}_p$-module X_{G_∞} (resp. Y_{G_∞}) est égal à $D_p(\mathbb{Z}_p(j))_{\mathcal{O}_F}$ (resp. $H^1_f(F_v, \mathbb{Z}_p(j))$), à un groupe fini près que l'on calcule exactement, et la flèche $F_v \otimes D_p(\mathbb{Z}_p(j)) \to H^1_f(F_v, \mathbb{Q}_p(j))$ qui se déduit de $X_{G_\infty} \cong Y_{G_\infty}$ est égale à $(j-1)! \exp_{F_v,\mathbb{Q}_p(j)}$.

1.6. La loi explicite de réciprocité s'énonce de la manière suivante : on note $[,]_{D_p(\mathbb{Q}_p(j))}$ la forme bilinéaire naturelle $D_p(\mathbb{Q}_p(j))_F \times D_p(\mathbb{Q}_p(1-j))_F \to F_p$ étendue par linéarité à

$$\mathbb{Q}_{n,p} \otimes_{\mathbb{Q}_p} D_p(\mathbb{Q}_p(j))_F \times \mathbb{Q}_{n,p} \otimes_{\mathbb{Q}_p} D_p(\mathbb{Q}_p(1-j))_F$$

et à

$$\Lambda \otimes D_p(\mathbb{Q}_p(j))_F \times \Lambda \otimes D_p(\mathbb{Q}_p(1-j))_F \,.$$

On note $<, >_{n,\mathbb{Q}_p(j)}$ l'application de dualité :

$$H^1(F_{n,p}, \mathbb{Q}_p(j)) \times H^1(F_{n,p}, \mathbb{Q}_p(1-j)) \to \mathbb{Q}_p$$

(somme des accouplements locaux pour $v|p$) et

$$<, >_{\infty,\mathbb{Q}_p(j)} \colon Z^1_{\infty,p}(F, \mathbb{Z}_p(j)) \times Z^1_{\infty,p}(F, \mathbb{Z}_p(1-j)) \to \Lambda$$

l'application définie par

$$< x, y >_{\infty, \mathbb{Q}_p(j)} = \left(\sum_{\tau \in G_n} < \tau^{-1} x_n, y_n >_{n, \mathbb{Q}_p(j)} \tau \right)_n .$$

Soit ι l'automorphisme de Λ induit par $\tau \mapsto \tau^{-1}$ pour $\tau \in G_\infty$.

THÉORÈME. (Réc($\mathbb{Q}_p(j)$)) *Si f est un élément de $\Lambda \otimes_{\mathbb{Z}_p} D_p(\mathbb{Q}_p(j))_F$ et g un élément de $\Lambda \otimes_{\mathbb{Z}_p} D_p(\mathbb{Q}_p(1-j))_F$, alors*

$$(-1)^{j-1} < \Omega^\epsilon_{\mathbb{Q}_p(j),j}(f), \Omega^{\epsilon^{-1}}_{\mathbb{Q}_p(1-j),1-j}(g) >_{\infty, \mathbb{Q}_p(j)} = Tr_{F/\mathbb{Q}}([f, \iota(g)]_{D_p(\mathbb{Q}_p(j))}) .$$

Ce théorème se déduit d'un résultat de Coleman et de l'invariance par torsion de la formule ([**Pa**, §4]).

1.7. Si S est un ensemble fini de places et L une extension finie de $\mathbb{Q}$, on note $G_{S,L}$ le groupe de Galois sur L de la plus grande extension galoisienne de L non ramifiée en dehors de S. Prenons $S = \{p, \infty\}$ et $S_f = \{p\}$. On pose

$$H^i_{\infty, \{p\}}(F, \mathbb{Z}_p(j)) = \varprojlim_n H^i(G_{S,F_n}, \mathbb{Z}_p(j))$$

pour $i \in \{1, 2\}$. La conjecture

$$\text{Leop}(\mathbb{Q}_p(j)) \quad : \quad H^2(G_{S,F_\infty}, \mathbb{Q}_p(j)/\mathbb{Z}_p(j)) = 0$$

est démontrée (théorème d'Iwasawa [**Iw69**], par invariance par torsion, il suffit de démontrer que $H^2(G_{S,F_\infty}, \mathbb{Q}_p/\mathbb{Z}_p) = 0$). On montre à l'aide des formules de caractéristique d'Euler-Tate que cela implique les propriétés suivantes : si ξ est un caractère d'ordre fini de $\text{Gal}(F_0/\mathbb{Q})$, $H^1_{\infty, \{p\}}(F, \mathbb{Z}_p(j))^{(\xi)}$ est un $\Lambda[G]^{(\xi)}$-module de rang 1 si $\epsilon(\xi\chi^j) = -1$ (resp. 0 si $\epsilon(\xi\chi^j) = 1$) et $H^2_{\infty, \{p\}}(F, \mathbb{Z}_p(j))$ est un $\Lambda[G]$-module de torsion. D'autre part, le sous-module de torsion de $H^1_{\infty, \{p\}}(F, \mathbb{Z}_p(j))$ est isomorphe à $\mathbb{Z}_p(j)$.

1.8. Soit $\omega \in H^1_{\infty, \{p\}}(F, \mathbb{Z}_p)$. Si $h = \chi(\gamma)\gamma - 1 \in \Lambda$ avec γ un générateur topologique de Γ, $h\omega$ appartient à l'image de $\Omega^\epsilon_{\mathbb{Q}_p, 0}$. On pose

$$L_{\omega, F} = (\Omega^\epsilon_{\mathbb{Q}_p, 0})^{-1}(\omega) =_{\text{déf}} h^{-1}(\Omega^\epsilon_{\mathbb{Q}_p, 0})^{-1}(h\omega) \in (\chi(\gamma)\gamma - 1)^{-1}\Lambda \otimes D_p(\mathbb{Q}_p)_F .$$

Prenons $F = \mathbb{Q}(\mu_m)$ et rappelons que α est une racine de l'unité d'ordre m. En utilisant l'isomorphisme $\mathbb{Q}_p[G] \to F$ induit par $\lambda \mapsto \lambda(\alpha)$, on peut voir $L_{\omega, F}$ comme l'image d'un élément $\underline{L}_{\omega, F}$ de $\Lambda[G]$. Si ξ est un caractère de G de conducteur m, on a le diagramme commutatif

$$
\begin{array}{ccc}
\mathbb{Q}[G] & \xrightarrow{\lambda \mapsto \xi(\lambda)} & \mathbb{Q}[\xi] \\
{\scriptstyle \lambda \mapsto \lambda(\alpha)} \downarrow & & \downarrow {\scriptstyle G(\xi^{-1}, \alpha)} \\
F & \xrightarrow[x \mapsto \sum_{\tau \in G} \xi(\tau)^{-1}\tau(x)]{} & \mathbb{Q}[\xi]
\end{array}
$$

Pour tout caractère ξ de G de conducteur m et pour $\rho \in \mathrm{Hom}_{\mathrm{cont}}(G_\infty, \mathbb{C}_p^\times)$, on a donc

$$\xi\rho(\underline{L}_{\omega,F}) = G(\xi^{-1}, \alpha)^{-1} \sum_{\tau \in G} \xi(\tau)^{-1} \rho(L_{\tau(\omega),F}) \,.$$

On note encore $\mathcal{L}_\omega(\xi.)$ la fonction sur $\mathrm{Hom}_{\mathrm{cont}}(G_\infty, \mathbb{C}_p^\times)$ telle que $\mathcal{L}_\omega(\xi\rho) = \xi\rho(\underline{L}_{\omega,F})$. Lorsque ξ est le caractère trivial, la fonction $\rho \mapsto \mathcal{L}_\omega(\rho)$ a éventuellement un pôle en χ^{-1}.

2. Quelques formules sur $\mathcal{L}_\omega$

2.1. Valeur de $\mathcal{L}_\omega$ en $\chi^{-j}\eta$ pour $j > 1$.

2.1.1. Soit j un entier > 1. Pour toute extension finie K de $\mathbb{Q}_p$, l'exponentielle est un isomorphisme de $K \otimes_{\mathbb{Q}_p} D_p(\mathbb{Q}_p(j))$ sur $H^1(K, \mathbb{Q}_p(j))$. On note

$$\log_{\mathbb{Q}_p(j)} : H^1(K, \mathbb{Q}_p(j)) \to K \otimes_{\mathbb{Q}_p} D_p(\mathbb{Q}_p(j))$$

l'application réciproque. On en déduit une application que l'on note de la même manière

$$\log_{\mathbb{Q}_p(j)} : H^1(F_{n,p}, \mathbb{Q}_p(j)) \to F_{n,p} \otimes_{\mathbb{Q}_p} D_p(\mathbb{Q}_p(j)) = \mathbb{Q}_{n,p} \otimes_{\mathbb{Q}_p} D_p(\mathbb{Q}_p(j))_F \,.$$

D'autre part, si $\omega \in H^1_{\infty,\{p\}}(F, \mathbb{Z}_p)$, on pose $\omega_j = \omega \otimes \epsilon^{\otimes j}$; si ρ est un caractère de $\mathrm{Gal}(F_\infty/\mathbb{Q})$ de conducteur mp^{n+1}, on pose

$$P^{(\rho)}(\omega_j) = \sum_{\tau \in \mathrm{Gal}(F_n/\mathbb{Q})} \rho^{-1}(\tau)\tau(\omega_{j,n})$$

(avec $\omega_{j,-1} = \mathrm{Tr}_{\mathrm{Gal}(F(\mu_p)/F)}\omega_{j,0} = \mathrm{Tr}_\Delta\omega_{j,0}$). Si ρ est le caractère trivial $\mathbf{1}$, on pose $P(\omega_j) = P^{(\mathbf{1})}(\omega_j)$. Si j est un entier ≥ 0, on pose $\Gamma^*(-j) = (-1)^j(j!)^{-1}$.

2.1.2.

PROPOSITION. *Soient ξ un caractère de G de conducteur m et j un entier > 1. Alors, on a*

$$(1 - p^{-j}\xi(\sigma))^{-1}(1 - p^{j-1}\xi(\sigma)^{-1})\mathcal{L}_\omega(\xi\chi^{-j}) \otimes e_{-j}$$
$$= -\Gamma^*(-j+1)G(\xi^{-1}, \alpha)^{-1} \log_{\mathbb{Q}_p(j)} P^{(\xi)}(\omega_j) \,;$$

si de plus η est un caractère de G_∞ de conducteur p^{n+1} avec $n \geq 0$, on a

$$(p^{1-j}\xi(\sigma))^{-(n+1)}\mathcal{L}_\omega(\xi\eta\chi^{-j}) \otimes e_{-j}$$
$$= -\Gamma^*(-j+1)G(\xi^{-1}\eta^{-1}, \alpha\zeta_n)^{-1} \log_{\mathbb{Q}_p(j)} P^{(\xi\eta)}(\omega_j) \,.$$

REMARQUE. Pour $\xi = \mathbf{1}$ (et donc $F = \mathbb{Q}$, $m = 1$), on obtient donc la formule

$$(1 - p^{-j})^{-1}(1 - p^{j-1})\mathcal{L}_\omega(\chi^{-j}) \otimes e_{-j} = -\Gamma^*(-j+1)\log_{\mathbb{Q}_p(j)} P(\omega_j) \,.$$

DÉMONSTRATION. On déduit du théorème 1.4, (B) que l'on a

$$\mathcal{L}_\omega(\xi\eta\chi^{-j}) \otimes e_{-j} = (-1)^j G(\xi^{-1}, \alpha)^{-1} \sum_{\tau \in G} \xi(\tau)^{-1}\eta((\Omega^\epsilon_{\mathbb{Q}_p(j),j})^{-1}(\tau\omega_j)).$$

En utilisant le diagramme commutatif du théorème 1.4 et en remarquant que

$$(j-1)!(1-p^{-j}\sigma)\log_{\mathbb{Q}_p(j)}\mathrm{Tr}_\Delta(\omega_{j,0}) = (1-p^{j-1}\sigma^{-1})\mathbf{1}(g) \otimes e_{-j}$$

si $\Omega^\epsilon_{\mathbb{Q}_p(j),j}(g \otimes e_{-j}) = \omega_j$, on obtient que pour $j \geq 0$ et ξ un caractère de G,

$$\sum_{\tau \in G} \xi^{-1}(\tau)(\Omega^\epsilon_{\mathbb{Q}_p(j),j}(L_{\tau\omega,F} \otimes e_{-j}))_{-1}$$
$$= (-1)^j(j-1)!\exp_{\mathbb{Q}_p(j)}((1-\xi(\sigma)p^{-j})^{-1}(1-\xi(\sigma^{-1})p^{j-1})$$
$$\sum_{\tau \in G} \xi^{-1}(\tau)\chi^{-j}(L_{\tau\omega,F}) \otimes e_{-j}),$$

d'où, en prenant le logarithme, la première formule. La deuxième se démontre de la même manière : on utilise que, si η est un caractère de G_∞ de conducteur p^{n+1}, on a avec $(1-\varphi)G = L_{\omega,F}.(1+T)$

$$\sum_{\tau \in \mathrm{Gal}(F_n/F)} \eta(\tau)^{-1}(p\sigma \otimes \varphi)^{-(n+1)}\tau G(\zeta_n - 1)$$
$$= (p\sigma \otimes \varphi)^{-(n+1)}\eta(L_{\omega,F}) \sum_{\tau \in \mathrm{Gal}(F_n/F)} \eta(\tau)^{-1}\tau(\zeta_n)$$
$$= (p\sigma \otimes \varphi)^{-(n+1)}\eta(L_{\omega,F})G(\eta^{-1}, \zeta_n).$$

□

2.2. Valeur de $\mathcal{L}_\omega$ en $\chi^{-j}\eta$ pour $j < 0$.

2.2.1. Soit j un entier < 0. Pour toute extension finie K de $\mathbb{Q}_p$, on note $\lambda_{\mathbb{Q}_p(j),K}$ l'application $H^1(K, \mathbb{Q}_p(j)) \to K \otimes_{\mathbb{Q}_p} D_p(\mathbb{Q}_p(j))$ dual de l'application

$$\exp_{\mathbb{Q}_p(1-j),K} : K \otimes_{\mathbb{Q}_p} D_p(\mathbb{Q}_p(1-j)) \to H^1(K, \mathbb{Q}_p(1-j)).$$

Si $\omega_j = \omega \otimes \epsilon^{\otimes j} \in H^1_{\infty,\{p\}}(F, \mathbb{Z}_p(j))$ et si ρ est un caractère de $\mathrm{Gal}(F_n/\mathbb{Q})$, on pose

$$R^{(\rho)}_{\mathbb{Q}_p(j)}(\omega_j) = (\#\mathrm{Gal}(\mathbb{Q}_n/\mathbb{Q}))^{-1} \sum_{\tau \in \mathrm{Gal}(F_n/\mathbb{Q})} \rho(\tau)^{-1}\tau(\lambda_{\mathbb{Q}_p(j),F_{n,p}}(\omega_{j,n})).$$

2.2.2.

PROPOSITION. *Soient ξ un caractère de G de conducteur m et j un entier < 0. Alors, on a,*

$$(1 - p^{-j}\xi(\sigma))^{-1}(1 - p^{j-1}\xi(\sigma)^{-1})\mathcal{L}_\omega(\xi\chi^{-j}) \otimes e_{-j}$$
$$= -\Gamma(-j+1)G(\xi^{-1}, \alpha)^{-1}R^{(\xi)}_{\mathbb{Q}_p(j)}(\omega_j);$$

si de plus η est un caractère de G_∞ de conducteur p^{n+1} avec $n \geq 0$, on a

$$(p^{1-j}\xi(\sigma))^{-(n+1)}\mathcal{L}_\omega(\xi\eta\chi^{-j}) \otimes e_{-j} = -\Gamma(-j+1).G(\xi^{-1}\eta^{-1}, \alpha\zeta_n)^{-1}R^{(\xi\eta)}_{\mathbb{Q}_p(j)}(\omega_j).$$

REMARQUE. Pour $\xi = 1$, on obtient la formule

$$(1-p^{-j})^{-1}(1-p^{j-1})\mathcal{L}_\omega(\chi^{-j}) \otimes e_{-j} = -\Gamma(-j+1)R_{\mathbb{Q}_p(j)}(\omega_j)$$

avec

$$R_{\mathbb{Q}_p(j)}(\omega_j) = \lambda_{\mathbb{Q}_p(j),\mathbb{Q}_p}(\omega_{j,-1}).$$

DÉMONSTRATION. Faisons la démonstration pour $\eta = 1$ (le cas général se démontre de manière analogue). On a

$$\mathcal{L}_\omega(\chi^{-j}) = Tw^{-j}(\mathcal{L}_\omega)(1) = (-1)^j \mathbf{1}((\Omega^\epsilon_{\mathbb{Q}_p(j),j})^{-1}(\omega_j)).$$

Posons $L = (1-p^{j-1}\sigma^{-1})(1-p^{-j}\sigma)^{-1}\mathcal{L}_\omega(\chi^{-j})$. Si $y \in D_p(\mathbb{Q}_p(1-j))$ avec $y = \Xi^{\epsilon^{-1}}_{0,\mathbb{Q}_p(1-j)}(f)$, on a

$$(-1)^j[L, \mathrm{Tr}_\Delta(y)]_{D_p(\mathbb{Q}_p(j))} = [\mathbf{1}((\Omega^\epsilon_{\mathbb{Q}_p(j),j})^{-1}(\omega_j)), \mathbf{1}(f)]_{D_p(\mathbb{Q}_p(j))}.$$

D'où en utilisant la loi de réciprocité Réc($\mathbb{Q}_p(j)$),

$$[L, \mathrm{Tr}_\Delta(y)]_{D_p(\mathbb{Q}_p(j))}] = - < \mathrm{Tr}_\Delta(\omega_{j,0}), \Omega^{\epsilon^{-1}}_{\mathbb{Q}_p(1-j),1-j}(f)_0 >_{0,\mathbb{Q}_p(j)}$$
$$= - < \mathrm{Tr}_\Delta(\omega_{j,0}), (-j)! \exp_{\mathbb{Q}_p(1-j)}(\mathrm{Tr}_\Delta(y)) >_{-1,\mathbb{Q}_p(j)}$$
$$= -(-j)![\lambda_{\mathbb{Q}_p(j),F_p}(\mathrm{Tr}_\Delta(\omega_{j,0})), \mathrm{Tr}_\Delta(y)]_{D_p(\mathbb{Q}_p(j))}.$$

D'où

$$L = -(-j)!\lambda_{\mathbb{Q}_p(j),F_p}(\mathrm{Tr}_\Delta(\omega_j)_0)$$

En appliquant $\sum_{\tau\in G}\xi^{-1}(\tau)\tau$, on en déduit la formule. $\square$

2.3. Valeur de $\mathcal{L}_\omega$ en $\chi^{-1}\eta$ et en η.

2.3.1. Commençons par étudier la valeur de $\mathcal{L}_\omega$ en $\chi^{-1}\eta\xi$ où $\xi\eta$ est d'ordre fini. L'exponentielle n'est plus un isomorphisme. Son image est $H^1_f(K, \mathbb{Q}_p(1))$. On note

$$\log_{\mathbb{Q}_p(1)} : H^1_f(K, \mathbb{Q}_p(1)) \to K \otimes_{\mathbb{Q}_p} D_p(\mathbb{Q}_p(1))$$

l'application réciproque. Les formules 2.1.2 sont valables pour $\omega \in \tilde{Z}^1_{\infty,p}(F, \mathbb{Z}_p)$, mais les deux termes sont nuls dès que $\xi(\sigma) = 1$. Soit $\omega \in \tilde{Z}^1_{\infty,p}(F, \mathbb{Z}_p)$ et $G_{\omega\otimes\epsilon}$ la série de Coleman associée à $\omega \otimes \epsilon$: c'est un élément de $F_p[[T]]$ vérifiant

$$(1-p^{-1}\sigma)G_{\omega\otimes\epsilon} \in F \otimes \mathbb{Z}_p[[T]]$$

et tel que

$$G^{\sigma^{-(n+1)}}_{\omega\otimes\epsilon}(\zeta_n - 1) = \log_{\mathbb{Q}_p(1)}\omega_{1,n} = \log_{\mathbb{Q}_p(1)}(\omega \otimes \epsilon)_n.$$

Posons

$$S(\omega \otimes \epsilon) = G_{\omega\otimes\epsilon}(0) \otimes e_{-1} \in D_p(\mathbb{Q}_p(1))_F.$$

Si $\omega \in Z^1_{\infty,p}(F, \mathbb{Z}_p)$,

$$S(\omega \otimes \epsilon) = (\log \chi(\gamma))^{-1} S((\gamma - 1)(\omega \otimes \epsilon))$$

ne dépend pas de γ. D'autre part, posons

$$\text{Res}_{\chi^{-1}}(\mathcal{L}_\omega) = (\log \chi(\gamma))^{-1} \mathbf{1}((\chi(\gamma)\gamma - 1)^{-1} L_{\omega, \mathbb{Q}})$$

(ce qui ne dépend pas non plus de γ). C'est le résidu de $\mathcal{L}_\omega$ en χ^{-1}. Lorsque $\omega \in \tilde{Z}^1_{\infty,p}(F, \mathbb{Z}_p)$, le résidu de $\mathcal{L}_\omega$ en χ^{-1} est nul. De manière générale, on a :

PROPOSITION. *Si ξ est non trivial,*

$$(1 - p^{-1}\xi(\sigma))^{-1} \mathcal{L}_\omega(\xi\chi^{-1}) \otimes e_{-1} = -\sum_{\tau \in G} \xi(\tau)^{-1} \tau(S(\omega \otimes \epsilon)).$$

Si $\xi(\sigma) = 1$,

$$(1 - p^{-1})^{-1} Res_{\chi^{-1}}(\mathcal{L}_\omega) \otimes e_{-1} = -S(\omega \otimes \epsilon).$$

2.3.2. On vérifie facilement que si $\omega \in \tilde{Z}^1_{\infty,p}(F, \mathbb{Z}_p)$, le calcul fait en 2.2 est encore valable à condition de multiplier par $(1 - p^{-j}\xi(\sigma))$ lorsque cela est nul. On obtient donc la proposition suivante :

PROPOSITION. *On a*

$$(1 - p^{-1}\xi(\sigma)^{-1})\mathcal{L}_\omega(\xi) = -(1 - \xi(\sigma))R^{(\xi)}_{\mathbb{Q}_p}(\omega)$$

et si η est un caractère de G_∞ de conducteur p^{n+1}, on a

$$(p\xi(\sigma))^{-(n+1)} \mathcal{L}_\omega(\xi\eta) = -G(\xi^{-1}\eta^{-1}, \alpha\zeta_n)^{-1} R^{(\xi\eta)}_{\mathbb{Q}_p}(\omega).$$

3. Fonction de Kubota-Leopoldt

3.1. Définition.

3.1.1. Dans tout ce qui suit, on suppose choisi un plongement de $\overline{\mathbb{Q}}$ dans $\mathbb{C}$ et un plongement de $\overline{\mathbb{Q}}$ dans $\mathbb{C}_p$; on en déduit pour tout entier M un choix naturel d'une racine de l'unité $\exp(2i\pi/M)$ d'ordre M (par exemple, $\zeta_n = \exp(2i\pi/p^{n+1})$ et $\alpha = \exp(2i\pi/m)$). Si ρ est un caractère d'ordre fini de $G_\mathbb{Q}$ à valeurs dans $\overline{\mathbb{Q}}^\times$ de conducteur M, on note $G(\rho) = G(\rho, \exp(2i\pi/M))$.

Soit $\zeta(s)$ la fonction de Riemann : $\zeta(s) = \sum_{n>0} n^{-s}$. Si ρ est un caractère de $G_\mathbb{Q}$ d'ordre fini à valeurs dans $\overline{\mathbb{Q}}^\times$ de conducteur M et si $\tilde{\rho}$ est le caractère de Dirichlet primitif associé (défini par $\tilde{\eta}(r) = \eta(\sigma_r)$ si r est premier à M et si σ_r est un élément de $\text{Gal}(\mathbb{Q}^{ab}/\mathbb{Q})$ tel que $\sigma_r(\beta) = \beta^r$ pour β racine de l'unité d'ordre premier à M), on pose

$$L(s, \rho) = L(s, \tilde{\rho}) = \sum_{\substack{n>0 \\ (n,M)=1}} \tilde{\rho}(n) n^{-s} = \prod_l (1 - \tilde{\rho}(l) l^{-s}).$$

On pose

$$L_{\{p\}}(s,\rho) = \prod_{l \neq p}(1 - \tilde{\rho}(l)l^{-s}).$$

3.1.2. Si ξ est un caractère d'ordre fini de $G_{\mathbb{Q}}$, rappelons que la fonction de Kubota-Leopoldt relative au caractère de Dirichlet primitif $\tilde{\xi}$ est l'unique fonction continue $L_p(s,\xi)$ de $\mathbb{Z}_p$ dans $\mathbb{C}_p$ (resp. $\mathbb{Z}_p - \{1\}$ dans $\mathbb{C}_p$ si ξ est le caractère trivial) telle que, pour tout $k \geq 1$,

$$L_p(1-k,\xi) = (1 - \tilde{\xi}_k(p)p^{k-1})L(1-k,\tilde{\xi}_k) = L_{\{p\}}(1-k,\tilde{\xi}_k)$$

où $\tilde{\xi}_k$ est le caractère primitif associé à $\xi\tilde{\omega}^{-k}$ (avec $\tilde{\omega}$ le caractère de Teichmüller). Si ξ est impair, la fonction $L_p(s,\xi)$ est identiquement nulle. Un des moyens de construire la fonction $L_p(s,\xi)$ ou ses variantes est d'exhiber un élément ω (élément cyclotomique) de $H^1_{\infty,\{p\}}(F,\mathbb{Z}_p)$ tel que $L_{\omega,F}$ soit lié à la fonction de Kubota-Leopoldt. Nous allons rappeler cette construction.

3.1.3. On fixe un caractère ξ de $G_{\mathbb{Q}}$ d'ordre fini de conducteur m premier à p et on prend pour simplifier $F = \mathbb{Q}(\mu_m)$. Les $(1 - \sigma^{-(n+1)}(\alpha)\zeta_n)$ forment un système projectif d'éléments de $F_n^\times$ relativement à la norme (on a $\sigma(\alpha) = \alpha^p$). On en déduit un élément de $H^1_{\infty,\{p\}}(F,\mathbb{Z}_p(1))$ que l'on note $\omega_{\alpha,1}$. On pose

$$\omega_\alpha = \omega_{\alpha,1} \otimes \epsilon^{\otimes -1} \in H^1_{\infty,\{p\}}(F,\mathbb{Z}_p).$$

On peut alors définir comme en 1.8 un élément $L_{\omega_\alpha,F}$ de $\Lambda \otimes D_p(\mathbb{Q}_p)_F = \Lambda \otimes_{\mathbb{Q}} F$ si $\alpha \neq 1$ (resp. de $(\chi(\gamma)\gamma - 1)^{-1}\Lambda \otimes D_p(\mathbb{Q}_p)$ si $\alpha = 1$). On pose pour $\epsilon(\xi\rho) = -1$

$$\mathcal{L}_{K-L}(\xi\rho) = -\mathcal{L}_{\omega_\alpha}(\xi\rho) = -G(\xi^{-1},\alpha)^{-1}\sum_{\tau \in G}\xi(\tau)^{-1}\rho(\dot{L}_{\tau(\omega_\alpha),F})$$

pour tout $\rho \in \mathrm{Hom}_{\mathrm{cont}}(G_\infty, \mathbb{C}_p^\times)$. Remarquons que $\mathcal{L}_{K-L}(\xi\rho)$ ne dépend pas du choix de α.

3.1.4.

PROPOSITION. (Iwasawa) *Soit ξ un caractère de conducteur m. Alors, pour tout caractère η de G_∞ d'ordre fini et pour tout entier $j \geq 0$, on a*

$$\mathcal{L}_{K-L}(\xi\eta\chi^j) = (1 - p^j\xi(\sigma)\eta(\sigma))L(-j,\xi\eta) = L_{\{p\}}(-j,\xi\eta)$$

où l'on pose $\eta(\sigma) = 0$ si η est non trivial.

REMARQUE. Lorsque $\epsilon(\rho) = 1$, la valeur de $\mathcal{L}_{\omega_\alpha}$ en ρ est nulle. On pose

$$\mathcal{L}_{K-L}(\rho) = \mathcal{L}_{K-L}(\chi^{-1}\rho^{-1})$$

pour $\rho \in \mathrm{Hom}_{\mathrm{cont}}(G_\infty, \mathbb{C}_p^\times)$ tel que $\epsilon(\rho) = 1$. Nous justifions en 3.3.5 et suivants cette définition.

DÉMONSTRATION. Elle est classique. Donnons-en une esquisse. Si

$$G_1(T) = \log(1 - \alpha(1 + T)) \otimes e_{-1} ,$$

on a

$$(1 \otimes p\varphi)^{-(n+1)}(G_1)(\zeta_n - 1) = \log(1 - \sigma^{-(n+1)}(\alpha)\zeta_n) \otimes e_{-1} .$$

On en déduit que

$$L_{\omega_{\alpha,1}}.(1 + T) = (\log(1 - \alpha(1 + T)) - p^{-1}\log(1 - \alpha^p(1 + T)^p)) \otimes e_{-1} .$$

Notons D l'opérateur de dérivation $(1 + T)d/dT$. On a

$$D(L_{\omega_{\alpha,1}}.(1 + T)) \otimes e_1 = -L_{\omega_\alpha}.(1 + T) .$$

D'où, en posant $X = 1 + T$,

$$L_{\omega_\alpha}.X = \frac{\alpha X}{\alpha X - 1} - \frac{\alpha^p X^p}{\alpha^p X^p - 1} .$$

Pour η de conducteur p^{n+1}, on a

$$\sum_{\tau \in \mathrm{Gal}(F_n/F)} \eta(\tau)^{-1}\tau(D^j(L_{\omega_\alpha}.X)_{|X=\zeta_n-1}) = \chi^j\eta(L_{\omega_\alpha})G(\eta^{-1}, \zeta_n) .$$

Lorsque ξ est le caractère trivial et donc $\alpha = 1$ et $j > 0$, on a

$$\chi^j(L_{\omega_1}) = (1 - p^j)(d/dt)^j(\frac{e^t}{e^t - 1}) ,$$

d'où par des formules classiques

$$\chi^j(L_{\omega_1}) = (1 - p^j)\frac{B_{j+1}}{j+1} = -(1 - p^j)\zeta(-j) .$$

En général, on a

$$G(\xi^{-1}, \alpha)^{-1} \sum_{\tau \in G} \xi(\tau)^{-1}L_{\omega_{\tau\alpha}}.X$$

$$= G(\xi^{-1}, \alpha)^{-1} \sum_{\tau \in G} \xi(\tau)^{-1}\frac{\tau(\alpha)X}{1 - \tau(\alpha)X}$$

$$= G(\xi^{-1}, \alpha)^{-1} \sum_{\tau \in G} \xi(\tau)^{-1}\frac{1}{1 - \tau(\alpha)X}$$

$$= \sum_{0 \le a < m} \tilde{\xi}(a)\frac{X^a}{1 - X^m}$$

où $\tilde{\xi}$ est le caractère de Dirichlet associé à ξ. Il est classique que

$$\sum_{\tau \in \mathrm{Gal}(F_n/F)} \eta(\tau)^{-1}\tau[D^j(\sum_{0 \le a < m} \tilde{\xi}(a)\frac{X^a}{1 - X^m})_{|X=\zeta_n-1}]$$

$$= -G(\eta^{-1}, \zeta_n)L(-j, \xi\eta) .$$

D'où le résultat. ☐

3.1.5. Faisons le lien avec la fonction de Kubota-Leopoldt. Notons $\omega(\chi)$ le composé de χ avec la projection $\mathbb{Z}_p \to \mu_{p-1}$ (on a donc $\chi = \omega(\chi) < \chi >$). On déduit de la proposition 3.1.3 que, si ξ est un caractère de conducteur m et η un caractère de G_∞ d'ordre fini et si $j \geq 1$

$$\mathcal{L}_{K-L}(\xi\eta\chi^j) = L_p(-j, \xi\eta\omega(\chi)^{1+j})$$

et donc que pour tout $s \in \mathbb{Z}_p - \{1\}$,

$$\mathcal{L}_{K-L}(\xi\eta < \chi >^s) = L_p(-s, \xi\eta\omega(\chi)).$$

3.2. Valeurs de $\mathcal{L}_{K-L}$ en $\xi\eta\chi^{-j}$.

3.2.1. Dans tout le paragraphe 3.2, on fixe un caractère ξ de G de conducteur m, un caractère η de G_∞ de conducteur p^{n+1} et un entier j tels que $\epsilon(\xi\eta\chi^j) = -1$ (les formules seraient encore vraies lorsque $\epsilon(\xi\eta\chi^j) = 1$ en remplaçant $\mathcal{L}_{K-L}$ par $\mathcal{L}_{\omega_\alpha}$, car identiquement nulles !). L'entier j sera négatif dans le paragraphe 3.2.2 et positif dans les paragraphes 3.2.3 et 3.2.4

Posons

$$C_j(\xi\eta) = \sum_{\tau \in \mathrm{Gal}(F_n/\mathbb{Q})} (\xi\eta)(\tau)^{-1}\tau((\omega_\alpha \otimes \epsilon^{\otimes j})_n)$$

$$= P^{(\xi\eta)}(\omega_\alpha \otimes \epsilon^{\otimes j}) \in H^1(F_n, \mathbb{Q}_p(j))^{(\xi\eta)}$$

où ω_α est l'élément cyclotomique défini en 3.1.3. Lorsque j est positif, $C_j(\xi\eta)$ est un élément de $H^1_f(F_n, \mathbb{Q}_p(j))^{(\xi\eta)}$; lorsque j est négatif, $C_j(\xi\eta)$ est seulement un élément de $H^1(G_{S,F_n}, \mathbb{Q}_p(j))^{(\xi\eta)}$. Lorsque le caractère est le caractère trivial $\mathbf{1}$, on pose simplement $C_j = C_j(\mathbf{1})$.

3.2.2.

PROPOSITION. *Supposons $j < 0$. Pour η trivial et $\epsilon(\xi\chi^{-j}) = -1$, on a*

$$(1 - p^{-j}\xi(\sigma))^{-1}(1 - p^{j-1}\xi(\sigma)^{-1})\mathcal{L}_{K-L}(\xi\chi^{-j}) \otimes e_{-j}$$
$$= \Gamma(-j+1)G(\xi^{-1}, \alpha)^{-1}\lambda_{\mathbb{Q}_p(j), F}(C_j(\xi)).$$

Si η est non trivial de conducteur p^{n+1} et $\epsilon(\xi\eta\chi^{-j}) = -1$, on a

$$(p^{1-j}\xi(\sigma))^{-(n+1)}\mathcal{L}_{K-L}(\xi\eta\chi^{-j}) \otimes e_{-j}$$
$$= \Gamma(-j+1)G(\eta^{-1}\xi^{-1}, \alpha\zeta_n)^{-1}\lambda_{\mathbb{Q}_p(j), F}(C_j(\eta\xi)).$$

Ici $\lambda_{\mathbb{Q}_p(j), F}C_j(\eta\xi) = \sharp(\mathrm{Gal}(\mathbb{Q}_n/\mathbb{Q}))^{-1}\lambda_{\mathbb{Q}_p(j), F_n}C_j(\eta\xi)$.

DÉMONSTRATION. Ce n'est qu'une retraduction de la proposition 2.2.2. ☐

En termes des valeurs de la fonction L complexe (ou des nombres de Bernoulli), cette formule devient pour $j < 0$ tel que $\epsilon(\xi\chi^{-j}) = -1$

$$(1 - p^{-1+j}\xi(\sigma)^{-1})L(j,\xi) \otimes e_{-j} = \Gamma(-j+1)G(\xi^{-1},\alpha)^{-1}\lambda_{\mathbb{Q}_p(j)}(C_j(\xi)).$$

3.2.3.

PROPOSITION. *Supposons $j > 0$. Si $\epsilon(\xi\chi^{-j}) = -1$, on a*

$$(1 - p^{-j}\xi(\sigma))^{-1}(1 - p^{j-1}\xi(\sigma)^{-1})\mathcal{L}_{K-L}(\xi\chi^{-j}) \otimes e_{-j}$$
$$= \Gamma^*(-j+1)G(\xi^{-1},\alpha)^{-1}\log_{\mathbb{Q}_p(j)} C_j(\xi).$$

Si η est non trivial de conducteur p^{n+1} et $\epsilon(\xi\eta\chi^{-j}) - -1$, on a

$$(p^{1-j}\xi(\sigma))^{-(n+1)}\mathcal{L}_{K-L}(\xi\eta\chi^{-j}) \otimes e_{-j}$$
$$= \Gamma^*(-j+1)G(\eta^{-1}\xi^{-1},\alpha\zeta_n)^{-1}\log_{\mathbb{Q}_p(j)} C_j(\xi\eta).$$

Cela se déduit de la proposition 2.1.2. Comme

$$\mathcal{L}_{K-L}(\xi\eta\chi^{-j}) = L_p(j,\xi\eta\omega(\chi)^{-j+1}),$$

on obtient ainsi une formule pour $L_p(j,\xi\eta\omega(\chi)^{-j+1})$ avec $\epsilon(\xi\eta\chi^j) = -1$:

$$(p^{1-j}\xi(\sigma))^{-(n+1)}(1 - p^{-j}\eta(\sigma)\xi(\sigma))^{-1}(1 - p^{j-1}\eta(\sigma)\xi(\sigma)^{-1})$$
$$L_p(j,\xi\eta\omega(\chi)^{-j+1}) \otimes e_{-j}$$
$$= \Gamma^*(-j+1)G(\eta^{-1}\xi^{-1},\alpha\zeta_n)^{-1}\log_{\mathbb{Q}_p(j)} C_j(\xi).$$

On retrouve donc dans le cas où η est trivial un résultat de Kurihara (démontré dans [G-K90] pour $j < p-1$) (notons que l'on traite ici le cas d'un caractère de conducteur premier ou non premier à p). En utilisant les propriétés des sommes de Gauss, la formule peut aussi s'écrire sous la forme : pour tout caractère d'ordre fini ρ de $G_{\mathbb{Q}}$ de conducteur $c(\rho)$ divisible par p et j entier > 0 tels que $\epsilon(\rho\chi^j) = -1$, on a

$$L_p(j,\rho\omega(\chi)^{-j+1}) \otimes e_{-j} = \mathcal{L}_{K-L}(\rho\chi^{-j}) \otimes e_{-j}$$
$$= c(\xi)^{-1}\Gamma^*(-j+1)(\chi^{-j}\eta)(\sigma_{c(\eta)}^{-1})c(\rho)^{-(j-1)}G(\rho^{-1})^{-1}\log_{\mathbb{Q}_p(j)} C_j(\rho)$$

où $\rho = \xi\eta$ avec η de conducteur une puissance $c(\eta)$ de p et ξ de conducteur $c(\xi)$ premier à p.

3.2.4. Supposons toujours $j > 0$. Remarquons que

$$(1 - p^{j-1}\sigma^{-1})^{-1}\log_{\mathbb{Q}_p(j)}(\omega_\alpha \otimes \epsilon^{\otimes j}) = \log_{\mathbb{Q}_p(j)}\left(\sum_{n\geq 0} p^{(j-1)n}\omega_{\sigma^{-n}(\alpha)}\right).$$

On pose

$$C_j^{Del}(\alpha) = \sum_{n\geq 0} p^{(j-1)n}\omega_{\sigma^{-n}(\alpha)}$$

et

$$C_j^{Del}(\xi) = \sum_{\tau \in G} \xi(\tau)^{-1} C_j^{Del}(\tau(\alpha)).$$

La première formule de 3.2.3 devient alors

$$\mathcal{L}_{K-L}(\xi\chi^{-j}) \otimes e_{-j} = \Gamma^*(-j+1)(1-p^{-j}\xi(\sigma))G(\xi^{-1}, \alpha)^{-1} \log_{\mathbb{Q}_p(j)} C_j^{Del}(\xi).$$

D'autre part, l'application $c_{j,2j-1}$ de [**G-K90**] est $(1-p^{-j}\sigma)\log_{\mathbb{Q}_p(j)}$. Je présume que l'application de [**De89**] (qui, avec ses notations, associe $\phi_p^{-1}(0)$ à x où ϕ_p est le Frobenius cristallin associé à x vu comme torseur, cf. §2.9) est

$$\pm(p^j - \sigma)\log_{\mathbb{Q}_p(j)} = \pm p^j(1 - p^{-j}\sigma)\log_{\mathbb{Q}_p(j)}.$$

On en déduirait alors (et c'est en tout cas le résultat qu'affirme Kurihara dans [**G-K90**]) que $C_j^{Del}(\xi)$ est l'image dans $\mathbb{Q}(\xi) \otimes H^1(F, \mathbb{Q}_p(j))$ de l'élément défini par Deligne comme extension de $\mathbb{Q}$ par $\mathbb{Q}(j)$ dans la catégorie des motifs mixtes de Deligne et dont l'image par l' "application régulateur à l'infini" (que l'on notera ici $\log_{\mathbb{Q}(j)}$) est $-(j-1)!L(j,\xi)$. De l'art et de la méthode de faire apparaître et disparaître les "facteurs d'Euler". Nous y reviendrons d'ailleurs dans les remarques de 5.2.

3.2.5. L'expression en termes de polylogarithmes se déduit facilement de la proposition 3.2.3 : d'après [**Co82**], si α est une racine de l'unité d'ordre m et si ξ est un caractère de conducteur m premier à p, on a

$$\mathcal{L}_{K-L}(\xi\chi^{-j}) = G(\xi^{-1}, \alpha)^{-1} \sum_{\tau \in G} \xi(\tau)^{-1} l_j^{(p)}(\tau(\alpha))$$

où $l_j^{(p)}$ est le polylogarithme p-adique : $l_j^{(p)}(x) = \sum_{\substack{n \geq 1 \\ (n,p)=1}} x^n/n^j$. On déduit que de la proposition 3.2.2 que

$$\mathcal{L}_{K-L}(\xi\chi^{-j}) \otimes e_{-j}$$
$$= \Gamma^*(-j+1)(1-p^{-j}\xi(\sigma))G(\xi^{-1}, \alpha)^{-1} \sum_{\tau \in G} \xi(\tau)^{-1} \log_{\mathbb{Q}_p(j)} C_j^{Del}(\tau\alpha)$$

et donc que

$$(1 - p^{-j}\sigma)\log_{\mathbb{Q}_p(j)} C_j^{Del}(\alpha) = (-1)^j(j-1)! l_j^{(p)}(\alpha)e_{-j}$$

ou encore si $l_{j,p}(\alpha) = (1-p^{-j}\sigma)^{-1} l_j^{(p)}(\alpha)$,

$$\log_{\mathbb{Q}_p(j)} C_j^{Del}(\alpha) = (-1)^j(j-1)! l_{j,p}(\alpha)e_{-j}$$

(on a d'autre part

$$\log_{\mathbb{Q}(j)} C_j^{Del}(\alpha) = (-1)^j(j-1)! l_{j,\infty}(\alpha)$$

où $l_{j,\infty}$ est le polylogarithme complexe : $l_{j,\infty}(x) = \sum_{n \geq 1} x^n/n^j$).

3.3. "Conjecture" principale.

3.3.1. Si $A = \Lambda$ ou $\Lambda[G] = \mathbb{Z}_p[[\Gamma]] \otimes \mathbb{Z}_p[G \times \Delta]$, on note $\mathrm{Frac}(A)$ l'anneau total des fractions de A. On note $A_\pm = (1\pm c)A$ où c est la conjugaison complexe de Δ ou de $G \times \Delta$. Si M est un A-module de type fini, on peut alors définir ([**K-M76**]) le A-module libre de rang 1 : $\det_A(M)$ et son dual : $\det_A(M)^{-1}$. Lorsque $M = 0$, $\det_A(0)$ est égal à $\mathrm{Frac}(A)$. Lorsque M est annulé par un élément non diviseur de 0, on a une application canonique de $\det_A(M)$ dans $\mathrm{Frac}(A)$ (par abus, on considère alors $\det_A(M)$ comme contenu dans $\mathrm{Frac}(A)$). Si M est un Λ-module sans torsion de rang r, tout Λ-homomorphisme de M dans un Λ-module libre L se prolonge en une application du Λ-bidual M^{**} de M dans N. En remarquant que $\det_\Lambda M \cong \det_\Lambda M^{**}$, on en déduit une application de $\det_\Lambda M$ dans la puissance extérieure r-ième de L. On utilise ici ce fait pour $r = 1$: si M est un Λ-module sans torsion de rang 1, on peut ainsi prolonger tout Λ-homomorphisme de M dans un module libre L en un Λ-homomorphisme de Λ-modules libres de $\det_\Lambda M$ dans L.

3.3.2. Considérons les $\Lambda[G]$-homomorphismes suivants :

$$H^1_{\infty,\{p\}}(F, \mathbb{Z}_p) \to Z^1_{\infty,p}(F, \mathbb{Z}_p)$$

$$(\Lambda[G] \otimes D_p(\mathbb{Q}_p))^{\tilde{\Delta}_0=0} \to (\Lambda \otimes D_p(\mathbb{Q}_p)_F)^{\tilde{\Delta}_0=0} \to Z^1_{\infty,p}(F, \mathbb{Z}_p)/ \oplus_{v|p} \mathbb{Z}_p ;$$

le premier homomorphisme est l'application naturelle de localisation en p (c'est une injection car $\mathrm{Leop}(\mathbb{Q}_p)$ est vrai) ; le deuxième homomorphisme est défini à partir de l'application $\Omega^\epsilon_{\mathbb{Q}_p,0,F}$.

Posons

$$\Delta^{glob}_{\infty,p}(F/\mathbb{Q}, \mathbb{Z}_p) = \otimes_{i\in\{1,2\}}(\det_{\Lambda[G]} H^i_{\infty,\{p\}}(F, \mathbb{Z}_p))^{(-1)^i}$$

$$\Delta^{loc}_{\infty,p}(F/\mathbb{Q}, \mathbb{Z}_p) = \otimes_{i\in\{1,2\}}(\det_{\Lambda[G]} Z^i_{\infty,p}(F, \mathbb{Z}_p))^{(-1)^i}$$

$$\Delta_{\infty,p}(F/\mathbb{Q}, \mathbb{Z}_p) = \Delta^{glob}_{\infty,p}(F/\mathbb{Q}, \mathbb{Z}_p) \otimes \Delta^{loc}_{\infty,p}(F/\mathbb{Q}, \mathbb{Z}_p)^{-1}.$$

On définit un isomorphisme de $\Lambda[G]$-modules $\lambda_{F/\mathbb{Q}}$:

$$\Delta_{\infty,p}(F/\mathbb{Q}, \mathbb{Z}_p)^{-1}_\pm \to \mathrm{Hom}_{\mathbb{Q}_p}(\wedge^{d_\pm} D(\mathbb{Q}_p), \mathrm{Frac}(\Lambda[G])_\pm)$$

avec $d_+ = 1$, $d_- = 0$: si ω_{glob} est une base de $\Delta^{glob}_{\infty,p}(F/\mathbb{Q}, \mathbb{Z}_p)_\pm$, si ω_{loc} est une base de $\Delta^{loc}_{\infty,p}(F/\mathbb{Q}, \mathbb{Z}_p)_\pm$, on définit $\lambda_{F/\mathbb{Q},\pm}$ par

$$\lambda_{F/\mathbb{Q},\pm}(\omega^{-1}_{glob} \otimes \omega_{loc})(s)\omega^{-1}_{loc} = \wedge^{d_\pm} \Omega^\epsilon_{\mathbb{Q}_p,0}(s) \wedge \omega^{-1}_{glob}$$

pour $s \in \wedge^{d_\pm} D(\mathbb{Q}_p)$. L'espace d'arrivée de λ_- est canoniquement $\mathrm{Frac}(\Lambda[G])$; celui de $\lambda_\pm$ s'identifie à $\mathrm{Frac}(\Lambda[G])$ en utilisant la base de $D(\mathbb{Q}_p)$. On identifiera donc en général $\mathrm{Hom}_{\mathbb{Q}_p}(\wedge^{d_\pm} D(\mathbb{Q}_p), \mathrm{Frac}(\Lambda[G])_\pm)$ avec $\mathrm{Frac}(\Lambda[G]_\pm)$; il est cependant intéressant de se souvenir qu'il s'agit d'une identification (cf. la remarque de 5.2).

3.3.3.

DÉFINITION. *Soit* $\mathbb{I}_{arith}(F/\mathbb{Q}, \mathbb{Z}_p)_{\pm}$ *l'image de* $\Delta_{\infty,p}(F/\mathbb{Q}, \mathbb{Z}_p)_{\pm}$ *dans*

$$\mathrm{Hom}_{\mathbb{Q}_p}(\wedge^{d\pm} D(\mathbb{Q}_p), Frac(\Lambda[G])_{\pm}) \cong Frac(\Lambda[G])_{\pm}$$

et $\mathbb{I}_{arith}(F/\mathbb{Q}, \mathbb{Z}_p)$ *le sous-$\Lambda[G]$-module libre de rang 1 dont les $\pm$-composantes sont les* $\mathbb{I}_{arith}(F/\mathbb{Q}, \mathbb{Z}_p)_{\pm}$. *Si ξ est un caractère de G de conducteur égal au conducteur de F, on note* $\mathbb{I}_{arith}(\xi, \mathbb{Z}_p)$ *le $\Lambda[G]^{(\xi)}$-module* $\mathbb{I}_{arith}(F/\mathbb{Q}, \mathbb{Z}_p)^{(\xi)}$.

On définit de la même manière pour tout entier j un $\Lambda[G]$-module

$$\Delta_{\infty,p}(F/\mathbb{Q}, \mathbb{Z}_p(j))$$

et une application $\lambda_{F/\mathbb{Q},j}$:

$$\Delta_{\infty,p}(F/\mathbb{Q}, \mathbb{Z}_p(j))_{\pm} \to \mathrm{Hom}_{\mathbb{Q}_p}(\wedge^{d\pm\epsilon_j} D(\mathbb{Q}_p(j)), Frac(\Lambda[G])_{\pm}) \cong Frac(\Lambda[G])_{\pm}$$

avec ϵ_j le signe de $(-1)^j$. Si $\mathbb{I}_{arith}(F/\mathbb{Q}, \mathbb{Z}_p(j))_{\pm}$ est l'image de

$$\Delta_{\infty,p}(F/\mathbb{Q}, \mathbb{Z}_p(j))_{\pm}$$

dans $Frac(\Lambda[G])_{\pm}$ et $\mathbb{I}_{arith}(F/\mathbb{Q}, \mathbb{Z}_p(j))$ le $\Lambda[G]$-module dont les $\pm$-composantes sont les $\mathbb{I}_{arith}(F/\mathbb{Q}, \mathbb{Z}_p(j))_{\pm}$, on vérifie facilement que

$$\mathbb{I}_{arith}(F/\mathbb{Q}, \mathbb{Z}_p(j)) = Tw^{-j}(\mathbb{I}_{arith}(F/\mathbb{Q}, \mathbb{Z}_p))$$

Si ξ est un caractère de G de conducteur le conducteur de $F/\mathbb{Q}$, on note encore

$$\mathbb{I}_{arith}(\xi, \mathbb{Z}_p(j)) = \mathbb{I}_{arith}(F/\mathbb{Q}, \mathbb{Z}_p(j))^{(\xi)} .$$

3.3.4. Dans les deux paragraphes qui suivent, nous allons écrire $\mathbb{I}_{arith}(\xi, \mathbb{Z}_p)$ de manière différente afin de retrouver les modules classiques qui interviennent dans les différentes formulations de la conjecture principale. La définition que nous venons de donner se généralise à des représentations p-adiques cristallines en p quelconques ([**Pb**]).

PROPOSITION. *Soit ω un élément du $\Lambda[G]_-$-module* $H^1_{\infty,\{p\}}(F, \mathbb{Z}_p)_-$ *engendrant un $\Lambda[G]_-$-module libre de rang 1. Alors,* $\mathbb{I}_{arith}(\xi, \mathbb{Q}_p)$ *est le sous-$\Lambda[G]^{(\xi)}$-module de $Frac(\Lambda[G]^{(\xi)})$ déterminé par*
  (i) *si* $\epsilon(\rho\xi) = -1$,

$$\rho(\mathbb{I}_{arith}(\xi, \mathbb{Z}_p)) =$$
$$\rho\xi((\det_{\Lambda[G]} H^2_{\infty,\{p\}}(F, \mathbb{Z}_p))^{-1} \otimes \det_{\Lambda[G]}(H^1_{\infty,\{p\}}(F, \mathbb{Z}_p)/\Lambda[G]\omega))\mathcal{L}_\omega(\rho\xi)$$

  (ii) *si* $\epsilon(\rho\xi) = 1$,

$$\rho(\mathbb{I}_{arith}(\xi, \mathbb{Z}_p)) = \rho\xi((\det_{\Lambda[G]} H^2_{\infty,\{p\}}(F, \mathbb{Z}_p))^{-1} \otimes \det_{\Lambda[G]} H^1_{\infty,\{p\}}(F, \mathbb{Z}_p)) .$$

Pour la définition de $\mathcal{L}_\omega$, voir 1.8. On identifie ici $D_p(\mathbb{Q}_p)$ avec $\mathbb{Q}_p$.

DÉMONSTRATION. La proposition se déduit de la définition de $\mathbb{I}_{arith}(\xi, \mathbb{Z}_p)$ et de ce que l'application

$$\Lambda[G] \otimes \det{}_{\mathbb{Z}_p} D_p(\mathbb{Z}_p) \to \otimes_{i \in \{1,2\}} (\det{}_{\Lambda[G]} Z^i_{\infty, \{p\}}(F, \mathbb{Z}_p))^{(-1)^i}$$

induite par

$$(\Lambda \otimes D_p(\mathbb{Z}_p)_{\mathcal{O}_F})^{\tilde{\Delta}_0 = 0} \to Z^1_{\infty, p}(F, \mathbb{Z}_p) / \oplus_{v|p} \mathbb{Z}_p$$

est un isomorphisme ([**Pa**], §4], se déduit du théorème (A)). $\square$

3.3.5. On note $X^i_{\infty, S}(F, \mathbb{Z}_p(j))$ le dual de Pontryagin de

$$H^i(G_{S, F_\infty}, \mathbb{Q}_p(j)/\mathbb{Z}_p(j)).$$

Le $\Lambda[G]$-module $X^1_{\infty, S}(F, \mathbb{Z}_p(1))$ est égal à $\mathrm{Gal}(M_\infty/F_\infty)$ où M_∞ est la plus grande p-extension abélienne de F_∞ non ramifiée en dehors de $S = \{\infty, p\}$. Soit A_∞ la limite projective des composantes p-primaires des groupes de classes d'idéaux des F_n.

Si ξ est un caractère de G de conducteur m, et si M est un $\Lambda[G]^{(\epsilon)}$-module de torsion, on note f_M une série caractéristique de M. La proposition suivante calcule $\mathbb{I}_{arith}(\xi, \mathbb{Z}_p(j))$ en termes plus classiques ([**C77**])

PROPOSITION. *Soit ξ un caractère de $\mathrm{Gal}(F/\mathbb{Q})$ de conducteur m et η un caractère de $\mathrm{Gal}(F_0/F)$. Alors,*

(i) *si $\epsilon(\xi \eta \chi^j) = 1$, on a*

$$\mathbb{I}_{arith}(\xi, \mathbb{Z}_p(j))^{(\eta)} = \Lambda[G]^{(\xi\eta)} f_{H^2_{\infty, \{p\}}(F, \mathbb{Z}_p(j))^{(\xi\eta)}} / f_{\mathbb{Z}_p(j)^{(\xi\eta)}}$$

$$= \Lambda[G]^{(\xi\eta)} f_{A_\infty(j-1)^{(\xi\eta)}} / f_{\mathbb{Z}_p(j)^{(\xi\eta)}}$$

(ii) *si $\epsilon(\xi \eta \chi^j) = -1$, on a*

$$\mathbb{I}_{arith}(\xi, \mathbb{Z}_p(j))^{(\eta)} = \Lambda[G]^{(\xi\eta)} f_{\mathrm{Gal}(M_\infty/F_\infty)(j-1)^{(\xi\eta)}} / f_{\mathbb{Z}_p(j-1)^{(\xi\eta)}}$$

$$= \Lambda[G]^{(\xi\eta)} f_{X^1_S(\mathbb{Z}_p(j))^{(\xi\eta)}} / f_{X^2_{\infty, S}(\mathbb{Z}_p(j))^{(\xi\eta)}}$$

DÉMONSTRATION. Il suffit de démontrer la proposition pour $j = 1$. On a le diagramme commutatif dont les lignes et les colonnes sont exactes

$$
\begin{array}{ccccccccc}
& & & & 0 & & & & \\
& & & & \uparrow & & & & \\
& & & & \oplus_{v \in S_p} \mathbb{Z}_p & & & & \\
& & & & \uparrow & & & & \\
0 \to & \mathcal{E}_{\infty, \{p\}} & \to & Z^1_{\infty, p}(F, \mathbb{Z}_p(1)) & \to & X^1_{\infty, S}(F, \mathbb{Z}_p(1)) & & & \\
& \uparrow & & \uparrow & & \| & & & \\
0 \to & \mathcal{E}_\infty & \to & U_\infty & \to & X^1_{\infty, S}(F, \mathbb{Z}_p(1)) & \to A_\infty \to 0 \\
& \uparrow & & \uparrow & & & & \\
& 0 & & 0 & & & & \\
\end{array}
$$

où $\mathcal{E}_{\infty, \{p\}} = H^1_{\infty, \{p\}}(F, \mathbb{Z}_p(1))$ est la limite projective des unités en dehors de p de F_n, $\mathcal{E}_\infty$ la limite projective des unités des F_n et U_∞ la limite projective des unités semi-locales de F_n en p. La suite exacte de Poitou-Tate donne d'autre part la suite exacte

$$0 \to H^1_{\infty,\{p\}}(F, \mathbb{Z}_p(1)) \to Z^1_{\infty,p}(F, \mathbb{Z}_p(1)) \to X^1_{\infty,S}(F, \mathbb{Z}_p(1))$$
$$\to H^2_{\infty,\{p\}}(F, \mathbb{Z}_p(1)) \to Z^2_{\infty,\{p\}}(F, \mathbb{Z}_p(1)) \to X^2_{\infty,S}(F, \mathbb{Z}_p(1)) \to 0.$$

Si $\epsilon(\xi\eta) = -1$, on a $Z^2_{\infty,p}(F, \mathbb{Z}_p(1))^{(\xi\eta)} = 0$, $X^2_{\infty,S}(F, \mathbb{Z}_p(1))^{(\xi\eta)} = 0$. On en déduit que $(A_\infty)^{(\xi\eta)} = H^2_{\infty,\{p\}}(F, \mathbb{Z}_p(1))^{(\xi\eta)}$. Le premier cas s'en déduit à l'aide de la proposition 3.3.2. Le deuxième cas se déduit de cette même suite exacte. $\square$

3.3.6.

PROPOSITION. (Iwasawa) *On a* $\mathbb{I}_{arith}(F/\mathbb{Q}, \mathbb{Z}_p(j)) = \mathbb{I}_{arith}(F/\mathbb{Q}, \mathbb{Z}_p(1-j))^\iota$ *où* ι *est l'involution de* $\Lambda[G]$ *induite par* $\tau \mapsto \tau^{-1}$ *pour* $\tau \in \mathrm{Gal}(F_\infty/\mathbb{Q})$.

Soit $I_{arith}(F/\mathbb{Q}, \mathbb{Z}_p)$ un générateur de $\mathbb{I}_{arith}(F/\mathbb{Q}, \mathbb{Z}_p)$; la proposition signifie qu'il existe une unité u de $\Lambda[G]$ telle que, si $\rho \in \mathrm{Hom}_{\mathrm{cont}}(\mathrm{Gal}(F_\infty/\mathbb{Q}), \mathbb{C}_p^\times)$, on a

$$\rho(I_{arith}(F/\mathbb{Q}, \mathbb{Z}_p)) = \rho(u)(\chi^{-1}\rho^{-1})(I_{arith}(F/\mathbb{Q}, \mathbb{Z}_p)).$$

DÉMONSTRATION. Il suffit de nouveau de démontrer la proposition pour $j = 1$. Iwasawa démontre qu'il existe un homomorphisme injectif de $\Lambda[G]$-modules

$$H^2_{\infty,\{p\}}(F, \mathbb{Z}_p(1)) \to \mathrm{Ext}^1_\Lambda(X^1_{\infty,S}(F, \mathbb{Z}_p), \Lambda)^\iota$$

à conoyau fini. Comme $f_{\mathbb{Z}_p(j)} = \iota(f_{X^2_{\infty,S}(F,\mathbb{Z}_p(1-j))})$, la proposition s'en déduit. Rappelons la construction de l'homomorphisme d'Iwasawa. La conjecture de Leopoldt est vraie pour F_n. On en déduit que $H^2(G_{S,F_n}, \mathbb{Z}_p(1))$ est égal à A_n et est fini pour tout entier n. Il en est de même de $t_\Lambda(X^1_{\infty,S}(F, \mathbb{Z}_p))_{\Gamma_n}$ où $t_\Lambda(M)$ désigne le sous-Λ-module de torsion d'un Λ-module M : en effet, pour tout caractère ξ de $\mathrm{Gal}(F_0/\mathbb{Q})$, si $s(\xi)$ est le rang de $X^1_{\infty,S}(F, \mathbb{Z}_p)^{(\xi)}$ comme $\Lambda[G]^{(\xi)}$-module, le rang sur $\mathbb{Z}_p$ de $X^1_{\infty,S}(F, \mathbb{Z}_p)^{(\xi)}_{\Gamma_n}$ est $[F_n : F_0]s(\xi)$: pour cela, on utilise de nouveau la conjecture de Leopoldt et le fait que $X^1_{\infty,S}(F, \mathbb{Z}_p)^{(\xi)}_{\Gamma_n}$ est égal à un groupe fini d'ordre borné près à $(H^1(G_{S,F_n}, \mathbb{Q}_p/\mathbb{Z}_p)^\smallfrown)^{(\xi)}$; un argument simple sur les $\mathbb{Z}_p[[\Gamma]]$-modules démontre la finitude de $t_\Lambda(X^1_{\infty,S}(F, \mathbb{Z}_p))_{\Gamma_n}$. On en déduit que $t_\Lambda(X^1_{\infty,S}(F, \mathbb{Z}_p))_{\Gamma_n}$ est égal à un groupe fini d'ordre borné près au sous-$\mathbb{Z}_p$-module de torsion de $X^1_{\infty,S}(F, \mathbb{Z}_p)_{\Gamma_n}$. D'autre part, on a la suite exacte

$$0 \to \mathbb{Q}_p/\mathbb{Z}_p \otimes H^1(G_{S,F_n}, \mathbb{Z}_p(1)) \to H^1(G_{S,F_n}, \mathbb{Q}_p(1)/\mathbb{Z}_p(1))$$
$$\to H^2(G_{S,F_n}, \mathbb{Z}_p(1)) \to 0$$

où la surjectivité se déduit de la finitude de $H^2(G_{S,F_n}, \mathbb{Z}_p(1))$. On obtient des homomorphismes de $\Lambda[G]$-modules dont les noyaux et conoyaux sont finis d'ordre borné par rapport à n

$$H^2(G_{S,F_n}, \mathbb{Z}_p(1)) \sim (t_\Lambda(X^1_{\infty,S}(F, \mathbb{Z}_p))_{\Gamma_n})^\smallfrown.$$

Par passage à la limite projective sur n, on obtient à gauche $H^2_{\infty,\{p\}}(F, \mathbb{Z}_p(1))$ et à droite $\mathrm{Ext}^1_\Lambda(t_\Lambda(X^1_{\infty,S}(F, \mathbb{Z}_p)), \Lambda)^\iota$ en remarquant que si M est un Λ-module de torsion et si $\omega_n = \gamma^{p^n} - 1$, avec γ générateur topologique de Γ, est premier à la série caractéristique de M pour tout entier n (ce qui est équivalent à ce que M_{Γ_n} soit fini pour tout entier n), on a

$$\mathrm{Ext}^1_\Lambda(M, \Lambda)^\iota = \varprojlim_n (M_{\Gamma_n})\widehat{} = (\varinjlim M_{\Gamma_n})\widehat{}$$

où la limite inductive est relative aux applications de transition

$$M_{\Gamma_{n+1}} \to M_{\Gamma_n}$$
$$x \mapsto (\omega_{n+1}/\omega_n).x.$$

La proposition s'en déduit. $\square$

3.3.7.

DÉFINITION. *Soit ξ un caractère de $\mathrm{Gal}(F/\mathbb{Q})$ de conducteur m. Notons $\mathbb{L}_{K-L}(\xi.)$ la fonction sur $\mathrm{Hom}(G_\infty, \mathbb{C}_p^\times)$ déterminée par*

$$\mathbb{L}_{K-L}(\xi\rho) = \begin{cases} \mathcal{L}_{K-L}(\xi\rho) & si\ \epsilon(\xi\rho) = -1, \\ \mathcal{L}_{K-L}(\chi^{-1}\xi^{-1}\rho^{-1}) & si\ \epsilon(\xi\rho) = 1 \end{cases}$$

pour $\rho \in \mathrm{Hom}_{cont}(G_\infty, \mathbb{C}_p^\times)$.

REMARQUE. Pour tout élément $\rho \in \mathrm{Hom}_{cont}(G_\infty, \mathbb{C}_p^\times)$, on a par définition même l'équation fonctionnelle

$$\mathbb{L}_{K-L}(\chi^{-1}\rho^{-1}) = \mathbb{L}_{K-L}(\rho).$$

3.3.8.

THÉORÈME. (Mazur-Wiles) *Il existe un générateur $I_{arith}(\xi, Z_p)$ du $\Lambda[G]^{(\xi)}$-module $\mathbb{I}_{arith}(\xi, \mathbb{Z}_p)$ tel que pour tout caractère ρ de $\mathrm{Hom}_{cont}(G_\infty, \mathbb{C}_p^\times)$, on a*

$$\rho(I_{arith}(\xi, \mathbb{Z}_p)) = \mathbb{L}_{K-L}(\xi\rho).$$

Nous n'avons fait bien sûr que réécrire la conjecture principale. Dans sa formulation même, le théorème 3.3.8 contient les deux versions classiques de la conjecture principale.

4. Cohomologie galoisienne

4.1. Notations. Soit toujours $S = \{\infty, p\}$. On définit $H^1_f(F, \mathbb{Q}_p(j))$ comme l'ensemble des éléments de $H^1(G_{S,F}, \mathbb{Q}_p(j))$ dont l'image par localisation en p appartient à $H^1_f(F_p, \mathbb{Q}_p(j))$ et $H^1_f(F, \mathbb{Z}_p(j))$ l'image réciproque de $H^1_f(F, \mathbb{Q}_p(j))$ dans $H^1(F, \mathbb{Z}_p(j))$. Pour $X = \mathbb{Z}_p$, $\mathbb{Q}_p$ ou $\mathbb{Q}_p/\mathbb{Z}_p$, notons $\mathrm{Loc}_{p,X(j)}$ l'application de localisation

$$H^1(G_{S,F}, X(j)) \to \oplus_{v|p} H^1(F_v, X(j)).$$

On vérifie facilement que pour $j < 0$, $H^1_f(F, \mathbb{Q}_p(j))$ est égal à $\ker \mathrm{Loc}_{p,\mathbb{Q}_p(j)}$; pour $j > 1$, $H^1_f(F, \mathbb{Q}_p(j)) = H^1(G_{S,F}, \mathbb{Q}_p(j))$.

Soit $H^1_f(F, \mathbb{Q}_p(j)/\mathbb{Z}_p(j))$ l'ensemble des éléments de $H^1(G_{S,F}, \mathbb{Q}_p(j)/\mathbb{Z}_p(j))$ dont l'image par localisation en p appartient à $\mathbb{Q}_p/\mathbb{Z}_p \otimes \oplus_{v|p} H^1_f(F_v, \mathbb{Z}_p(j))$. Pour $j < 0$, $H^1_f(F, \mathbb{Q}_p(j)/\mathbb{Z}_p(j))$ est égal à $\ker \mathrm{Loc}_{p,\mathbb{Q}_p(j)/\mathbb{Z}_p(j)}$; pour $j > 1$, c'est l'ensemble des éléments de $H^1(G_{S,F}, \mathbb{Q}_p(j)/\mathbb{Z}_p(j))$ dont l'image dans $\oplus_{v|p} H^2(F_v, \mathbb{Z}_p(j))$ est nulle. Définissons comme dans [**B-K90**] le groupe de Shafarevich-Tate $\mathrm{III}(\mathbb{Z}_p(j))$ comme le quotient de $H^1_f(F, \mathbb{Q}_p(j)/\mathbb{Z}_p(j))$ par son sous-groupe divisible maximal. On déduit de la suite exacte

$$0 \to \mathbb{Q}_p/\mathbb{Z}_p \otimes_{\mathbb{Z}_p} H^1(G_{S,F}, \mathbb{Z}_p(j)) \to H^1(G_{S,F}, \mathbb{Q}_p(j)/\mathbb{Z}_p(j))$$
$$\to H^2(G_{S,F}, \mathbb{Z}_p(j))_{\mathrm{tors}} \to 0$$

que pour $j > 1$

$$\mathrm{III}(\mathbb{Z}_p(j)) = \ker(H^2(G_{S,F}, \mathbb{Z}_p(j))_{\mathrm{tors}} \to \oplus_{v|p} H^2(F_v, \mathbb{Z}_p(j))).$$

Lorsque $j < 0$, on déduit du théorème de dualité de Poitou-Tate que le groupe $H^1_f(F, \mathbb{Q}_p(j)/\mathbb{Z}_p(j))$ est le dual de Pontryagin du noyau de l'application de localisation

$$H^2(G_{S,F}, \mathbb{Z}_p(j)) \to \oplus_{v|p} H^2(F_v, \mathbb{Z}_p(j)).$$

On en déduit une dualité de groupes finis

$$\mathrm{III}(\mathbb{Z}_p(j)) \times \mathrm{III}(\mathbb{Z}_p(1-j)) \to \mathbb{Q}_p/\mathbb{Z}_p,$$

ce qui est connu de manière beaucoup plus générale ([**Fl90**]), [**F-P92**]).

4.2. Cohomologie galoisienne et dimensions. Le théorème qui suit se déduit de théorèmes dûs à C. Soulé ([**So80**], [**So83**], [**So87**]). Rappelons que $d_+ = 1$, $d_- = 0$

4.2.1.

THÉORÈME. *Soit ξ un caractère de $\mathrm{Gal}(F_0/\mathbb{Q})$.*

(A) $H^2(G_{S,F_0}, \mathbb{Q}_p(j))$ *est nul pour $j \geq 2$.*

(B) $H^2(G_{S,F_0}, \mathbb{Q}_p(j))^{(\xi)}$ *est nul pour $j < 0$ et $\epsilon(\xi\chi^j) = -1$.*

(C)   (i) $H^1(G_{S,F_0}, \mathbb{Q}_p(j))^{(\xi)}$ *est de dimension $d_{-\epsilon(\xi\chi^j)}$ pour $j \geq 2$.*

     (ii) $C_j(\xi)$ *engendre dans $H^1(G_{S,F_0}, \mathbb{Q}_p(j))^{(\xi)}$ un sous-$\mathbb{Z}_p[\xi]$-module libre de rang 1 pour $j \geq 2$ et $\epsilon(\xi\chi^j) = -1$.*

(iii) $(ker\ Loc_{p,\mathbb{Q}_p(j)})^{(\xi)}$ *est nul si et seulement si* $\log_{\mathbb{Q}_p(j)} C_j(\xi)$ *est non nul, ce qui est encore équivalent à la non nullité de* $\mathcal{L}_{K-L}(\xi\chi^{-j}) = L_p(j, \xi\omega(\chi)^{-j+1})$.

(D) $H^1(G_{S,F_0}, \mathbb{Q}_p(j))^{(\xi)}$ *est nul pour* $j \geq 2$, $\epsilon(\xi\chi^j) = 1$.

(E) $H^1_f(F_0, \mathbb{Q}_p(j))$ *est nul pour* $j < 0$.

(F) *Si* j *est un entier* < 0 *tel que* $\epsilon(\xi\chi^j) = 1$, $H^2(G_{S,F_0}, \mathbb{Q}_p(j))^{(\xi)}$ *est nul si et seulement si* $\mathcal{L}_{K-L}(\xi\chi^{j-1})$ *est non nul.*

4.2.2. Faisons quelques commentaires rapides sur la démonstration de ce théorème (voir aussi [**I-S87**]). Le point crucial et très profond que démontre Soulé est l'assertion (A). La démonstration généralisant une démonstration de Tate pour j—2 utilise le théorème de Borel sur la dimension des groupes de K-théorie $K_{2j-2}(F_0)$. Soulé montre alors qu'il existe un isomorphisme

$$c_{j,2} : \mathbb{Q}_p \otimes K_{2j-2}(F_0) \cong H^2(G_{S,F_0}, \mathbb{Q}_p(j)).$$

Les assertions (C,i) et (D) sont équivalentes à (A) grâce aux formules de caractéristique d'Euler de Tate. Comme $H^2(G_{S,F_0}, \mathbb{Q}_p(j))$ est le dual de

$$ker\ Loc_{p,\mathbb{Q}_p(1-j)} \subset H^1(G_{S,F_0}, \mathbb{Q}_p(1-j)),$$

(B) se déduit de (D). Comme $H^1_f(F_0, \mathbb{Q}_p(j)) = ker\ Loc_{p,\mathbb{Q}_p(j)}$ pour $j < 0$ est le dual du noyau de $H^2(G_{S,F_0}, \mathbb{Q}_p(1-j)) \to \oplus_{v|p} H^2(F_{0,v}, \mathbb{Q}_p(1-j))$, (E) est équivalent à (A). Soulé déduit (C,ii) de la conjecture principale (théorème de Mazur-Wiles et de Wiles) et de la nullité de $H^2(G_{S,F_0}, \mathbb{Q}_p(j))$ pour $j > 0$. Remarquons (comme le fait Kurihara) qu'un résultat récent de Beilinson montre que les C_j proviennent d'un élément non nul de $\mathbb{Q} \otimes_{\mathbb{Z}} K_{2j-1}(F_0)$ par l'application de Chern $c_{j,2}$. Cette application étant injective (Soulé), on en déduit que $C_j(\xi)$ est non nul.

Montrons maintenant (C,iii) (Soulé). Comme $H^1(G_{S,F_0}, \mathbb{Q}_p(j))^{(\xi)}$ est de dimension 1 et que $C_j(\xi)$ est non nul, celui-ci l'engendre. Comme $\log_{\mathbb{Q}_p(j)}$ est injective sur $\oplus_{v|p} H^1(F_{0,v}, \mathbb{Q}_p(j))$, $\log_{\mathbb{Q}_p(j)} C_j(\xi) \neq 0$ si et seulement si le noyau de localisation sur $H^1(G_{S,F_0}, \mathbb{Q}_p(j))^{(\xi)}$ est nul. Par la proposition 3.2.3, cela est équivalent à la non-nullité de $\mathcal{L}_{K-L}(\xi\chi^{-j}) = L_p(j, \xi\omega(\chi)^{-j+1})$.

REMARQUE. Le théorème précédent est suffisant pour la partie de la conjecture de Bloch-Kato donnant l'ordre du zéro de la fonction L du motif $\mathbb{Q}(1-j)$ en termes de $r(\xi, \mathbb{Q}_p(j)) = \dim_{\mathbb{Q}_p} H^1_f(F_0, \mathbb{Q}_p(j))^{(\xi)} - \dim_{\mathbb{Q}_p} H^0(F_0, \mathbb{Q}_p(j))^{(\xi)}$ (dans la version [**F-P91**], [**F-P92**], [**F92**]). On a en effet :

$$r(\xi, \mathbb{Q}_p(j)) = \begin{cases} 0 & \text{si } j < 0 \text{ ou } j = 0,\ \xi \text{ non trivial} \\ -1 & \text{si } j = 0,\ \xi \text{ trivial} \\ 1 & \text{si } j = 1,\ \xi \text{ non trivial} \\ 0 & \text{si } j = 1,\ \xi \text{ trivial} \\ 1 & \text{si } j \geq 2,\ \epsilon(\xi\chi^j) = -1 \\ 0 & \text{si } j \geq 2,\ \epsilon(\xi\chi^j) = 1 \end{cases}$$

4.3. Démonstrations à la "Kolyvagin".

4.3.1. Nous allons démontrer la proposition suivante sans supposer le théorème 4.2.2 en suivant les idées de Kolyvagin (cf. Kurihara [**Ku92**]). Posons

$$\tau_p(k) = \sharp(\mathbb{Q}_p(k)/\mathbb{Z}_p(k))^{G_\mathbb{Q}}.$$

PROPOSITION.   (i) *Si j est un entier impair et si $\mathcal{L}_{K-L}(\chi^{-j}) \neq 0$, alors $H^1(G_{S,\mathbb{Q}}, \mathbb{Z}_p(1-j))$ est fini et annulé par $\tau_p(1-j)\mathcal{L}_{K-L}(\chi^{-j})$.*

(ii) *Si j est un entier impair > 0 et si $C_j = C_j(\mathbf{1})$ est non nul, alors $H^1_f(\mathbb{Q}, \mathbb{Q}_p(1-j)/\mathbb{Z}_p(1-j))$ est fini et annulé par l'indice de C_j dans $H^1(G_{S,\mathbb{Q}}, \mathbb{Z}_p(j))$.*

La démonstration est donnée dans 4.3.3 et suivants.

4.3.2. La proposition implique une partie du théorème 4.2.2 pour $F = \mathbb{Q}$ et ξ caractère trivial. Comme $\mathcal{L}_{K-L}(\chi^{-j})$ est non nul pour j impair et négatif, on en déduit la nullité de $H^1(G_{S,\mathbb{Q}}, \mathbb{Q}_p(1-j))$ et donc de $H^2(G_{S,\mathbb{Q}}, \mathbb{Q}_p(1-j))$ et de $H^2(G_{S,\mathbb{Q}}, \mathbb{Q}_p(j))$, c'est-à-dire (A) pour j pair positif, (D) et (B).

Pour j positif, impair, la proposition implique que si $\mathcal{L}_{K-L}(\chi^{-j})$ est non nul, $H^1(G_{S,\mathbb{Q}}, \mathbb{Q}_p(1-j))$ et $H^2(G_{S,\mathbb{Q}}, \mathbb{Q}_p(1-j))$ sont nuls (moitié de (F)). (ii) implique que si C_j est non nul, $H^1_f(\mathbb{Q}, \mathbb{Q}_p(1-j))$, $H^2(G_{S,\mathbb{Q}}, \mathbb{Q}_p(1-j))$ et $H^2(G_{S,\mathbb{Q}}, \mathbb{Q}_p(j))$ sont nuls (j impair, positif). Enfin, la non-nullité de $\mathcal{L}_{K-L}(\chi^{-j})$ pour j impair, positif implique la non nullité de C_j.

La proposition 4.3.1 n'utilise pas la K-théorie. Par contre, il ne semble pas y avoir jusqu'à présent de moyen d'éviter la K-théorie pour montrer que $C_j \neq 0$ pour j impair, positif.

REMARQUE. Montrons que $\mathrm{III}(\mathbb{Z}(j)) = \oplus_p \mathrm{III}(\mathbb{Z}_p(j))$ est fini. Pour $j < 0$ et impair, il est en effet annulé par $\prod \tau_p(1-j)L(j, \omega(\chi)^{j+1})$ qui est un rationnel. Si $j > 0$ et impair, il est annulé par le produit des indices de C_j dans $H^1(G_{S,\mathbb{Q}}, \mathbb{Z}_p(j))$. Pour p assez grand, les $\mathbb{Z}_p$-modules $\mathbb{Z}_p \otimes_\mathbb{Z} K_{2j-1}(\mathbb{Z})$ et $H^1(G_{S,\mathbb{Q}}, \mathbb{Z}_p(j))$ sont des $\mathbb{Z}_p$-modules libres de même rang. Or l'application

$$\mathbb{Z}_p \otimes_\mathbb{Z} K_{2j-1}(\mathbb{Z}) \to H^1(G_{S,\mathbb{Q}}, \mathbb{Z}_p(j))$$

est surjective ([**D-F85**, theorem 8.7 et remark 8.8]). On en déduit facilement que l'indice de C_j dans $H^1(G_{S,\mathbb{Q}}, \mathbb{Z}_p(j))$ est égal à 1 pour presque tout p. Le cas où j est pair se déduit du cas où j est impair par la dualité décrite dans 4.1.

4.3.3. Nous donnons ici une démonstration de la proposition 4.3.1 due essentiellement à Kurihara au moins pour le (ii) (d'après les idées de Kolyvagin).

Soit j un entier impair, M une puissance de p et $S(M)$ l'ensemble des entiers naturels sans facteurs carrés et dont les facteurs premiers sont congrus à 1 mod M. Si $L \in S(M)$, posons $S = \{p, \infty\}$ et $S \cup \{L\} = S \cup \{l|L\}$. Choisissons un système de racines de l'unité α_L d'ordre L pour tout $L \in S(M)$ tel que $\alpha_L^l = \alpha_{L/l}$ pour tout $l|L$. Choisissons un générateur τ_l de $\mathrm{Gal}(\mathbb{Q}(\mu_L)/\mathbb{Q}(\mu_{L/l}))$

pour tout nombre premier $l|L$. Si L est un élément de $S(M)$, on définit un élément $P_j(L)$ de $H^1(G_{S\cup\{L\},\mathbb{Q}}, (\mathbb{Z}/M\mathbb{Z})(j))$ de la manière suivante : Soient

$$D_L = \prod_{l|L} \sum_{i=1}^{l-2} i\tau_l^i$$

l'opérateur de Kolyvagin et $P(\omega_{\alpha_L}) \in H^1(G_{S\cup\{L\},\mathbb{Q}(\mu_L)}, \mathbb{Z}_p(j))$ défini (cf. 3.3.1) par

$$P(\omega_{\alpha_L,j}) = \sum_{\tau\in\mathrm{Gal}(\mathbb{Q}_n(\mu_L)/\mathbb{Q})} \tau((\omega_{\alpha_L}\otimes\epsilon^{\otimes j})_n)\,.$$

Alors, l'image de $D_L(P(\omega_{\alpha_L,j}))$ dans $H^1(G_{S\cup\{L\},\mathbb{Q}(\mu_L)}, (\mathbb{Z}/M\mathbb{Z})(j))$ est invariant sous l'action de $\mathrm{Gal}(\mathbb{Q}(\mu_L)/\mathbb{Q})$. La nullité de $(\mathbb{Z}/M\mathbb{Z})(j)^{G_{\mathbb{Q}(\mu_L)}}$ (j est impair) implique par la suite inflation-restriction qu'il existe un unique élément $P_j(L)$ de $H^1(G_{S\cup\{L\},\mathbb{Q}}, (\mathbb{Z}/M\mathbb{Z})(j))$ dont l'image dans $H^1(G_{S\cup\{L\},\mathbb{Q}(\mu_L)}, (\mathbb{Z}/M\mathbb{Z})(j))$ est $D_L(P(\omega_{\alpha_L,j}))$; on a $P_j(1) = P(\omega_j) = C_j(\mathbf{1})$. L'intérêt de ces éléments vient de leurs propriétés locales. Comme μ_M est contenu dans $\mathbb{Q}_l$ pour $l \equiv 1 \mod M$, on a des isomorphismes non canoniques

$$H^1(\mathbb{Q}_l^{nr}/\mathbb{Q}_l, (\mathbb{Z}/M\mathbb{Z})(j)) = \mathrm{Hom}_{\mathbb{Z}}(\mathrm{Gal}(\mathbb{Q}_l^{nr}/\mathbb{Q}_l), (\mathbb{Z}/M\mathbb{Z})(j)) \cong (\mathbb{Z}/M\mathbb{Z})(j)$$

et

$$H^1(\mathbb{Q}_l, (\mathbb{Z}/M\mathbb{Z})(j))/H^1(\mathbb{Q}_l^{nr}/\mathbb{Q}_l, (\mathbb{Z}/M\mathbb{Z})(j))$$
$$= H^1(\mathbb{Q}_l^{nr}, (\mathbb{Z}/M\mathbb{Z})(j))^{\mathrm{Gal}(\mathbb{Q}_l^{nr}/\mathbb{Q}_l)}$$
$$= \mathrm{Hom}_{\mathbb{Z}}(\mathbb{Z}_p(1), (\mathbb{Z}/M\mathbb{Z})(j))^{\mathrm{Gal}(\mathbb{Q}_l^{nr}/\mathbb{Q}_l)}$$
$$\cong (\mathbb{Z}/M\mathbb{Z})(j).$$

Soit u_l un générateur de $H^1(\mathbb{Q}_l^{nr}/\mathbb{Q}_l, (\mathbb{Z}/M\mathbb{Z})(j))$. Si $L \in S(M)$ et $l|L$, l'image $P_j(L/l)_l$ de $P_j(L/l)$ dans $H^1(\mathbb{Q}_l, (\mathbb{Z}/M\mathbb{Z})(j))$ est dans $H^1(\mathbb{Q}_l^{nr}/\mathbb{Q}_l, (\mathbb{Z}/M\mathbb{Z})(j))$ et on pose

$$P_j(L/l)_l = m_l(L/l)u_l$$

où u_l est un générateur de $H^1(\mathbb{Q}_l^{nr}/\mathbb{Q}_l, (\mathbb{Z}/M\mathbb{Z})(j))$ avec $m_l(L/l) \in \mathbb{Z}/M\mathbb{Z}$.

4.3.4.

LEMME. *Si v est une place divisant L, on peut écrire*

$$P_j(L)_l \equiv m_l(L/l)U_l \mod H^1(\mathbb{Q}_l^{nr}/\mathbb{Q}_l, (\mathbb{Z}/M\mathbb{Z})(j))$$

avec U_l un générateur de $H^1(\mathbb{Q}_l, (\mathbb{Z}/M\mathbb{Z})(j))/H^1(\mathbb{Q}_l^{nr}/\mathbb{Q}_l, (\mathbb{Z}/M\mathbb{Z})(j)))$.

Le lemme se déduit de la proposition 2.4 de [**Ru90**] (remarquons que pour l premier divisant L, $\mathbb{Q}_l$ contient $(\mathbb{Z}/M\mathbb{Z})(1)$).

4.3.5. Reprenons la démonstration de 4.3.1. On suppose toujours j impair (lorsque j est pair, les C_j sont nuls). Remarquons que $H^1(G_{S,\mathbb{Q}}, \mathbb{Z}_p(j))$ et $H^1(\mathbb{Q}_p, \mathbb{Z}_p(j))$ sont sans torsion (la torsion est égale à $H^0(G_{S,\mathbb{Q}}, \mathbb{Q}_p(j)/\mathbb{Z}_p(j))$ et à $H^0(\mathbb{Q}_p, \mathbb{Q}_p(j)/\mathbb{Z}_p(j)))$ respectivement. Soit m la plus grande puissance de p par laquelle $P_j(1) = C_j$ est divisible dans $H^1(G_{S,\mathbb{Q}}, \mathbb{Z}_p(j))$. Ecrivons $C_j = mz$ avec $z \in H^1(G_{S,\mathbb{Q}}, \mathbb{Z}_p(j))$. De même, $C_j = m_p u_p$ où u_p est un générateur de $H^1(\mathbb{Q}_p, \mathbb{Z}_p(j))$ avec $m | m_p$. Ce qui suit n'a d'intérêt que si m_p est non nul. Nous le supposons et calculons m_p plus tard. On a $z = m^{-1} m_p u_p$. Soit $x \in H^1(G_{S,\mathbb{Q}}, (M^{-1}\mathbb{Z}_p/\mathbb{Z}_p)(1-j))$. Par la dualité globale et comme z est non ramifié en dehors de p, on a $< z, x >_p = 0$. Donc

$$< u_p, m^{-1} m_p x >_p = 0.$$

On en déduit que l'image de $m^{-1} m_p x$ dans $H^1(\mathbb{Q}_p, (\mathbb{Q}_p/\mathbb{Z}_p)(1-j))$ est nulle et donc de $m_p x$ aussi.

Si $l \in S(M)$ et si $m(l)$ est la plus grande puissance de p telle que $P(l) = m(l) z(l)$ avec $z(l) \in H^1(G_{S \cup \{l\}, \mathbb{Q}}, (\mathbb{Z}/M\mathbb{Z})(j))$, on a

$$< z(l), x >_l + < z(l), x >_p = 0 .$$

D'où

$$< z(l), m^{-1} m_p x >_l = 0 .$$

D'autre part, l'image de $z(l)$ dans $H^1(\mathbb{Q}_l, (\mathbb{Z}/M\mathbb{Z})(j))/H^1(\mathbb{Q}_l^{nr}/\mathbb{Q}_l, (\mathbb{Z}/M\mathbb{Z})(j))$ est égale à $m(l)^{-1} m_l(1) U_l$. Comme x, étant non ramifié en l, est orthogonal à $H^1(\mathbb{Q}_l^{nr}/\mathbb{Q}_l, (\mathbb{Z}/M\mathbb{Z})(j))$, l'image dans $H^1(\mathbb{Q}_l, (\mathbb{Z}/M\mathbb{Z})(1-j))$ de l'élément $m(l)^{-1} m_l(1) m^{-1} m_p x$ est orthogonale à $H^1(\mathbb{Q}_l, (\mathbb{Z}/M\mathbb{Z})(j))$ et est donc nulle. En particulier, l'image de $m_l(1) m^{-1} m_p x$ dans $H^1(\mathbb{Q}_l, (\mathbb{Z}/M\mathbb{Z})(1-j))$ est nulle.

L'application restriction

$$H^1(G_{S,\mathbb{Q}}, (\mathbb{Z}/M\mathbb{Z})(j)) \to H^1(G_{S,\mathbb{Q}(\mu_M)}, (\mathbb{Z}/M\mathbb{Z})(j))$$
$$= \operatorname{Hom}_{\mathbb{Z}_p}(G_{S,\mathbb{Q}(\mu_M)}, (\mathbb{Z}/M\mathbb{Z})(j))$$

est injective. Soit H l'extension cyclique d'ordre M/m de $\mathbb{Q}(\mu_M)$, galoisienne sur $\mathbb{Q}$, définie par C_j : l'image de C_j dans $\operatorname{Hom}_{\mathbb{Z}_p}(G_{S,\mathbb{Q}(\mu_M)}, (\mathbb{Z}/M\mathbb{Z})(j))$ se factorise donc exactement par $\operatorname{Gal}(H/\mathbb{Q}(\mu_M))$. Si $l \in S(M)$ est tel que le Frobenius en l engendre $\operatorname{Gal}(H/\mathbb{Q}(\mu_M))$, on a $m_l(1) = m$. Pour un tel l, l'image de $m_p x$ dans $H^1(\mathbb{Q}_l, (\mathbb{Z}/M\mathbb{Z})(1-j))$ est nulle et l'extension $K/\mathbb{Q}(\mu_M)$ définie par $m_p x$ est totalement décomposée en l. Ce qui est impossible par le théorème de Chebotarev sauf si $K = \mathbb{Q}(\mu_M)$. En effet, nécessairement, $H \cap K = \mathbb{Q}(\mu_M)$; soit h un élément de $\operatorname{Gal}(H.K/\mathbb{Q}(\mu_M))$ se projetant sur un générateur de $\operatorname{Gal}(H/\mathbb{Q}(\mu_M))$ et sur un générateur de $\operatorname{Gal}(K/\mathbb{Q}(\mu_M))$. Il existe un nombre premier l tel que le Frobenius en l dans $\operatorname{Gal}(HK/\mathbb{Q})$ soit h ; ainsi l appartient nécessairement à $S(M)$ et l'extension $K/\mathbb{Q}(\mu_M)$ doit être totalement décomposée en l, ce qui est contradictoire avec le choix de h. L'image de $m_p x$ dans $H^1(G_{S,\mathbb{Q}(\mu_M)}, (M^{-1}\mathbb{Z}_p/\mathbb{Z}_p)(1-j))$ est donc nulle. Ainsi, $m_p x$ appartient à $H^1(\mathbb{Q}(\mu_M)/\mathbb{Q}, (M^{-1}\mathbb{Z}_p/\mathbb{Z}_p)(1-j))$.

Comme l'image de $m_p x$ dans $H^1(\mathbb{Q}_p, (M^{-1}\mathbb{Z}_p/\mathbb{Z}_p)(1-j))$ est nulle, on en déduit que $m_p x = 0$.

On a donc montré que m_p annule $H^1(G_{S,\mathbb{Q}}, (M^{-1}\mathbb{Z}_p/\mathbb{Z}_p)(1-j))$ au moins si m_p est non nul. Calculons m_p en termes de $\log_{\mathbb{Q}_p(j)} C_j$ ou de $\lambda_{\mathbb{Q}_p(j)} C_j$ selon le signe de j, puis de $\mathcal{L}_{K-L}(\chi^{-j})$. Supposons d'abord que $j > 1$. Ecrivons

$$\log_{\mathbb{Q}_p(j)} C_j = (\mathrm{Log}_{\mathbb{Q}_p(j)} C_j) e_j$$

avec $\mathrm{Log}_{\mathbb{Q}_p(j)} C_j \in \mathbb{Q}_p$ et e_j la base canonique de $D_p(\mathbb{Z}_p(j))$; l'indice généralisé $[H^1(\mathbb{Q}_p, \mathbb{Z}_p(j))/\mathrm{tors} : D_p(\mathbb{Z}_p(j))]$ est égal à $(1 - p^{-j})^{-1} \mathrm{Tam}_p(\mathbb{Z}_p(j))^{-1}$. Comme $\mathrm{Tam}_p(\mathbb{Z}_p(1-j)) \sim \tau_p(1-j)$, on obtient en utilisant 1.5 que

$$\mathrm{Tam}_p(\mathbb{Z}_p(j))^{-1} \sim \tau_p(1-j)^{-1}(j-1)! \,.$$

Donc,

$$m_p \sim (1 - p^{-j})(j-1)!^{-1} \tau_p(1-j) \mathrm{Log}_{\mathbb{Q}_p(j)} C_j \sim \tau_p(1-j) \mathcal{L}_{K-L}(\chi^{-j})$$

en utilisant 2.1.2 et $\tau_p(j) \sim 1$

Supposons $j < 0$. On a maintenant $\mathrm{Tam}_p(\mathbb{Z}_p(j)) \sim \tau_p(j) \sim 1$. Donc, avec $\lambda_{\mathbb{Q}_p(j)}(C_j) = (\tilde{\lambda}_{\mathbb{Q}_p(j)}(C_j)) e_j$ et en utilisant 1.5, on trouve

$$\begin{aligned}
m_p &\sim \tau_p(1-j)(1 - p^{j-1}) \mathrm{Tam}_p(\mathbb{Z}_p(1-j))^{-1} \tilde{\lambda}_{\mathbb{Q}_p(j)}(C_j) \\
&\sim \tau_p(1-j)(-j)! \mathrm{Tam}_p(\mathbb{Z}_p(j))^{-1} \tilde{\lambda}_{\mathbb{Q}_p(j)}(C_j) \\
&\sim \tau_p(1-j)(-j)! \tilde{\lambda}_{\mathbb{Q}_p(j)}(C_j) \\
&\sim \tau_p(1-j) \mathcal{L}_{K-L}(\chi^{-j}).
\end{aligned}$$

D'où l'assertion (i) de la proposition.

4.3.6.   Montrons rapidement (ii). Soit $x \in H^1_f(\mathbb{Q}, \mathbb{Q}_p(1-j)/\mathbb{Z}_p(1-j))$, c'est-à-dire dans $\ker \mathrm{Loc}_{p, \mathbb{Q}_p(1-j)/\mathbb{Z}_p(1-j)}$. En reprenant la démonstration faite en 4.3.5, on voit qu'il n'est pas nécessaire de multiplier x par m_p mais uniquement par $m(l)$. On obtient alors que m annule x, c'est-à-dire que l'indice de C_j dans $H^1(G_{S,\mathbb{Q}}, \mathbb{Z}_p(j))$ annule $H^1_f(\mathbb{Q}, \mathbb{Q}_p(1-j)/\mathbb{Z}_p(1-j))$.

5. Conjectures p-adiques

Les conjectures qui suivent sont inspirées de nombreux travaux, en particulier ceux de Schneider, Gros, Bloch et Kato...

5.1.   Si α est toujours une racine de l'unité d'ordre m premier à p et F de conducteur m, Soulé ([So83]) définit un élément $C_j^{Sou}(\alpha)$ de $\mathbb{Z}_p(\alpha) \otimes K_{2j-1}(\mathcal{O}_F)$ comme la limite des

$$N_{\mathbb{Q}(\alpha)(\mu_{p^{n+1}})/F}(1 - \alpha^{1/p^{n+1}} \zeta_n) B(\zeta_n)^{j-1} \in K_{2j-1}(\mathcal{O}_F; \mathbb{Z}/p^{n+1}\mathbb{Z})$$

où $B : \mu_{p^{n+1}} \to K_2(\mathcal{O}_{F_n}; \mathbb{Z}/p^{n+1}\mathbb{Z})$ est l'inverse du morphisme de Bockstein. Avec les notations de ce texte, l'image de $C_j^{Sou}(\alpha)$ dans $H^1(F, \mathbb{Z}_p(j))$ est

$$(j-1)! Tr_{\mathbb{Q}(\alpha)/F} P(\omega_\alpha \otimes \epsilon^{\otimes j}).$$

Si ξ est un caractère de conducteur m, posons

$$C_j^{Sou}(\xi) = \sum_{\tau \in G} \xi(\tau)^{-1} C_j^{Sou}(\tau\alpha);$$

l'image de $C_j^{Sou}(\xi)$ dans $H^1(F, \mathbb{Z}_p(j))(\xi)$ est donc $(j-1)! C_j(\xi)$. Les éléments $C_j(\xi)$ proviennent donc de $\mathbb{Q}_p(\xi) \otimes K_{2j-1}(\mathcal{O}_F) = K_{2j-1}(\mathcal{O}_F, \mathbb{Q}_p(\xi))$.

Plus précisément, Beilinson ([**Be**]) montre que $C_j^{Del}(\alpha)$ défini en 3.2.4 provient d'un élément de $K_{2j-1}(\mathcal{O}_F)$, le cas $j = 2$ avait été montré par Soulé dans [**So87**]. Rappelons enfin que $C_j^{Del}(\xi) = (1 - p^{j-1}\xi(\sigma)^{-1})^{-1} C_j(\xi)$.

5.2. Si ξ est un caractère de Dirichlet, posons

$$\Gamma^*(j,\xi) = \begin{cases} \Gamma^*(j) & \text{si } \epsilon(\xi\chi^j) = 1, \\ 1 & \text{si } \epsilon(\xi\chi^j) = -1. \end{cases}$$

La proposition suivante a déjà été vue. Nous la réécrivons afin de comparer les formules avec celle de la conjecture qui les suit.

PROPOSITION. *Soit ξ un caractère de Dirichlet de conducteur m.*

(i) *Si j est un entier négatif et si $\epsilon(\xi\chi^j) = -1$, on a*

$$\mathbb{L}_{K-L}(\xi\chi^{-j}) = L_{\{p\}}(j,\xi)$$

(ii) *Si j est un entier positif et si $\epsilon(\xi\chi^j) = 1$, on a*

$$(1 - p^{j-1}\xi(\sigma)^{-1})^{-1}(1 - p^{-j}\xi(\sigma))\mathbb{L}_{K-L}(\xi\chi^{-j})$$

$$= 2\Gamma^*(j,\xi) G(\xi^{-1})^{-1} \frac{L_{\{p\}}(j,\xi)}{(2i\pi)^j}.$$

CONJECTURE. *(théorème pour $j=2$ [**G-K90**]) Soit ξ un caractère de Dirichlet de conducteur m. Si j est un entier positif et si $\epsilon(\xi\chi^j) = -1$, pour tout élément B non nul de $\mathbb{Q} \otimes K_{2j-1}(\mathcal{O}_F)^{(\xi^{-1})}$, on a*

$$\mathbb{L}_{K-L}(\xi\chi^{-j}) = \Gamma^*(j,\xi) \frac{L_{\{p\}}(j,\xi)}{\log_{\mathbb{Q}(j)} B} (\log_{\mathbb{Q}_p(j)} B)/e_{-j}).$$

Ici, $\log_{\mathbb{Q}(j)}$ désigne l'application régulateur complexe et $(\log_{\mathbb{Q}_p(j)} B)/e_{-j}$ est l'élément de $\mathbb{Q}_p$ tel que

$$\log_{\mathbb{Q}_p(j)} B = (\log_{\mathbb{Q}_p(j)} B)/e_{-j})e_{-j}.$$

REMARQUES.   (i) Si l'on revient à la définition de $\mathbb{I}_{arith}(F, \mathbb{Z}_p(j))_{-\epsilon_j}$ où
ϵ_j est le signe de $(-1)^j$, celui-ci est dans $\wedge^0 D_p(\mathbb{Q}_p(1-j)) \otimes \Lambda[G]_{-\epsilon_j}$, ce
qui explique l'absence de facteurs du type $(1 - \varphi)^{-1}(1 - p^{-1}\varphi^{-1})$ dans
les formules de (i) et de la conjecture : la puissance extérieure 0-ième
d'un endomorphisme est 1.

(ii) Par contre, $I_{arith}(F, \mathbb{Z}_p(j))_{\epsilon_j}$ est un élément de $\wedge^1 D_p(\mathbb{Q}_p(1-j)) \otimes \Lambda[G]_{\epsilon_j}$,
ce qui est égal à $D_p(\mathbb{Q}_p(1-j)) \otimes \Lambda[G]_{\epsilon_j}$. Dans la formule (ii), on reconnait
dans $(1-p^{j-1}\xi(\sigma)^{-1})^{-1}(1-p^{-j}\xi(\sigma))$ l'action de l'opérateur $(1-\varphi)^{-1}(1-p^{-1}\varphi^{-1})$ sur $D_p(\mathbb{Q}_p(1-j)) \otimes \Lambda[G]^{(\xi^{-1})}$.

(iii) Toutes ces formules peuvent se généraliser à un "motif quelconque"
([**Pb**]) et on peut en donner alors une formulation unifiée.

(iv) Les formules 5.2 peuvent être écrites sous la forme de conjectures de
type Bloch-Kato p-adiques sans faire appel aux valeurs complexes de la
fonction L.

RÉFÉRENCES

[B-K90]   S. Bloch et K. Kato, *L functions and Tamagawa numbers of motives*, The Grothendieck Festschrift, vol. 1, Prog. in Math. 86, Birkhaüser, Boston, 1990, pp. 333–400.

[Be]   A. A. Beilinson, *Polylogarithm and cyclotomic elements*, prépublication.

[C77]   J. Coates, *p-adic L functions and Iwasawa theory*, Algebraic Number Fields, (A. Fröhlich, ed) Academic Press, 1977, pp. 269–353.

[Co82]   R. F. Coleman, *Dilogarithms, regulators and p-adic L-functions*, Invent. Math. 69 (1982), 171-208.

[De89]   P. Deligne, *Le groupe fondamental de la droite projective moins trois points*, dans Galois Groups over $\mathbb{Q}$, (Y. Ihara, K. Ribet, J.-P. Serre eds.), MSRI publications 16, Springer Berlin (1989), pp. 79-293.

[D-F85]   W. Dwyer et E. Friedlander, *Algebraic and étale K-theory*, Trans. Am. Math. Soc. 292 (1985), 247-280.

[Fl90]   M. Flach, *A generalisation of the Cassels pairing*, J. reine. angew. Math 412 (1990), 113-127.

[F92]   J.-M. Fontaine, *Valeurs spéciales de fonctions L des motifs*, Séminaire Bourbaki, exposé 751, février 1992.

[F-P91]   J.-M. Fontaine et B. Perrin-Riou, *Autour des conjectures de Bloch et Kato*, C.R. Acad. Sci. Paris, 313, série I (1991), *I - Cohomologie galoisienne*, 189-196, *II - Structures motiviques f-closes*, 349-356; *III - Le cas général*, 421-428.

[F-P92]   J.-M. Fontaine et B. Perrin-Riou, *Autour des conjectures de Bloch et Kato : cohomologie galoisienne et valeurs de fonctions L*, à paraître dans les Proceedings d'une conférence sur les motifs à Seattle (AMS)

[G-K90]   M. Gros, *Régulateurs syntomiques et valeurs de fonctions L p-adiques I*, avec un appendice de M. Kurihara : *Computation of the syntomic regulator in the cyclotomic case*, Invent. Math. 99 (1990), 293-320.

[I-S87]   H. Ichimura et K. Sakaguch, *The non-vanishing of a certain Kummer character χ_m (after Soulé) and some related topics*, Galois representations and arithmetic algebraic geometry, Kyoto 1985/ Tokyo 1986 Adv. on Math 12 (1987), pp. 53-64.

[Iw69]   K. Iwasawa, *On p-adic L functions*, Ann. of Math. 89 (1969), 198-205.

[Iw73]   K. Iwasawa, *On $\mathbb{Z}_l$-extensions of algebraic number fields*, Ann. of Math. 98 (1973), 246-326.

[K-M76]   F. Knudsen et D. Mumford, *The projectivity of the moduli space of stable curves I : Preliminaries on "det" and "Div"*, Math. Scand. 39 (1976), 19-55.

[Ko90]   V.A. Kolyvagin, *Euler systems*, The Grothendieck Festschrift, vol. 2, Prog. in Math. 87, Birkhaüser, Boston, 1990, pp. 436–483.

[Ku92]   M. Kurihara, *Some remarks on conjectures about cyclotomic fields and K-groups of* $\mathbb{Z}$, Comp. Math. 81 (1992), 223-236.

[M-W84]   B. Mazur et A. Wiles, *Class fields of abelian extensions of* $\mathbb{Q}$, Invent. Math. 76 (1984), 179-330.

[N93]   W. Niziol,*Cohomology of crystalline representations*, Duke Math. J. 71 (1993),747-791.

[P92]   B. Perrin-Riou, *Théorie d'Iwasawa des représentations p-adiques : le cas local*, C.R. Acad. Sci. Paris, 315, série I (1992), 629-632.

[Pa]   B. Perrin-Riou, *Théorie d'Iwasawa des représentations p-adiques sur un corps local*, (à paraître dans Invent. Math.).

[Pb]   B. Perrin-Riou, *Fonctions L p-adiques des représentations p-adiques*, en préparation.

[Pc]   B. Perrin-Riou, *Fonctions L p-adiques d'une courbe elliptique et points rationnels*, Ann. Inst. Fourier, 43, (1993).

[Ru90]   K. Rubin, *The main conjecture*, Appendix to Cyclotomic fields (seconde édition) par S. Lang, Graduate Texts in Math. 121, Springer-Verlag (1990).

[S89]   P. Schneider, *Motivic Iwasawa Theory*, Adv. Stud. Pure Math. 17 (1989), 421-456.

[So80]   C. Soulé, *On higher p-adic regulators*, Algebraic K-theory, Evanston 1980, Lecture Notes in Math., vol 854, Springer-Verlag, Berlin et New-York, pp. 372-401.

[So83]   C. Soulé, *K-théorie et zéros aux points entiers de fonctions zéta*, Proc. ICM. Warszawa (1983).

[So87]   C. Soulé, *Eléments cyclotomiques en K-theorie*, Astérisque 147-148 (1987), 225-257.

[Wi90]   A. Wiles, *The Iwasawa conjecture for totally real field*, Ann. Math. 131 (1990), 493-540.

UNIVERSITÉ PIERRE ET MARIE CURIE, LMF, UFR 21, TOUR 45-46, 4 PLACE JUSSIEU, F-75252 PARIS CEDEX 05, FRANCE

E-mail address: bpr@ccr.jussieu.fr

Contemporary Mathematics
Volume **174**, 1994

Supersingular p-adic Height Pairings
on Elliptic Curves

ANDREW PLATER

ABSTRACT. We study two possible techniques for defining p-adic height pairings on supersingular elliptic curves, one using Iwasawa theory and the other using p-adic Hodge theory, and show that they give the same result.

0. Introduction

Let E be an elliptic curve defined over $\mathbb{Q}$ having good supersingular reduction at an odd prime p. The problem of constructing a p-adic height pairing for E can be reduced to the problem of splitting certain extensions associated to E. There are a number of possible ways to do this and the purpose of this paper is to show that two of these methods give rise to the same splitting and so to the same height pairing.

More specifically, start with an extension

$$0 \to \mathcal{G}_m \to \mathcal{G}_X \to \mathcal{G}_E \to 0$$

of the formal group $\mathcal{G}_E$ of E by the formal multiplicative group $\mathcal{G}_m$. We shall compare two splittings of the extension of points

$$(1) \qquad 0 \to \mathcal{G}_m(\mathbb{Z}_p) \otimes \mathbb{Q}_p \to \mathcal{G}_X(\mathbb{Z}_p) \otimes \mathbb{Q}_p \to \mathcal{G}_E(\mathbb{Z}_p) \otimes \mathbb{Q}_p \to 0.$$

The first splitting we consider depends upon a choice of splitting of the Hodge filtration of the Dieudonné module of $\mathcal{G}_E$ and is due to Nekovář. The second splitting is defined using Iwasawa theory and a generalisation of the notion of universal norms due to Perrin-Riou.

The paper is arranged as follows: in section 1, we recall some definitions in the theory of formal groups over rings of Witt vectors. In section 2, we show

1991 *Mathematics Subject Classification.* Primary 11G07, 11S31; Secondary 11R23, 14L05.
The author was supported by SERC Fellowship #B/91/RFO/1149.
This paper is in final form, and no version of it will be submitted for publication elsewhere

how splittings of the Hodge filtration induce splittings of the local extensions
of points. In section 3, we construct the sequences of points in $\mathbb{Z}_p$-extensions
satisfying certain compatibility conditions which we shall use to construct the
Iwasawa theoretic splitting. In section 4, we construct this splitting, and show
that it is the same as that constructed in section 2.

I am grateful to Jan Nekovář for explaining his work to me and to the referee
for a number of useful comments.

1 Formal groups over rings of Witt vectors

1.1. In this section, we shall define the Dieudonné module of a formal group
and give enough of the basic theory to allow us to define our splittings. We use
the model for the Dieudonné module given in [1] and more details concerning
the basic definitions can be found in [1].

1.2. Let H be a finite unramified extension of $\mathbb{Q}_p$ with ring of integers W,
residue field k, and absolute Frobenius σ. Let $\mathcal{P}_d$ be the subset of $H[[T_1, \ldots T_d]]$
of those formal power series f satisfying $\partial f/\partial T_i \in W[[T_1, \ldots T_d]]$ for all i and let

$$\overline{\mathcal{P}}_d = \mathcal{P}_d/pW[[T_1, \ldots T_d]].$$

For $d = 1$, we write $\mathcal{P}$ for $\mathcal{P}_d$. Following Fontaine, we identify $CW(k[[T_1, \ldots T_d]])$
with $\overline{\mathcal{P}}_d$ via the map

$$(\ldots, a_{-1}, a_0) \mapsto \sum (\tilde{a}_{-n})^{p^n}/p^n$$

where $\tilde{a}_{-n}$ is any lift of a_{-n} to $W[[T_1, \ldots T_d]]$. With this identification, the endomorphisms

$$F : (\ldots, a_{-1}, a_0) \mapsto (\ldots, a_{-1}^p, a_0^p)$$
$$V : (\ldots, a_{-1}, a_0) \mapsto (\ldots, a_{-2}, a_{-1})$$

of $CW(k[[T]])$ correspond to endomorphisms F and V of $\overline{\mathcal{P}}$ given by the formulae

$$F\left(\sum a_n T^n\right) = \sum \sigma(a_n) T^{pn}$$
$$V\left(\sum a_n T^n\right) = \sum p\sigma^{-1}(a_{pn}) T^n.$$

1.3. Let $\mathcal{G}$ be a formal group over W of finite dimension d. Choosing coordinates, we can identify the affine algebra $\mathcal{R}$ of $\mathcal{G}$ with a formal power series ring
$W[[T_1, \ldots T_d]]$. We define the Dieudonné module M of $\mathcal{G}$ by

$$M = \{f \in \overline{\mathcal{P}}_d : f(X \oplus Y) = f(X) + f(Y)\}$$

where $\oplus$ denotes the formal group law of $\mathcal{G}$. Define $\mathcal{L} \subset \mathcal{P}_d$ by

$$\mathcal{L} = \{f \in \mathcal{P}_d : f(X \oplus Y) = f(X) + f(Y)\}$$

and let $L \subset M$ be the image of $\mathcal{L}$ mod $pW[[T_1, \ldots T_d]]$.

1.4. We now define a logarithm map for $\mathcal{G}$. We denote by $\operatorname{Hom}_W^{alg}(\ ,\ )$ and $\operatorname{Hom}_W(\ ,\ )$ homomorphisms in the categories of p-adic W-algebras and p-adic W-modules respectively. Let A be any commutative W-algebra and define $A_H = A \otimes_W H$. Then, by definition, $\mathcal{G}(A) = \operatorname{Hom}_W^{alg}(\mathcal{R}, A)$. We identify $\operatorname{Hom}_W(VL, A_H)$ with the tangent space of $\mathcal{G}(A)$. Let $\varphi \in \mathcal{G}(A)$. Then φ extends uniquely to a homomorphism, also denoted by φ, from $\mathcal{L}$ to A_H. We define the logarithm map $\log_{\mathcal{G}} : \mathcal{G}(A) \to \operatorname{Hom}_W(VL, A_H)$ by

$$\log_{\mathcal{G}}(\varphi)(V\bar{l}) = \varphi(l)$$

where l is the unique lift of $\bar{l}$ to $\mathcal{L}$. We can extend $\log_{\mathcal{G}}$ by linearity to define an isomorphism from $\mathcal{G}(A) \otimes \mathbb{Q}_p$ to $\operatorname{Hom}_W(VL, A_H)$.

2. Analytic splittings

2.1. In this section, we use the approach of Zarhin and Nekovář to associate a splitting of (1) to any splitting of the Hodge filtration of the Dieudonné module of $\mathcal{G}_E$. The reader should consult [**3**] for the same theory in a more general setting.

2.2. We keep the same notation as in 1.2. Let E be an elliptic curve over W with formal group $\mathcal{G}_E$ and let

$$(2) \qquad 0 \to \mathcal{G}_m \to \mathcal{G}_X \to \mathcal{G}_E \to 0$$

be an extension of $\mathcal{G}_E$ by the formal multiplicative group. Let M_E, M_X, and M_m be the corresponding Dieudonné modules, with canonical subgroups L_E, L_X, and L_m, as defined in 1.3. So we have extensions

$$(3) \qquad 0 \to M_E \to M_X \to M_m \to 0$$

and

$$(4) \qquad 0 \to L_E \to L_X \to L_m \to 0.$$

For a $\mathbb{Z}_p$-module M, we shall add a subscript $\mathbb{Q}_p$ to denote the $\mathbb{Q}_p$-vector space $M_{\mathbb{Q}_p} = M \otimes \mathbb{Q}_p$ obtained by extending scalars to $\mathbb{Q}_p$. In order to construct a splitting of the extension of points

$$(5) \qquad 0 \to \mathcal{G}_m(W) \otimes \mathbb{Q}_p \to \mathcal{G}_X(W) \otimes \mathbb{Q}_p \to \mathcal{G}_E(W) \otimes \mathbb{Q}_p \to 0$$

it is enough, thanks to the logarithm map, to construct a splitting of the corresponding sequence of tangent spaces

$$(6) \qquad 0 \to \operatorname{Hom}_W(VL_m, H) \to \operatorname{Hom}_W(VL_X, H) \to \operatorname{Hom}_W(VL_E, H) \to 0$$

obtained by applying V to (4) and dualising. To do this, it is enough to construct a splitting of the extension

$$(7) \qquad 0 \to VL_{E,\mathbb{Q}_p} \to VL_{X,\mathbb{Q}_p} \to VL_{m,\mathbb{Q}_p} \to 0,$$

or equivalently, to construct a complementary subspace to $VL_{E,\mathbb{Q}_p}$ in $VL_{X,\mathbb{Q}_p}$.

2.3. Because F acts on M_E and M_m with different slopes, the sequence

$$0 \to M_{E,\mathbb{Q}_p} \to M_{X,\mathbb{Q}_p} \to M_{m,\mathbb{Q}_p} \to 0$$

splits uniquely as an extension of $H[F]$-modules, i.e. there is a complementary H-subspace $\hat{M}_m$ to $M_{E,\mathbb{Q}_p}$ in $M_{X,\mathbb{Q}_p}$ which is stable under F. In the case $H = \mathbb{Q}_p$, this is just linear algebra; F is a linear map and acts on $M_{E,\mathbb{Q}_p}$ with characteristic polynomial $X^2 - a_p X + p$, on $M_{m,\mathbb{Q}_p}$ with characteristic polynomial $X - p$, and on $M_{X,\mathbb{Q}_p}$ with characteristic polynomial $(X^2 - a_p X + p)(X - p)$. So in this case, $\hat{M}_m$ is just the p-eigenspace of the action of F on $M_{X,\mathbb{Q}_p}$. We write $\alpha : M_{m,\mathbb{Q}_p} \to M_{X,\mathbb{Q}_p}$ for the splitting of (3) associated to this decomposition. Because the resultant of $X^2 - a_p X + p$ and $X - p$ has p-adic valuation equal to one, α does not map M_m to M_X in general, but only to $p^{-1} M_X$. Because of this, we must use the map $p\alpha$ rather than α when dealing with $\mathbb{Z}_p$-modules. I am grateful to the referee for pointing out this fact.

2.4. The Hodge filtration of $M_{E,\mathbb{Q}_p}$ is given by $Fil^2 M_{E,\mathbb{Q}_p} = 0$, $Fil^1 M_{E,\mathbb{Q}_p} = VL_{E,\mathbb{Q}_p}$, $Fil^0 M_{E,\mathbb{Q}_p} = M_{E,\mathbb{Q}_p}$. Let N be any splitting of this filtration, i.e. any H-submodule of $M_{E,\mathbb{Q}_p}$ such that

$$M_{E,\mathbb{Q}_p} = N \oplus VL_{E,\mathbb{Q}_p}.$$

This splitting, together with that of 2.3, allows us to write $M_{X,\mathbb{Q}_p}$ as a direct sum

$$M_{X,\mathbb{Q}_p} = M_{E,\mathbb{Q}_p} \oplus \hat{M}_m = N \oplus VL_{E,\mathbb{Q}_p} \oplus \hat{M}_m.$$

This decomposition induces a splitting of (7), the complementary subgroup of $VL_{E,\mathbb{Q}_p}$ in $VL_{X,\mathbb{Q}_p}$ being given by $VL_{X,\mathbb{Q}_p} \cap \{N \oplus \hat{M}_m\}$, and this splitting induces a splitting of (5), the complementary subspace of $\mathcal{G}_m(W) \otimes \mathbb{Q}_p$ in $\mathcal{G}_X(W) \otimes \mathbb{Q}_p$ consisting of those φ in $\mathcal{G}_X(W) \otimes \mathbb{Q}_p$ satisfying $\log_X(\varphi)(m) = 0$ for all m in $VL_X \cap \{N \oplus \hat{M}_m\}$.

2.5. We have now shown that any splitting of the Hodge filtration of M_E allows us to construct splittings of the extensions (5) for all possible extensions (2) of $\mathcal{G}_E$ by $\mathcal{G}_m$, and these splittings (or rigidifications) are compatible with taking Baer sums (see [5] for the definition of the Baer sum of two rigidified extensions). Conversely, any such collection of compatible splittings come from a splitting of the Hodge filtration in this way; this is immediate from the identification of $M_E \otimes \mathbb{Q}_p$ with the group of rigidified extensions of $\mathcal{G}_E$ by $\mathcal{G}_m$. See [2],[5] or [3] for this identification.

3. Compatible sequences of points

3.1. In this section, we shall follow Perrin-Riou [4] to construct sequences of points of formal groups lying in $\mathbb{Z}_p$-extensions. These sequences will take the place played by norm compatible sequences for curves with ordinary reduction. All the techniques of this section are taken from [4] and the reader should consult that paper for more details.

3.2. For all $n \geq 0$, fix a primitive p^{n+1}-th root of unity ζ_n related by $\zeta_{n+1}^p = \zeta_n$. Define $K_n = H(\zeta_n)$ with ring of integers $\mathcal{O}_{K_n}$ and let K_∞ be the union of the K_n, with Galois group $G_\infty \cong \mathbb{Z}_p^\times$. Let H_n be the sub-extension of K_n/H of degree p^n with ring of integers $\mathcal{O}_{H_n}$ and let H_∞ be the union of the H_n, with Galois group Γ. So H_∞ is the cyclotomic $\mathbb{Z}_p$-extension of H.

3.3. Define $\varphi(T) = (1+T)^p - 1 \in W[[T]]$ and consider φ as an endomorphism of $\mathcal{P}$ with action

$$\varphi\left(\sum a_n T^n\right) = \sum \sigma(a_n)\varphi(T)^n.$$

Let ψ be the unique σ^{-1}-linear endomorphism of $\mathcal{P}$ satisfying

$$\psi \circ \varphi = p \qquad \text{and} \qquad \varphi \circ \psi(f)(T) = \sum_\zeta f\left(\zeta(1+T) - 1\right)$$

where the sum is over the p-th roots of unity. So φ and ψ reduce to F and V on $\overline{\mathcal{P}}$.

3.4. Let $\mathcal{G}$ be any connected formal group over W, and also write $\mathcal{G}$ for the reduction of $\mathcal{G}$ to a formal group over k. Given $\rho \in \mathcal{G}(k[[T]]) \cong \mathrm{Hom}_{W[F]}(M, \overline{\mathcal{P}})$, let $\hat{\rho}$ be the unique lift to $\mathrm{Hom}_W(M, \mathcal{P})$ satisfying

$$\psi[\varphi(\hat{\rho}(m)) - \hat{\rho}(Fm)] = 0 \qquad \text{for all } m \in M.$$

Then $\hat{\rho}$ also satisfies

$$P(\psi^f/p^f)(\hat{\rho}(m)) = 0 \qquad \text{for all } m \in M$$

where

$$P(X) = \det_W\left(1 - F^f X | M\right)$$

and f is the degree of the local field extension $H/\mathbb{Q}_p$.

3.5. For any commutative p-adic W-algebra A, let $x \in \mathcal{G}(A) \cong \mathrm{Hom}_W^{alg}(\mathcal{R}, A)$. Then x extends uniquely to a homomorphism $x_{\mathcal{L}} : \mathcal{L} \to A_H$. Reducing x mod p gives a map $\tilde{x} : \mathcal{R}/p\mathcal{R} \to A/pA$, which induces a map

$$CW(\tilde{x}) : CW(\mathcal{R}/p\mathcal{R}) \to CW(A/pA).$$

Let x_M be the restriction of $CW(\tilde{x})$ to $M \subseteq CW(\mathcal{R}/p\mathcal{R})$. Then it is proven in [1] (IV.1.3) that the map $x \mapsto (x_M, x_{\mathcal{L}})$ defines an isomorphism from $\mathcal{G}(A)$ to the fibre product $\mathrm{Hom}_{W[F]}(M, CW(A/pA)) \times \mathrm{Hom}_W(\mathcal{L}, A_H)$. Using this isomorphism, we define a homomorphism

$$\mathcal{G}(k[[T]]) \to \mathcal{G}(W[[T]])$$

$$\rho \mapsto (\rho, \hat{\rho}|_{\mathcal{L}})$$

where we abuse notation and write $\hat{\rho}|_{\mathcal{L}}(l) = \hat{\rho}(\bar{l})$, where $l \mapsto \bar{l}$ is the reduction map from $\mathcal{L}$ to L.

3.6. For $n \geq 0$, let $\Sigma_n : \mathcal{G}(k[[T]]) \to \mathcal{G}(\mathcal{O}_{K_n})$ be the composite of the homomorphism $\mathcal{G}(k[[T]]) \to \mathcal{G}(W[[T]])$ of 3.5 with the specialisation $T \mapsto \zeta_n - 1$. For a point $P \in \mathcal{G}(k[[T]])$, we write P'_n for $\Sigma_n(P)$ and $P_n = Tr_{K_n/H_n}(P'_n)$. It is easy to check that the specialisation maps and the logarithm maps are related as follows; let $P \in \mathcal{G}(W[[T]]) \cong \mathrm{Hom}_W^{alg}(\mathcal{R}, W[[T]])$ with $\log_{\mathcal{G}}(P) \in \mathrm{Hom}_W(VL, H[[T]])$, with specialisations $P'_n \in \mathcal{G}(\mathcal{O}_{K_n})$ and $P_n \in \mathcal{G}(\mathcal{O}_{H_n})$. Then

$$\log_{\mathcal{G}}(P'_n)(m) = \left(\log_{\mathcal{G}}(P)(m)\right)(\zeta_n - 1)$$

and

$$\log_{\mathcal{G}}(P_n)(m) = Tr_{K_n/H_n}\left(\log_{\mathcal{G}}(P)(m)\right)(\zeta_n - 1).$$

Recall from [4] Lemme 1.4, that the operators φ and ψ are related to the specialisation maps by the formulae

$$(8) \quad \varphi(x)(\zeta_{n+1} - 1) = x^\sigma(\zeta_n - 1), \qquad \psi(x^\sigma)(\zeta_{n-1} - 1) = Tr_{K_n/K_{n-1}} x(\zeta_n - 1),$$

for all $x \in \mathcal{P}$. So points $P \in \mathcal{G}(k[[T]])$ parametrise sequences of points in $\mathcal{G}(\mathcal{O}_{K_n})$ and $\mathcal{G}(\mathcal{O}_{H_n})$ satisfying certain relations. We will now work out these relations for $\mathcal{G}_m$, $\mathcal{G}_X$, and $\mathcal{G}_E$.

3.7. Assume from now on that $W = \mathbb{Z}_p$ and $H = \mathbb{Q}_p$. We begin with the case $\mathcal{G} = \mathcal{G}_E$. So F acts on M_E with inverse characteristic polynomial

$$P_E(X) = pX^2 - a_p X + 1.$$

Let P be an element of $\mathcal{G}_E(k[[T]])$, which we identify with a homomorphism $\rho \in \mathrm{Hom}_{W[F]}(M_E, \overline{\mathcal{P}})$. By 3.4, $\hat{\rho}(m)$ is killed by $P_E(\psi/p)$ for all $m \in M_E$. So, setting $f = \hat{\rho}(m)$ and specialising to $T = \zeta_n - 1$, we have

$$0 = (\psi^2 - a_p\psi + p)f(\zeta_n - 1)$$

$$= Tr_{K_{n+2}/K_n} f(\zeta_{n+2} - 1) - a_p Tr_{K_{n+1}/K_n} f(\zeta_{n+1} - 1) + pf(\zeta_n - 1)$$

On the level of points in $\mathcal{G}_E(\mathcal{O}_{K_n})$, this means

$$Tr_{K_{n+2}/K_n}(P'_{n+2}) - a_p Tr_{K_{n+1}/K_n}(P'_{n+1}) + pP'_n = 0$$

and taking traces from K_n to H_n gives

$$(9) \quad Tr_{H_{n+2}/H_n}(P_{n+2}) - a_p Tr_{H_{n+1}/H_n}(P_{n+1}) + pP_n = 0.$$

3.8. For $\mathcal{G} = \mathcal{G}_m$, F acts on M_m with inverse characteristic polynomial

$$P_m(X) = pX - 1.$$

Let R be a point of $\mathcal{G}_m(k[[T]])$ identified with $\rho \in \mathrm{Hom}_{W[F]}(M_m, \overline{\mathcal{P}})$. If we set $f = \hat{\rho}(m)$, then f is killed by $\psi - 1$, so we get

$$Tr_{K_{n+1}/K_n} f(\zeta_{n+1} - 1) - f(\zeta_n - 1) = 0,$$

which implies that

$$(10) \quad Tr_{H_{n+1}/H_n}(R_{n+1}) - R_n = 0.$$

So R specialises to a norm compatible sequence $\{R_n\}$ with $R_n \in \mathcal{G}_m(\mathcal{O}_{H_n})$.

3.9. For $\mathcal{G} = \mathcal{G}_X$, F acts on M_X with inverse characteristic polynomial

$$P_X(X) = (pX - 1)(pX^2 - a_p X + 1).$$

Let Q be a point of $\mathcal{G}_X(k[[T]])$ identified with $\rho \in \text{Hom}_{W[F]}(M_X, \overline{\mathcal{P}})$. If we set $f = \hat{\rho}(m)$, then f is killed by

$$(\psi - 1)(\psi^2 - a_p\psi + p) = \psi^3 - (a_p + 1)\psi^2 + (a_p + p)\psi - p$$

which implies that the specialisations Q_n are related by

(11) $$\begin{aligned} Tr_{H_{n+3}/H_n}(Q_{n+3}) - (a_p + 1)Tr_{H_{n+2}/H_n}(Q_{n+2}) \\ + (a_p + p)Tr_{H_{n+1}/H_n}(Q_{n+1}) - pQ_n = 0. \end{aligned}$$

3.10. We have now seen that the specialisation maps allow us to parametrise sequences of points $P_n \in \mathcal{G}(\mathcal{O}_{H_n})$ satisfying certain relation by elements of $\mathcal{G}(k[[T]])$. We shall now restrict to certain subgroups of $\mathcal{G}(k[[T]])$, on which the points parametrised will satisfy stronger relations. For any connected formal group $\mathcal{G}$ and any W-submodule N of M, define

$$Z_N = \{\rho \in \mathcal{G}(k[[T]]) : \varphi(\hat{\rho}(m)) - \hat{\rho}(Fm) = 0 \text{ for all } m \in N\}.$$

When it is unclear which formal group we are using, we will add another subscript and write $Z_{X,N}$ or $Z_{E,N}$ for Z_N.

3.11. Take $\mathcal{G} = \mathcal{G}_E$ and $N = L_E \subset M_E$. Then by [4] Lemme 4.10, for all $l \in L_E$, $\hat{\rho}(l)$ is killed by $\varphi - a_p + \psi$, which implies that points $P \in Z_{L_E}$ parametrise sequences of points P_n satisfying

(12) $$Tr_{H_{n+2}/H_{n+1}}(P_{n+2}) - a_p P_{n+1} + P_n = 0$$

as an element of $\mathcal{G}_E(\mathcal{O}_{H_{n+1}})$. Taking the trace from $\mathcal{G}_E(\mathcal{O}_{H_{n+1}})$ to $\mathcal{G}_E(\mathcal{O}_{H_n})$, we recover the weaker relation (9).

LEMMA 3.12. *Given $\rho_X \in \mathcal{G}_X(k[[T]])$, define $\rho_E \in \mathcal{G}_E(k[[T]])$ and $\rho_m \in \mathcal{G}_m(k[[T]])$ by*

$$\rho_E(m_E) = \rho_X(m_E), \qquad \rho_m(m_m) = \rho_X(p\alpha(m_m)).$$

Then

$$\hat{\rho}_E(m_E) = \hat{\rho}_X(m_E) \qquad and \qquad \hat{\rho}_m(m_m) = \hat{\rho}_X(p\alpha(m_m)).$$

PROOF. Define $\tilde{\rho}_E \in \text{Hom}_W(M_X, \mathcal{P})$ by $\tilde{\rho}_E(m_E) = \hat{\rho}_X(m_E)$. Then $\tilde{\rho}_E(m_E)$ is a lift of ρ_E to $\text{Hom}_W(M_X, \mathcal{P})$ which satisfies the property uniquely defining $\hat{\rho}_E$, namely that

$$\psi[\varphi(\tilde{\rho}_E(m)) - \tilde{\rho}_E(Fm)] = 0$$

for all $m \in M_E$. So $\tilde{\rho}_E = \hat{\rho}_E$. Similarly, define $\tilde{\rho}_m$ by $\tilde{\rho}_m(m_m) = \hat{\rho}_X(p\alpha(m_m))$. Then $\tilde{\rho}_m$ satisfies the property uniquely defining $\hat{\rho}_m$. So $\tilde{\rho}_m = \hat{\rho}_m$.

LEMMA 3.13. *Let A and B be W-submodules of M_E and M_m respectively. Let $< A, p\alpha(B) >$ be the W-submodule of M_X generated by A and $p\alpha(B)$. Given $\rho_X \in \mathcal{G}_X(k[[T]])$, define $\rho_E \in \mathcal{G}_E(k[[T]])$ and $\rho_m \in \mathcal{G}_m(k[[T]])$ as in Lemma 3.12. Then $\rho_X \in Z_{X,<A,p\alpha(B)>}$ if and only if $\rho_E \in Z_{E,A}$ and $\rho_m \in Z_{m,B}$.*

PROOF. Suppose $\rho_E \in Z_{E,A}$ and $\rho_M \in Z_{m,B}$. We must show that ρ_X lies in $Z_{X,<A,p\alpha(B)>}$. Take $m_X = a + p\alpha(b) \in< A, p\alpha(B) >$, with $a \in A, b \in B$. Then

$$\varphi(\hat{\rho}_X(a + p\alpha(b))) - \hat{\rho}_X(F(a + p\alpha(b)))$$
$$= \{\varphi(\hat{\rho}_X(a)) - \hat{\rho}_X(Fa)\} + \{\varphi(\hat{\rho}_X(p\alpha(b))) - \hat{\rho}_X(Fp\alpha(b))\}$$
$$= \{\varphi(\hat{\rho}_E(u)) - \hat{\rho}_E(Fu)\} + \{\varphi(\hat{\rho}_m(b)) - \hat{\rho}_m(Fb)\}$$
$$= 0.$$

The converse is just as easy.

LEMMA 3.14. *The image of the homomorphism $Z_{X,p\alpha(M_m)} \to \mathcal{G}_E(k[[T]])$ contains $p\mathcal{G}_E(k[[T]])$.*

PROOF. Denote by β the map $M_X \to M_m$ in (3). For any $m_X \in M_X$ we have $pm_X - p\alpha(\beta(m_X)) \in M_E$. Given $\rho_E \in \mathcal{G}_E(k[[T]]) = \operatorname{Hom}_{W[F]}(M_E, \overline{\mathcal{P}})$, define $\rho_X \in \mathcal{G}_X(k[[T]]) = \operatorname{Hom}_{W[F]}(M_X, \overline{\mathcal{P}})$ by

$$\rho_X(m_X) = \rho_E(pm_X - p\alpha(\beta(m_X))).$$

Then ρ_X maps to $p\rho_E$ in $\mathcal{G}_E(k[[T]])$ and ρ_X lies in $Z_{X,p\alpha(M_m)}$ by 3.13.

COROLLARY 3.15. *The image of the map $Z_{X,<L_E,p\alpha(M_m)>} \to Z_{E,L_E}$ contains pZ_{E,L_E}.*

PROOF. The subgroup Z_{X,L_E} is the inverse image of Z_{E,L_E} in $\mathcal{G}_X(k[[T]])$, and $Z_{X,<L_E,p\alpha(M_m)>}$ is the intersection of Z_{X,L_E} and $Z_{X,p\alpha(M_m)}$.

LEMMA 3.16. *Take $\mathcal{G} = \mathcal{G}_X$ and $N = p\alpha(M_m)$. Then for all $\rho_X \in Z_{X,N}$, $\hat{\rho}_X(m)$ is killed by $(\varphi - p)\psi^2(\varphi^2 - a_p\varphi + p)$ for all $m \in M_X$.*

PROOF. In the extension

$$0 \to M_{E,\mathbb{Q}_p} \to M_{X,\mathbb{Q}_p} \overset{\alpha}{\underset{\beta}{\rightleftarrows}} M_{m,\mathbb{Q}_p} \to 0,$$

$\alpha(M_{m,\mathbb{Q}_p})$ is equal to $(F^2 - a_pF + p)M_{X,\mathbb{Q}_p}$. It follows that, for any $m_X \in M_X$, $(F^2 - a_pF + p)m_X$ lies in $\alpha(M_m)$ (explicitly, $(F^2 - a_pF + p)m_X = \alpha(m_m)$ where $m_m = (F^2 - a_pF + p)m_X$.) Let $m_X \in M_X$. Recall from 3.8 that ψ acts as the identity on $\hat{\rho}_m(m_m)$ for any $m_m \in M_m$, and therefore also on $\hat{\rho}_X(\alpha(m_m))$. So for $\rho_X \in Z_{X,p\alpha(M_m)}$ we have

$$
\begin{aligned}
(\varphi - p)\psi^2(\varphi^2 - a_p\varphi + p)\hat{\rho}_X(m) &= (\varphi - p)\psi^2\hat{\rho}_X((F^2 - a_pF + p)m) \\
&= (\varphi - p)\psi^2\hat{\rho}_X(\alpha(m_m)) \quad \text{for some } m_m \in M_m \\
&= p^{-1}(\varphi - p)\hat{\rho}_X(p\alpha(m_m)) \\
&= p^{-1}\hat{\rho}_X((F - P)p\alpha(m_m)) \\
&= 0.
\end{aligned}
$$

COROLLARY 3.17. *A point $Q \in Z_{X,\hat{M}_m}$ specialises to a sequence of points $Q_n \in \mathcal{G}_X(\mathcal{O}_{H_n})$ which satisfy the relation*

$$
\begin{aligned}
(13) \quad &\left\{ Tr_{H_{n+2}/H_n}(Q_{n+2}) - a_pTr_{H_{n+1}/H_n}(Q_{n+1}) + pQ_n \right\} \\
&- p\left\{ Tr_{H_{n+3}/H_{n+1}}(Q_{n+3}) - a_pTr_{H_{n+2}/H_{n+1}}(Q_{n+2}) + pQ_{n+1} \right\} = 0.
\end{aligned}
$$

Taking traces from $\mathcal{G}_X(\mathcal{O}_{H_{n+1}})$ to $\mathcal{G}_X(\mathcal{O}_{H_n})$ of this relation, we get the weaker relation (11).

4. Algebraic splittings

4.1. In this section we will use the sequences of points constructed in section 3 to construct splittings of (1) analogous to the construction using norm compatible sequences in the case of ordinary reduction.

4.2. We begin with the exact sequence

$$
0 \to \mathcal{G}_m(\mathbb{Z}_p) \to \mathcal{G}_X(\mathbb{Z}_p) \to \mathcal{G}_E(\mathbb{Z}_p) \to 0
$$

of free $\mathbb{Z}_p$-modules. Let P be a point of $\mathcal{G}_E(\mathbb{Z}_p)$. Our aim is to construct a canonical lift $\tilde{P}$ of P to $\mathcal{G}_X(\mathbb{Z}_p)$. Recall from [4] Lemme 4.11 that the map $Tr_{K_0/\mathbb{Z}_p}\Sigma_0 : Z_{E,L_E} \to \mathcal{G}_E(\mathbb{Z}_p)$ is surjective (for odd p), and so we can choose a sequence of points $P_n \in \mathcal{G}_E(\mathcal{O}_{H_n})$ with $P_0 = P$ specialising a point in Z_{E,L_E}, and therefore satisfying the relations

$$
Tr_{H_{n+2}/H_{n+1}}(P_{n+2}) - a_pP_{n+1} + P_n = 0.
$$

These points will take the place of a norm compatible sequence for a curve with ordinary reduction. For all n, let Q_n be any lift of P_n to $\mathcal{G}_X(\mathcal{O}_{H_n})$ and let $\check{Q}_n = Tr_{H_n/\mathbb{Q}_p}(Q_n)$. Then the $\check{Q}_n$ form a sequence in $\mathcal{G}_X(\mathbb{Z}_p)$ satisfying

$$
\check{Q}_{n+2} - a_p\check{Q}_{n+1} + p\check{Q}_n \equiv 0 \quad \mod Tr_{H_{n+1}/\mathbb{Q}_p}\mathcal{G}_m(\mathcal{O}_{H_{n+1}}).
$$

So the $\check{Q}_n$ almost satisfy a recurrence relation.

LEMMA 4.3. *There is a unique sequence $\{\theta_n\}$ in $\mathcal{G}_X(\mathbb{Z}_p)$ satisfying $\theta_n - \check{Q}_n \in Tr_{H_n/\mathbb{Q}_p}\mathcal{G}_m(\mathcal{O}_{H_n})$ and*

$$
(14) \qquad\qquad \theta_{n+2} - a_p\theta_{n+1} + p\theta_n = 0.
$$

PROOF. Both existence and uniqueness are easy linear algebra.

Uniqueness: Suppose that $\{\theta_n\}$ and $\{\theta'_n\}$ are two such sequences and let $\phi_n = \theta_n - \theta'_n$. So $\phi_n \equiv 0 \mod Tr_{H_n/\mathbb{Q}_p}\mathcal{G}_m(\mathcal{O}_{H_n})$. We claim that

$$\phi_n \equiv 0 \qquad \mod Tr_{H_{n+k}/\mathbb{Q}_p}\mathcal{G}_m(\mathcal{O}_{H_{n+k}})$$

for all $k \geq 0$. Assuming the result for k, (14) shows that

$$p\phi_n \equiv 0 \qquad \mod Tr_{H_{n+k+2}/\mathbb{Q}_p}\mathcal{G}_m(\mathcal{O}_{H_{n+k+2}})$$

and so $\phi_n \equiv 0 \mod Tr_{H_{n+k+1}/\mathbb{Q}_p}\mathcal{G}_m(\mathcal{O}_{H_{n+k+1}})$ and we are done by induction.

Existence: We will refine the sequence $\{\check{Q}_n\}$ to get a sequence of sequences $\{\theta^1_n\}, \{\theta^2_n\}, \{\theta^3_n\} \ldots$ which tend to $\{\theta_n\}$. Define $\theta^1_n = \check{Q}_n$ and define $\{\theta^{i+1}_n\}$ from $\{\theta^i_n\}$ by

$$\theta^i_{n+2} - a_p\theta^i_{n+1} + p\theta^{i+1}_n = 0$$

(It is easy to see that $\theta^i_{n+2} - a_p\theta^i_{n+1}$ is divisible by p in $\mathcal{G}_X(\mathbb{Z}_p)$ so we can choose such a θ^i_{n+1}). The θ^i_n are related by

$$\theta^{i+1}_n \equiv \theta^i_n \mod Tr_{H_{n+i-1}/\mathbb{Q}_p}\mathcal{G}_m(\mathcal{O}_{H_{n+i-1}})$$

and $\qquad \theta^i_{n+2} - a_p\theta^i_{n+1} + p\theta^i_n \equiv 0 \mod Tr_{H_{n+i-1}/\mathbb{Q}_p}\mathcal{G}_m(\mathcal{O}_{H_{n+i-1}})$

so the θ^i_n converge to a sequence $(\theta_n)_{n\geq 0}$ with the required properties.

LEMMA 4.4. *The sequence $(\theta_n)_{n\geq 0}$ is independent of the choice of P_n.*

PROOF. Because Z_{E,L_E} has Λ-rank one, two choices (P_n) and (P'_n) are related by $P_n = f(T)P'_n$, for some $f \in \Lambda$ with $f(0) = 1$. If Q'_n is a lifting of P'_n then we can take $Q_n = f(T)Q'_n$ as a lifting of P_n. Then we have $\check{Q}_n = \check{Q}'_n$, and $\theta_n = \theta'_n$.

4.5. In particular, θ_0 gives a canonical lift of P to $\mathcal{G}_X(\mathbb{Z}_p)$ and we define $\tilde{P} = \theta_0$. In order to relate this lifting to the splitting of section 2, we now give another characterisation of $\tilde{P}$ as an image of the specialisation maps. For any n, define $I'_{N,n}$ to be the image of the specialisation map $\Sigma_n : Z_N \to \mathcal{G}(\mathcal{O}_{K_n})$ and define $I_{N,n} = Tr_{K_n/H_n}\left(I'_{N,n}\right)$ and $I_N = I_{N,0} \subseteq \mathcal{G}(\mathbb{Z}_p)$. To make clear which formal group we are in, we shall add the subscript X or E to I_N.

LEMMA 4.6. *There is a unique sequence $\{\theta_n\}$ in $\mathcal{G}_X(\mathbb{Z}_p)$ satisfying $\theta_n - \check{Q}_n \in Tr_{H_n/\mathbb{Q}_p}\mathcal{G}_m(\mathcal{O}_{H_n})$ and*

(15) $\{\theta_{n+3} - a_p\theta_{n+2} + p\theta_{n+1}\} - \{\theta_{n+2} - a_p\theta_{n+1} + p\theta_n\} = 0.$

PROOF. The sequence $\{\theta_n\}$ constructed in 4.3 satisfies (15). The proof of uniqueness is identical to that in 4.3.

LEMMA 4.7. *Given $P_0 \in p\mathcal{G}_E(\mathbb{Z}_p)$, there is a unique lift $\tilde{P}_0$ to $\mathcal{G}_X(\mathbb{Z}_p)$ which lies in $I_{X,<L_E,p\alpha(M_m)>}$, and this $\tilde{P}_0$ is the same as the $\tilde{P}$ constructed in 4.5.*

PROOF. Choose $P \in pZ_{E,L_E}$ which specialises to a sequence $\{P_n\}$ starting with P_0 and let $Q \in Z_{X,<L_E,p\alpha(M_m)>}$ map to P (such a Q exists by 3.14). Let $\{Q_n\}$ be the specialisations of Q to $\mathcal{G}_X(\mathcal{O}_{H_n})$. Then Q_n maps to P_n in $\mathcal{G}_E(\mathcal{O}_n)$

and the traces $\check{Q}_n = Tr_{H_n/\mathbb{Q}_p}(Q_n)$ satisfy (15). So $\check{Q}_n = \theta_n$. In particular, $\check{Q}_0 = \tilde{P}_0$ lies in $I_{X,<L_E,p\alpha(M_m)>}$.

COROLLARY 4.8. *The subspace* $I_{X,<L_E,p\alpha(M_m)>} \otimes \mathbb{Q}_p = I_{X,L_E} \otimes \mathbb{Q}_p$ *is a complementary subspace to* $\mathcal{G}_m(\mathbb{Z}_p) \otimes \mathbb{Q}_p$ *in* $\mathcal{G}_X(\mathbb{Z}_p) \otimes \mathbb{Q}_p$.

THEOREM 4.9. *The splitting of (1) constructed in this section is the same as that constructed in Section 2 using* $N = L_E$.

PROOF. We must show that if Q is an element of $Z_{X,<L_E,p\alpha(M_m)>}$ which specialises to $Q_0 \in \mathcal{G}_X(\mathbb{Z}_p)$, then $\log_{\mathcal{G}_X}(Q_0) \in \text{Hom}_W(VL_X, \mathbb{Q}_p)$ is identically zero on $VL_X \cap < L_E, p\alpha(M_m) >$. Identify Q with $\rho_X \in \text{Hom}_{W[F]}(M_X, \overline{\mathcal{P}})$ which we use to define a pair (ρ_E, ρ_m) as in 3.13. Then ρ_E lies in Z_{E,L_E}, which implies that $\hat{\rho}_E$ is in $\text{Hom}_W(M_E/L_E, \mathcal{P})$ by [4] Lemme 4.9. Also, ρ_m lies in Z_{m,M_m}, which implies that $\hat{\rho}_m$ is in $\text{Hom}_W(M_m, \mathbb{Z}_p)^{V=1} \log(1+T)$ by [4] Théorème 3.1. Now write $Vl \in VL_X \cap < L_E, p\alpha(M_m) >$ as $m_E + p\alpha(m_m)$, with $m_E \in L_E$ and $m_m \in M_m$. Then we have

$$\hat{\rho}_X(Vl) = \hat{\rho}_E(m_E) + \hat{\rho}_m(m_m) \in \mathbb{Z}_p \log(1+T).$$

Now $\log_{\mathcal{G}_X}(\hat{\rho}_X)$ is the map given by $Vl \mapsto \hat{\rho}_X(l)$. So we have

$$\begin{aligned}
\log_{\mathcal{G}_X}(\hat{\rho}_X)(Vl) = \hat{\rho}_X(l) &= p^{-1}\hat{\rho}_X(FVl)\\
&= p^{-1}\varphi\left(\hat{\rho}_X(Vl)\right)\\
&= p^{-1}\varphi(c.\log(1+T)) \qquad \text{for some } c \in W\\
&= c.\log(1+T).
\end{aligned}$$

So finally, using the relations between specialising and taking logarithms given in 3.6, we get

$$\log(Q_0)(Vl) = Tr_{K_0/\mathbb{Q}_p}(c.\log(1+(\zeta_0-1))) = Tr_{K_0/\mathbb{Q}_p}(c.\log(\zeta_0)) = 0$$

as required.

REFERENCES

1. J.-M. Fontaine, *Groupes p-divisible sur les corps locaux*, Astérisque **47-48** (1977).

2. N. Katz, *Crystalline cohomology, Dieudonné modules and Jacobi sums.*, Automorphic forms, representation theory and arithmetic., Tata Inst. Fund. Res. Studies in Math., vol. 10, Bombay, 1981, pp. 165–246.

3. J. Nekovář, *On p-adic height pairings*, Séminaire de Théorie des Nombres, Paris 1990-91.

4. B. Perrin-Riou., *Théorie d'Iwasawa p-adique locale et globale*, Invent. Math. **99** (1990), 247–292.

5. Yu. G. Zarhin, *p-adic height pairings on abelian varieties.*, Séminaire de Théorie des Nombres, Paris 1987-88, Progress in Math., vol 81, Birkhäuser, 1990.

UNIVERSITY OF CAMBRIDGE, DEPARTMENT OF PURE MATHEMATICS AND MATHEMATICAL STATISTICS, 16 MILL LANE, CAMBRIDGE, CB2 1SB, ENGLAND.

E-mail address: A.J.Plater@pmms.cam.ac.uk

Fields of definition of abelian varieties with real multiplication

KENNETH A. RIBET

1. Introduction

Let K be a field, and let $\overline{K}$ be a separable closure of K. Let C be an elliptic curve over $\overline{K}$. For each g in the Galois group $G := \mathrm{Gal}(\overline{K}/K)$, let $^g C$ be the elliptic curve obtained by conjugating C by g. One says that C is an elliptic K-curve if all the elliptic curves $^g C$ are $\overline{K}$-isogenous to C.

Recall that a subfield L of $\overline{K}$ is said to be a $(2,\dots,2)$-extension of K if L is a compositum of a finite number of quadratic extensions of K in $\overline{K}$. The extension L/K is then Galois, and $\mathrm{Gal}(L/K)$ is an elementary abelian 2-group. Recently, N. Elkies proved:

(1.1) THEOREM (Elkies, [**2**]). *Let C be an elliptic K-curve over $\overline{K}$ with no complex multiplication. Then C is $\overline{K}$-isogenous to an elliptic curve defined over a $(2,\dots,2)$-extension of K.*

In this article, we present an approach to (1.1) which seems different from that of Elkies. At the same time, we generalize (1.1) to include higher-dimensional analogues of elliptic K-curves with no complex multiplication. These are abelian varieties A over $\overline{K}$ whose endomorphism algebras are totally real fields of dimension $\dim(A)$.

For lack of a better term, we borrow the phrase "Hilbert-Blumenthal abelian varieties" to refer to abelian varieties whose endomorphism algebras are totally real fields of maximal dimension. Our use of this expression is a bit unusual. Indeed, in standard parlance, a Hilbert-Blumenthal abelian variety relative to

1991 AMS Subject Classification: 11F11, 11G10 (primary); 14K10, 14K15. This manuscript is in final form. It was written primarily during the program on L-functions at the Isaac Newton Institute for Mathematical Sciences. It is a pleasure to thank the INIMS for its hospitality. The author wishes to thank N. Elkies as well as W. Raskind and J.-L. Colliot-Thélène for helpful e-mail correspondence. This work was partially supported by NSF Grants DMS 88-06815 and DMS 93-06898.

a totally real number field F is an abelian variety A over $\overline{K}$ which is furnished
with an action of the ring of integers $\mathcal{O}$ of F. One requires that the Lie algebra
$\mathrm{Lie}(A/\overline{K})$ be free of rank one over $\mathcal{O} \otimes \overline{K}$, which acts on $\mathrm{Lie}(A/\overline{K})$ by functoriality; in particular, this requirement forces the dimension of A and the degree
of F to be equal. In this article, we insist that $\mathrm{End}(A) \otimes \mathbf{Q}$ be equal to (and not
bigger than) a totally real field of dimension $\dim(A)$. On the other hand, we do
not require the full ring of integers of this field to act on A.

Suppose that A is a Hilbert-Blumenthal abelian variety in our sense, and let
F be the totally real number field $\mathrm{End}(A) \otimes \mathbf{Q}$. We say that A is a K-Hilbert-
Blumenthal abelian variety (or "K-HBAV") if $^g A$ is F-equivariantly isogenous
to A for all $g \in G$. The equivariance refers to the evident isomorphism $\varphi \mapsto {}^g\varphi$
between the endomorphism algebras of A and of $^g A$: we demand that there be
for each $g \in \mathrm{Gal}(\overline{K}/K)$ an isogeny $\mu_g : {}^g A \to A$ which satisfies $\varphi \circ \mu_g = {}^g\varphi \circ \mu_g$ for
all $\varphi \in F$.

(1.2) THEOREM. *Suppose that A is a K-HBAV. Then A is F-equivariantly
isogenous to a Hilbert-Blumenthal abelian variety over a finite $(2, \dots, 2)$ extension of K.*

One motivation for proving Theorem 1.2 is the study of Jacobians of modular
curves. Indeed, let f be a weight-two newform on the group $\Gamma_1(N)$, and let X_f
be the abelian variety associated to f be Shimura's construction [**7**, Th. 7.14].
Thus X_f is a $\mathbf{Q}$-simple factor of the abelian variety $J_1(N)$. If f is a cusp for
with complex multiplication, then X_f becomes isogenous to a power of a CM
elliptic curve over $\overline{\mathbf{Q}}$. In the opposite case, Propositions 2.1–2.2 below show that
the $\overline{\mathbf{Q}}$-simple factors of X_f are then either $\mathbf{Q}$-HBAVs or quaterionic analogues
of $\mathbf{Q}$-HBAVs. (It should be possible to prove a version of Theorem 1.2 in the
quaternionic case as well.) The absolute decomposition of X_f is controlled by
coincidences between the Galois conjugates of f and twists of f by Dirichlet
characters (see [**4**]). From the point of view of [**4**], one sees that this absolute
decomposition is achieved over the abelian extension of $\mathbf{Q}$ cut out by the set
of Dirichlet characters which intervene. This extension is a $(2, \dots, 2)$-extension
of $\mathbf{Q}$ if f has trivial Nebentypus character, but not in general. Nevertheless,
Theorem 1.2 tells us that the absolute "building blocks" of X_f are defined over
a $(2, \dots, 2)$-extension of $\mathbf{Q}$ in all cases.

We turn now to a discussion of the proof of Theorem 1.2 and some associated
results. First of all, let us indicate how Theorem 1.2 follows immediately from a
series of results in §3. As we will see, if A is a K-HBAV, then A defines a class γ in
the cohomology group $\mathrm{H}^2(G, F^*)$ made from locally constant cocycles $G \times G \longrightarrow$
F^* and the trivial action of G on F^*. Proposition 3.1 shows that the class γ
represents the obstruction to finding a Hilbert-Blumenthal abelian variety over
K which is isogenous over $\overline{K}$ to the given one. Therefore, to prove (1.2) is to show
that γ becomes trivial under the cohomological restriction map corresponding
to a base extension $K \rightsquigarrow K'$, where K' is a $(2, \dots, 2)$-extension of K.

Now, by Proposition 3.2, γ lies in the subgroup $\mathrm{H}^2(G, F^*)[2]$ of $\mathrm{H}^2(G, F^*)$ consisting of classes of order at most two. On the other hand, Theorem 3.3 asserts that each element of $\mathrm{H}^2(G, F^*)[2]$ becomes trivial under the restriction map corresponding to a base extension $K \rightsquigarrow K'$ of the desired type. This completes our discussion of (1.2).

Secondly, we wish to highlight a technical point that arises in applying Theorem 3.3 to γ. Namely, let P be the quotient $F^*/\{\pm 1\}$ so that we have an exact sequence of abelian groups

$$0 \to \{\pm 1\} \to F^* \to P \to 0.$$

This sequence is *split*, since the abelian group P is free (Lemma 3.5). Consequently, $\mathrm{H}^2(G, F^*)[2]$ is a split extension of $\mathrm{H}^2(G, P)[2]$ by $\mathrm{H}^2(G, \{\pm 1\})$. As will be seen in §3, there is an elementary isomorphism $\mathrm{Hom}(G, P/P^2) \xrightarrow{\sim} \mathrm{H}^2(G, P)[2]$. Hence we have a split exact sequence

$$0 \to \mathrm{H}^2(G, \{\pm 1\}) \to \mathrm{H}^2(G, F^*)[2] \to \mathrm{Hom}(G, P/P^2) \to 0.$$

Call $\overline{\gamma}$ the image of γ in $\mathrm{Hom}(G, P/P^2)$. Then $\overline{\gamma}$ cuts out a $(2, \dots, 2)$-extension of K. This is the extension K_P of K in $\overline{K}$ such that $\mathrm{Gal}(\overline{K}/K_P)$ is the kernel of $\overline{\gamma}$. Certainly, any extension of K in $\overline{K}$ which trivializes γ must contain K_P.

In the case of elliptic curves (i.e., the case $F = \mathbf{Q}$), Elkies shows that γ is trivialized by the field K_P. This point, although admittedly technical, seems very striking to us. The present article may be viewed as an attempt to find a generalization of this phenomenon to the Hilbert-Blumenthal case.

Such a generalization is presented in §4, where we introduce the presumably superfluous requirement that K has characteristic zero. (This hypothesis does not intervene in [2].) We show (in Corollary 4.5 below) that γ is trivialized by K_P whenever $[F : \mathbf{Q}]$ is odd, and more generally whenever there is an embedding $F \hookrightarrow \overline{K}$ for which the degree $[FK : K]$ is odd. It would be interesting to determine whether this hypothesis is necessary.

In the case where $[F : \mathbf{Q}]$ is odd, F^* is canonically a product $P \times \{\pm 1\}$. Indeed, P may be identified with the *subgroup* of F^* consisting of elements with positive norm to $\mathbf{Q}^*$. Hence we have canonically

$$\mathrm{H}^2(G, F^*)[2] = \mathrm{H}^2(G, \{\pm 1\}) \times \mathrm{Hom}(G, P/P^2).$$

This suggests a study of the image $\gamma_\pm$ of γ in the first factor $\mathrm{H}^2(G, \{\pm 1\})$. One may be tempted to think that $\gamma_\pm$ is trivial, which would certainly explain the vanishing of γ over K_P. Although $\gamma_\pm$ is trivial for the $\mathbf{Q}$-elliptic curves constructed by Shimura in [8], there seem to be examples where $\gamma_\pm$ can be non-trivial. One such example was communicated to the author by E. Pyle of Berkeley, California; here, $K = \mathbf{Q}$ and A is a $\mathbf{Q}$-elliptic curve.

2. K-Hilbert-Blumenthal abelian varieties

Let K, $\overline{K}$ and G be as above, and let F be a totally real number field.

We consider pairs (A, ι), where A is an abelian variety of dimension $[F : \mathbf{Q}]$ over $\overline{K}$, and where ι is an isomorphism $F \xrightarrow{\sim} \mathbf{Q} \otimes \mathrm{End}(A)$; such a pair will be called a Hilbert-Blumenthal abelian variety. As mentioned above, it would be more standard to allow ι to be an *injection* $F \hookrightarrow \mathbf{Q} \otimes \mathrm{End}(A)$; however, the more restrictive definition seems to be convenient in what follows. It should be stressed that our Hilbert-Blumenthal abelian varieties are, in particular, non-CM abelian varieties.

Abusing notation, we will generally write A for the pair (A, ι). Moreover, we will frequently view A as an object in the category of abelian varieties *up to isogeny* over $\overline{K}$. In this category, isogenies of abelian varieties become isomorphisms, and the endomorphism algebra usually denoted $\mathbf{Q} \otimes \mathrm{End}(A)$ can be written more simply as $\mathrm{End}(A)$.

Suppose that g is an element of G. Then ${}^g A$ admits a natural multiplication ${}^g \iota$ by F, so that ${}^g A$ is again a Hilbert-Blumenthal abelian variety. As mentioned in §1, we say that A is a K-HBAV if there is an F-equivariant isomorphism $\mu_g \colon {}^g A \xrightarrow{\sim} A$ of abelian varieties up to isogeny for each $g \in G$. Notice that a K-HBAV of dimension one is an elliptic K-curve with no complex multiplication.

To motivate the study of K-HBAVs, we record some facts concerning $\overline{\mathbf{Q}}$-simple factors of abelian varieties over $\mathbf{Q}$ with many endomorphisms. We begin with some terminology: an abelian variety A over $\overline{\mathbf{Q}}$ is said to be a "fake" Hilbert-Blumenthal abelian variety if its endomorphism algebra is a quaternion division algebra over a totally real field F and if $\dim(A) = 2 \cdot [F : \mathbf{Q}]$. Also, an abelian variety C over $\mathbf{Q}$ is said to be of $\mathbf{GL}_2$-type if $\mathbf{Q} \otimes \mathrm{End}_{\mathbf{Q}}(C)$ is a number field of degree $\dim(C)$. One knows that the $\mathbf{Q}$-simple factors of the Jacobian of a modular curve $X_1(N)$ are of $\mathbf{GL}_2$-type. Moreover, it is reasonable to conjecture that *all* abelian varieties of $\mathbf{GL}_2$-type over $\mathbf{Q}$ are $\mathbf{Q}$-simple factors of the Jacobian of some $X_1(N)$, cf. [5, 4.4]. (Such a conjecture may be viewed as a higher-dimensional analogue of the conjecture of Taniyama and Shimura to the effect that all elliptic curves over $\mathbf{Q}$ are modular.)

(2.1) PROPOSITION. *Suppose that C is an abelian variety over $\mathbf{Q}$ of $\mathbf{GL}_2$-type. Then $C_{/\overline{\mathbf{Q}}}$ is "isotypical": it is isogenous to a product $A \times \cdots \times A$, where A is a simple abelian variety over $\overline{\mathbf{Q}}$. Further, A must be one of the following: (i) an elliptic curve with complex multiplication; (ii) a Hilbert-Blumenthal abelian variety (for some totally real field F); (iii) a "fake" Hilbert-Blumenthal abelian variety.*

PROOF. Proposition 2.1 is implicit in the discussion of §5 of [5] (which is based principally on results of G. Shimura). For completeness, we shall deduce the proposition from some results included in [5].

First of all, if $C_{/\overline{\mathbf{Q}}}$ contains any non-zero abelian subvariety with complex multiplication, then a result of Shimura (Proposition 1.5 of [8]) implies that

$C_{/\overline{\mathbf{Q}}}$ is a power of an elliptic curve with complex multiplication; thus, we are in case (i). Assume instead that $C_{/\overline{\mathbf{Q}}}$ has no non-zero abelian subvariety with complex multiplication. The number field $E := \mathbf{Q} \otimes \mathrm{End}_{\mathbf{Q}}(C)$ is then its own commutant in $\mathcal{X}$, the algebra of all endomorphisms of C over $\overline{\mathbf{Q}}$. This implies that the center of $\mathcal{X}$ is contained in E: it is therefore a subfield F of E. One shows that F is in fact a *totally real* number field [**5**, 5.4]. Since the center of $\mathcal{X}$ is a single field (as opposed to a product of several fields), C is isotypical, as we claimed. In fact, if we write $\mathcal{X}$ as $\mathrm{M}(n, D)$, where D is a division algebra with center F, then $C_{/\overline{\mathbf{Q}}}$ is isogenous to the nth power of a simple abelian variety A over $\overline{\mathbf{Q}}$ whose endomorphism algebra is D.

Let t be such that t^2 is the rank of D over F. Then a short calculation, based on the fact that E is a maximal commutative semisimple subalgebra of $\mathrm{M}(n, D)$, establishes the formula $\dim(A) = t \cdot [F : \mathbf{Q}]$. An argument exploiting the action of D on $\mathrm{H}^1(A(\mathbf{C}), \mathbf{Q})$ shows that t is at most 2. (See the proof of Proposition 5.2 of [**5**] for these facts.) If $t = 1$, we are in case (ii), while if $t = 2$ we are in case (iii). ∎

(2.2) PROPOSITION. *Suppose that C is an abelian variety over $\mathbf{Q}$ of $\mathbf{GL}_2$-type. Let A be a simple $\mathbf{Q}$-quotient of C whose endomorphism algebra is a totally real field. Then A is a $\mathbf{Q}$-HBAV.*

PROOF. From what we have seen, the abelian variety $C_{/\overline{\mathbf{Q}}}$ is isogenous to a product $A \times \cdots \times A$ (with, say, n factors), and we are in case (ii). Using the analysis of [**5**], §5 again, we see that the center of the endomorphism algebra $\mathcal{X}$ of $C_{/\overline{\mathbf{Q}}}$ is a subalgebra of the algebra of $\mathbf{Q}$-endomorphisms of C. If this center is F, then $\mathcal{X}$ is isomorphic to $\mathrm{M}(n, F)$, and the endomorphism algebra of A is F. After fixing an isomorphism $\mathcal{X} \approx \mathrm{M}(n, F)$, we may view A as the image of the matrix whose upper left-hand corner entry is 1 and whose other entries are 0. In this model, there is an obvious F-equivariant isogeny $\lambda \colon C_{\overline{\mathbf{Q}}} \to A^n$, given by the n different matrices with a single 1 in the first column and 0's elsewhere.

Let g be an element of G and take $d \in F$. Then we have a commutative diagram

$$
\begin{array}{ccccccc}
A^n & \xleftarrow{\lambda} & C & = & {}^g C & \xrightarrow{{}^g\lambda} & {}^g A^n \\
\downarrow d & & \downarrow d & & \downarrow {}^g d & & \downarrow {}^g d \\
A^n & \xleftarrow{\lambda} & C & = & {}^g C & \xrightarrow{{}^g\lambda} & {}^g A^n
\end{array}
$$

in which the central square expresses the fact that d is defined over $\mathbf{Q}$. Contracting it, we get a diagram

$$
\begin{array}{ccc}
A^n & \xrightarrow{\sim} & {}^g A^n \\
\downarrow d & & \downarrow {}^g d \\
A^n & \xrightarrow{\sim} & {}^g A^n
\end{array}
$$

in which the isomorphism $A^n \xrightarrow{\sim} {}^g A^n$ is ${}^g\lambda \circ \lambda^{-1}$.

For each pair of integers (i, j) with $1 \le i, j \le n$, we get a map $A \to {}^g A$ by composing the following maps: the inclusion $A \hookrightarrow A^n$ which uses the ith coordinate, the isomorphism $A^n \approx {}^gA^n$, and the projection ${}^gA^n \to {}^gA$ which uses the jth coordinate. For some pair (i,j), this map is non-zero, hence an isogeny. (Recall that A is simple.) Let κ be this isogeny. Then we have

$$
\begin{array}{ccc}
A & \overset{\kappa}{\to} & {}^gA \\
\downarrow d & & \downarrow {}^gd \\
A & \overset{\kappa}{\to} & {}^gA
\end{array}
$$

as desired.   ■

3. The class γ

Let A be a K-HBAV. For each g, let μ_g be an F-equivariant isomorphism up to isogeny ${}^gA \to A$. We can, and do, assume that the collection (μ_g) has been constructed from a model A_o of A over a finite extension L of K in such a way that μ_g and $\mu_{g'}$ are the same map ${}^gA \to A$ whenever g and g' coincide on L. Thus the association $g \mapsto \mu_g$ is in an obvious sense locally constant. For each pair $\sigma, \tau \in G$, we note that $\mu_\sigma \circ {}^\sigma\mu_\tau \circ \mu_{\sigma\tau}^{-1}$ is an automorphism of A up to isogeny, and consequently of the form $\iota(c(\sigma,\tau))$ with $c(\sigma,\tau) \in F^*$. The map $(\sigma,\tau) \mapsto c(\sigma,\tau)$ is a continuous two-cocycle on G with values in F^*, with F^* being regarded as a trivial G-module. The image γ of c in $\mathrm{H}^2(G,F^*)$ is independent of the choices of the μ_σ.

The following proposition is a mild generalization of [**5**, Th. 8.2].

(3.1) PROPOSITION. *Let E be an extension of K in $\overline{K}$; let $H = \mathrm{Gal}(\overline{K}/E)$ be the corresponding closed subgroup of G. Then the Hilbert-Blumenthal abelian variety A is F-equivariantly isogenous to a Hilbert-Blumenthal abelian variety over E if and only if the class γ lies in the kernel of the restriction map $\mathrm{H}^2(G,F^*) \to \mathrm{H}^2(H,F^*)$.*

PROOF. It suffices to prove the proposition in the case $E = K$, in which case the assertion to be proved is that $\gamma = 1$ if and only if A is (F-equivariantly) isogenous to a Hilbert-Blumenthal abelian variety over K.

Suppose first that there is a Hilbert-Blumenthal abelian variety B over K, together with an isomorphism $\lambda \colon A \overset{\sim}{\to} B_{/\overline{K}}$ of Hilbert-Blumenthal abelian varieties up to isogeny over $\overline{K}$. For each $g \in G$, we obtain an isomorphism ${}^g\lambda \colon {}^gA \overset{\sim}{\to} B$. After setting $\mu_g := \lambda^{-1} \circ {}^g\lambda$, we find that c is identically 1, so that γ is trivial.

Conversely, suppose that $\gamma = 1$. Let L and A_o be as above; to simplify notation, we shall write A for A_o. After replacing L by a finite extension of L, we can, and do, assume that the μ_σ are defined over L. Further, the hypothesis that $\gamma = 1$ means that there is a locally constant function $\alpha \colon G \to F^*$ so that $c(\sigma,\tau) = \alpha(\sigma)\alpha(\tau)/\alpha(\sigma\tau)$; we enlarge L, if necessary so that α is defined modulo $\mathrm{Gal}(\overline{K}/L)$, and so that L is a Galois extension of K. We then replace μ_g by $(1/\alpha(g)) \cdot \mu_g$ for each $g \in G$. The new μ's satisfy $\mu_{\sigma\tau} = \mu_\sigma{}^\sigma\mu_\tau$, and μ_g depends only on the image of g in the finite group $\Delta = \mathrm{Gal}(L/K)$. Finally, μ_g may be viewed as an F-equivariant isomorphism up to isogeny ${}^g\!A \xrightarrow{\sim} A$ which is defined over L.

Let R be the abelian variety $\mathrm{Res}_{L/K} A$, where "Res" denotes Weil's "Restriction of scalars" functor. In other words, R represents the functor $C \mapsto \mathrm{Hom}(C_{/L}, A)$ from the category of abelian varieties up to isogeny over K to the category of $\mathbf{Q}$-vector spaces. Thus R is an abelian variety over K which is furnished with a structural homomorphism (up to isogeny) $\lambda : R_{/L} \to A$. Given C over K and a homomorphism (of abelian varieties up to isogeny) $\varphi : C_{/L} \to A$, there is a unique homomorphism $\theta : C \to R$ over K such that $\lambda \circ \theta = \varphi$.

One constructs R by considering the product $\prod_{g \in \mathrm{Gal}(L/K)} {}^g\!A$ and using obvious patching data to descend this product to K. In particular, R has dimension $[L : K] \cdot \dim(A)$. From this point of view, the map $\lambda : R_{/L} \to A$ is the projection of $\prod_{g \in \mathrm{Gal}(L/K)} {}^g\!A$ onto its factor A. Alternatively, given R with its structural map λ, and an element g of $\mathrm{Gal}(L/K)$, we can view ${}^g\!\lambda$ as a map $R_{/L} \to {}^g\!A$. One shows that the map $R_{/L} \to \prod_{g \in \mathrm{Gal}(L/K)} {}^g\!A$ induced by the family of ${}^g\!\lambda$ is an isomorphism.

By functoriality, F acts on R over K. Thus we have $F \subseteq \mathcal{X}$, where $\mathcal{X} = \mathrm{End}_K(R)$. Further, the universal property satisfied by R makes it easy to compute $\mathrm{End}_K(R)$ as a $\mathbf{Q}$-vector space. Indeed, we have

$$\mathrm{End}_K(R) = \mathrm{Hom}_K(R, R) = \mathrm{Hom}_L(R_{/L}, A) = \underset{g}{\oplus} \mathrm{Hom}_L({}^g\!A, A) = \underset{g}{\oplus} F \cdot \mu_g.$$

Let $[g]$ be the element of $\mathrm{End}_K(R)$ which corresponds to μ_g in the gth factor of the direct sum. Then $[g]$ is the unique element in $\mathrm{End}_K(R)$ such that $\lambda \circ [g] = \mu_g \circ {}^g\!\lambda$. The unicity implies that each $[g]$ commutes with the action of F on R. A short computation, based on the unicity and the formula $\mu_{\sigma\tau} = \mu_\sigma {}^\sigma\!\mu_\tau$, shows that $[\sigma][\tau] = [\sigma\tau]$ for $\sigma, \tau \in \Delta$. Also, the $[g]$ commute with F. Hence $\mathcal{X}$ is in the end the group algebra $F[\Delta]$.

The field F is a direct summand of the algebra $\mathcal{X} = F[\Delta]$ via the inclusion $a \in F \mapsto a \cdot [1]$ and the augmentation map $[g] \mapsto 1$. Let B be the subvariety of R corresponding to the direct summand F of $\mathcal{X}$. Then B is an abelian variety over K with an induced action of F. We claim that B is a Hilbert-Blumenthal abelian variety over K, i.e., that $\dim(B) = \dim(A)$, and that in fact A and B are isomorphic Hilbert-Blumenthal abelian varieties up to isogeny over L. To see this, we remark that $R_{/L}$ is isogenous to a product of copies of A (since the ${}^g\!A$ are each isogenous to A), so that $B_{/L}$ is certainly isogenous to a product of some number of copies of A. To determine the dimension of B, we remark that

$$F = \mathrm{End}_K(B) = \mathrm{Hom}_K(B, R) = \mathrm{Hom}_L(B_{/L}, A).$$

Since F is the endomorphism algebra of A, $B_{/L}$ is isogenous to A. ∎

Suppose now that A is a K-HBAV over $\overline{K}$. Fix a polarization $\theta_A : A \to A^\vee$ of A as an abelian variety. The associated Rosati involution of the endomorphism algebra of A may be viewed as a positive involution of the totally real field F and therefore is forced to be the identity.

Let B again be a Hilbert-Blumenthal abelian variety over $\overline{K}$, and fix a polarization θ_B of B. Then for each F-equivariant map $\mu: B \to A$, we define the "degree" $\deg(\mu)$ by the formula

$$\deg(\mu) = \mu \circ \theta_B^{-1} \circ \mu^\vee \circ \theta_A,$$

so that $\deg(\mu)$ is an element of the endomorphism algebra of A. Identifying this algebra with F, we consider that $\deg(\mu)$ is an element of F. If $\overline{K}$ is a subfield of $\mathbf{C}$, then θ_A and θ_B identify $\bigwedge_F^2 H_1(A(\mathbf{C}), \mathbf{Q})$ and $\bigwedge_F^2 H_1(B(\mathbf{C}), \mathbf{Q})$ with F. The number $\deg(\mu)$ is the element of F describing the map $\bigwedge_F^2 H_1(B(\mathbf{C}), \mathbf{Q}) \to \bigwedge_F^2 H_1(A(\mathbf{C}), \mathbf{Q})$ induced by μ.

If $\mu: A \to A$ is given by an element c of F, then $\deg(\mu) = c^2$. Also, it is easy to check that deg is multiplicative in the following sense. Suppose that (C, θ_C) is a third polarized Hilbert-Blumenthal abelian variety, and that $\lambda: C \to B$ is F-equivariant. Then $\deg(\mu \circ \lambda) = \deg(\mu) \deg(\lambda)$, provided of course that the "degrees" are computed with respect to a fixed set of polarizations.

(3.2) PROPOSITION. *Let A be a K-HBAV over $\overline{K}$. Then the order of the associated cohomology class $\gamma \in \mathrm{H}^2(G, F^*)$ is at most two.*

PROOF. Fix a polarization $\theta: A \to A^\vee$ of A and for each $g \in G$ let $^g\theta$ denote the associated polarization of ^gA. Choose a family (μ_g) as above, and let c be the F^*-valued two-cocycle on G defined by this family. For each g, let

$$d_g := \mu_g \circ {}^g\theta^{-1} \circ \mu_g^\vee \circ \theta$$

be the degree of μ_g, calculated with respect to the polarizations θ and $^g\theta$. This element of F is canonical in the following sense: if we replace θ by another polarization θ' of A, then d_g will remain unchanged provided that we replace $^g\theta$ by $^g\theta'$.

By construction,

$$c(\sigma, \tau) = \mu_\sigma \circ {}^\sigma\mu_\tau \circ \mu_{\sigma\tau}^{-1}$$

for $\sigma, \tau \in G$. On taking degrees, we find the formula

$$c(\sigma, \tau)^2 = \frac{d_\sigma d_\tau}{d_{\sigma\tau}}$$

which expresses the square of c as a coboundary.   ∎

(3.3) THEOREM. *Let γ be an element of order dividing two in $\mathrm{H}^2(G, F^*)$. Then there is an open normal subgroup H of G such that G/H is an elementary abelian two-group and such that the image of γ in $\mathrm{H}^2(H, F^*)$ is trivial. In other words, γ becomes trivial after K is replaced by a finite $(2, \ldots, 2)$-extension of K.*

PROOF. Let $P = F^*/\{\pm 1\}$, so that there is a tautological exact sequence

(3.4) $$0 \to \{\pm 1\} \to F^* \to P \to 0.$$

In the case where $[F: \mathbf{Q}]$ is odd, there is a natural splitting of this exact sequence: we may view P as the *subgroup* of F^* consisting of elements with positive norm to $\mathbf{Q}^*$. In the general case, (3.4) is still split, but apparently in no natural way. This follows directly from:

(3.5) LEMMA. *The group P is a free abelian group, of countable rank.*

To prove the lemma, we consider the map $\phi\colon a \mapsto (a)$ which takes an element a of P to the fractional ideal of F generated by a lift of a to F^*. The image of ϕ is a subgroup (of finite index) of the group of all fractional ideals of F; this latter group is the free abelian group on the set of non-Archimedean places of F. Hence the image of ϕ is a free abelian group, so that P is abstractly isomorphic to the direct sum of the image of ϕ and the kernel of ϕ. On the other hand, $\ker(\phi)$ is the group $U/\{\pm 1\}$, where U is the group of units of F. According to Dirichlet's theorem, $\ker(\phi)$ is a free abelian group of rank $n-1$, where $n = [F: \mathbf{Q}]$. It follows that P is the direct sum of two free abelian groups, one finitely generated and one countably generated. This proves the lemma.

Returning to the proof of (3.3), we fix a splitting of (3.4). We shall assume that this is the indicated natural splitting if $[F: \mathbf{Q}]$ is odd. The splitting fixes an isomorphism of abelian groups $F^* \approx \{\pm 1\} \times P$. This isomorphism induces in turn a decomposition

$$\mathrm{H}^2(G, F^*)[2] \approx \mathrm{H}^2(G, \{\pm 1\}) \times \mathrm{H}^2(G, P)[2],$$

where the notation "[2]" indicates the kernel of multiplication by 2. The element γ of $\mathrm{H}^2(G, F^*)$ is then the product of its two projections $\gamma_\pm \in \mathrm{H}^2(G, \{\pm 1\})$ and $\overline{\gamma} \in \mathrm{H}^2(G, P)[2]$. The factor $\overline{\gamma}$ is independent of the chosen splitting of (3.4): it is the image of γ under the map on cohomology induced by the quotient map $F^* \to P$. On the other hand, the "sign" component $\gamma_\pm$ of γ depends on the splitting. We thus consider that $\gamma_\pm$ is defined intrinsically only in the case where $[F: \mathbf{Q}]$ is odd.

To prove (3.3), we must show that both $\gamma_\pm$ and $\overline{\gamma}$ become trivial when K is replaced by a $(2, \dots, 2)$-extension of K.

To treat $\overline{\gamma}$, we consider the exact sequence

$$1 \to P \xrightarrow{x \mapsto x^2} P \to P/P^2 \to 1;$$

note that P is torsion free. The associated long exact cohomology sequence then gives an isomorphism

$$\mathrm{Hom}(G, P/P^2) \xrightarrow{\sim} \mathrm{H}^2(G, P)[2].$$

Suppose that $\overline{\gamma}$ corresponds to the homomorphism $\varphi\colon G \to P/P^2$. Then $\ker(\varphi)$ is a subgroup of G which corresponds to a $(2, \dots, 2)$-extension K_P of K. It is clear that $\overline{\gamma}$ becomes trivial over this extension.

It remains to split $\gamma_\pm$. It is known that $\mathrm{H}^2(G, \{\pm 1\}) = 0$ if K has characteristic 2, cf. [6, p. II-5]. Assume, then, that the characteristic of K is different from 2. The group $\mathrm{H}^2(G, \{\pm 1\})$ may then be identified with $\mathrm{Br}(K)[2]$, where $\mathrm{Br}(K)$ is the Brauer group of K. A theorem of S. A. Merkur'ev [3] states that $\mathrm{Br}(K)[2]$ is generated by the classes of quaternion algebras over K. Since each quaternion algebra over K is split by a quadratic extension of K, it follows that $\gamma_\pm$ is split by a $(2, \dots, 2)$-extension of K, as required. ∎

4. The odd-dimensional situation

In this section, we assume that K has characteristic 0. (An assumption concerning the parity of $[F : \mathbf{Q}]$ will be made later.)

We begin with some preliminary comments concerning the class γ defined by a K-HBAV. First, we note that the tensor product $F \otimes_{\mathbf{Q}} K$ decomposes as some direct sum of fields $\oplus_{\omega \in \Omega} K_\omega$, where the K_ω are finite extensions of K. The index set Ω is the set $\mathrm{Hom}(F, \overline{K})$ of field embeddings $F \hookrightarrow \overline{K}$, modulo the action of G on $\mathrm{Hom}(F, \overline{K})$.

Next, suppose that L is a finite extension of K, and consider $M := L \otimes_K \overline{K}$ as a G-module, with G acting trivially on the first factor. Choose an embedding σ of L into $\overline{K}$, and let $H = \mathrm{Gal}(\overline{K}/\sigma(L))$. Then $L \otimes_K \overline{K}$ is the induced representation $\mathrm{Ind}_H^G \overline{K}$. Indeed, M may be written as the group of functions $f : \Sigma \to \overline{K}$, where Σ is the set of embeddings $L \hookrightarrow \overline{K}$. In this optic, G acts via $(g \cdot f)(\tau) = g(f(g^{-1}\tau))$ for $g \in G$ and $\tau \in \Sigma$. It is clear that $M = \oplus_{\tau \in \Sigma} M_\tau$, where M_τ consists of those functions which vanish outside of τ. Further, the subgroups M_τ of M are permuted transitively by G, since G acts transitively on Σ. The formula $M = \mathrm{Ind}_H^G \overline{K}$ then follows by a well known criterion (see, e.g., [1, Ch. III, 5.4]).

Finally, we consider $F \otimes_{\mathbf{Q}} \overline{K}$ as a G-module, with G acting trivially on the first factor and in the natural way on the second. Then

$$F \otimes_{\mathbf{Q}} \overline{K} = (F \otimes_{\mathbf{Q}} K) \otimes_K \overline{K} = \bigoplus_\omega (K_\omega \otimes_K \overline{K}).$$

Therefore,

$$F \otimes_{\mathbf{Q}} \overline{K} = \bigoplus_\omega \mathrm{Ind}_{H_\omega}^G \overline{K}.$$

In this latter formula, we have chosen for each ω a K-embedding $K_\omega \hookrightarrow \overline{K}$ and have put $H_\omega := \mathrm{Gal}(\overline{K}/K_\omega)$.

In a similar vein, one proves

$$(F \otimes_{\mathbf{Q}} \overline{K})^* = \bigoplus_\omega \mathrm{Ind}_{H_\omega}^G \overline{K}^*.$$

Shapiro's lemma [1, Ch. III, 6.2] provides an identification

$$\mathrm{H}^i(G, \mathrm{Ind}_{H_\omega}^G \overline{K}^*) = \mathrm{H}^i(H_\omega, \overline{K}^*)$$

for each $i \geq 0$. (The induced module $\mathrm{Ind}_{H_\omega}^G \overline{K}^*$ may be regarded as a co-induced module because the index of H_ω in G is finite.) Thus

$$\mathrm{H}^i(G, (F \otimes_{\mathbf{Q}} \overline{K})^*) = \bigoplus_\omega \mathrm{H}^i(H_\omega, \overline{K}^*)$$

for each i. Consequently,

$$H^1(G, F \otimes_{\mathbf{Q}} \overline{K})^*) = 0$$

by Hilbert's Theorem 90. Similarly,

(4.1) $$H^2(G, (F \otimes_{\mathbf{Q}} \overline{K})^*) = \bigoplus_{\omega} \operatorname{Br}(K_{\omega})$$

because $H^2(H_{\omega}, \overline{K}^*)$ is the Brauer group of K_{ω}.

Now consider the exact sequence of G-modules

(4.2) $$0 \to F^* \to (F \otimes \overline{K})^* \to (F \otimes \overline{K})^*/F^* \to 0,$$

where G acts in the usual way on $\overline{K}$ and the indicated tensor products are taken over $\mathbf{Q}$. Exploiting the vanishing of $H^1(G, (F \otimes \overline{K})^*)$, we obtain

$$0 \to H^1(G, (F \otimes \overline{K})^*/F^*) \xrightarrow{\delta} H^2(G, F^*) \to H^2(G, (F \otimes \overline{K})^*) \to \cdots .$$

Here, δ is the indicated connecting homomorphism in the long cohomology sequence arising from (4.2).

(4.3) LEMMA. *Let A be a K-HBAV. Then the element γ of $H^2(G, F^*)$ defined by A lies in the image of δ. Equivalently, the image of γ in $H^2(G, (F \otimes \overline{K})^*)$ is zero.*

PROOF. We must exhibit an element β of $H^1(G, (F \otimes \overline{K})^*/F^*)$ such that $\gamma = \delta(\beta)$. Let $V = \operatorname{Lie}(A/\overline{K})$. For each $g \in G$, the map $\mu_g : {}^g A \to A$ induces a $(F \otimes \overline{K})$-linear homomorphism $\operatorname{Lie}({}^g A/\overline{K}) \to \operatorname{Lie}(A/\overline{K})$, or equivalently an F-linear homomorphism $\lambda_g : V \to V$ which is g-linear in the sense that it satisfies $\lambda(a \cdot v) = g(a)\lambda(v)$ for $a \in \overline{K}$ and $v \in V$. Now it is well known, and easy to verify, that the Lie algebra $\operatorname{Lie}(A/\overline{K})$ is free of rank one over $F \otimes \overline{K}$. Let v be a basis of V, considered as a free rank-one $F \otimes \overline{K}$-module. Then one has $\lambda_g(v) = a_g \cdot v$ for some element a_g in $(F \otimes \overline{K})^*$. The relations among the μ_g provide the formula $c(\sigma, \tau)a_{\sigma\tau} = a_{\sigma}{}^{\sigma}a_{\tau}$ for $\sigma, \tau \in G$. It follows that the function $G \to (F \otimes \overline{K})^*/F^*$ induced by $g \mapsto a_g$ is a 1-cocycle, and that the corresponding class β in $H^1(G, (F \otimes \overline{K})^*)$ maps to γ under δ. ∎

(4.4) PROPOSITION. *Suppose that γ is the element of $H^2(G, F^*)$ arising from a K-HBAV and that $\overline{\gamma} = 0$. Then $\gamma = 0$ provided that no element of order two in the Brauer group of K is split by all extensions of K of the form K_{ω}.*

PROOF. In view of the hypothesis $\overline{\gamma} = 0$, we have $\gamma = \gamma_{\pm} \in H^2(G, \{\pm 1\})$. According to (4.3), γ lies in the kernel of the map

$$j \colon H^2(G, \{\pm 1\}) \to H^2(G, (F \otimes \overline{K})^*)$$

induced by the inclusion of $\{\pm 1\}$ in $(F \otimes \overline{K})^*$. This map may be viewed as the natural map

$$\operatorname{Br}(K)[2] \to \bigoplus_{\omega} \operatorname{Br}(K_{\omega}),$$

which is injective by hypothesis. Thus γ is indeed 0. ∎

(4.5) COROLLARY. *Suppose that $\gamma \in H^2(G, F^*)$ arises from a K-HBAV. Suppose that $[F : \mathbf{Q}]$ is odd. Then the class γ becomes zero after K is replaced by the extension K_P of K defined by $\bar{\gamma}$. In particular, if $\bar{\gamma} = 0$, then $\gamma = 0$.*

PROOF. The two assertions of the corollary are equivalent, since $\bar{\gamma}$ becomes trivial after K is replaced by K_P. Because of (4.4), to prove the second assertion it is enough to prove that the map j which occurs in the proof of (4.4) is injective whenever $[F : \mathbf{Q}]$ is odd.

However, $[F : \mathbf{Q}] = \sum_\omega [K_\omega : K]$. Thus, if $[F : \mathbf{Q}]$ is odd, then there is at least one index ω for which $[K_\omega : K]$ is odd. Further, if $[K_\omega : K]$ is odd, then it is evident that the map $\mathrm{Br}(K)[2] \to \mathrm{Br}(K_\omega)$ is injective since there is a corestriction map cor: $\mathrm{Br}(K_\omega) \to \mathrm{Br}(K)$ whose composition with the natural map $\mathrm{Br}(K) \to \mathrm{Br}(K_\omega)$ is multiplication by $[K_\omega : K]$. ∎

REFERENCES

1. K. S. Brown, *Cohomology of groups*, Graduate Texts in Math., vol. 87, Springer-Verlag, New York, Heidelberg and Berlin, 1982.
2. N. Elkies, *Remarks on elliptic K-curves*, Preprint, 1993.
3. A. S. Merkur'ev, *On the norm residue symbol of degree 2*, Dokl. Akad. Nauk. SSSR **261** (1981), 542–547; Soviet Math. Dokl. **24** (1981), 546–551.
4. K. A. Ribet, *Twists of modular forms and endomorphisms of abelian varieties*, Math. Ann. **253** (1980), 43–62.
5. ———, *Abelian varieties over $\mathbf{Q}$ and modular forms*, 1992 Proceedings of KAIST Mathematics Workshop, Korea Advanced Institute of Science and Technology, Taejon, 1992, pp. 53–79.
6. J-P. Serre, *Cohomologie galoisienne*, Lecture Notes in Math., vol. 5, 4ᵉ édition, Springer-Verlag, Berlin and New York, 1973.
7. G. Shimura, *Introduction to the arithmetic theory of automorphic functions*, Princeton University Press, Princeton, 1971.
8. ———, *Class fields over real quadratic fields and Hecke operators*, Ann. of Math. **95** (1972), 131–190.

UC MATHEMATICS DEPARTMENT, BERKELEY, CA 94720 USA

E-mail address: ribet@math.berkeley.edu

Contemporary Mathematics
Volume **174**, 1994

p-adic Interpolation of Half-Integral Weight Modular Forms

ADRIANA SOFER

ABSTRACT. We construct a p-adic interpolation of a half-integral weight p-adic modular form and its "derivatives". We obtain it as the Gamma transform of a p-adic measure.

CONTENTS

1. Introduction

The p-adic interpolation of modular forms on congruence subgroups of $SL_2(\mathbb{Z})$ has been successfully used in the past to interpolate values of L-series.

In [**12**], Serre interpolated the values at negative integers of the ζ-series of a totally real number field (in fact of L-series of powers of the Teichmüller character) by interpolating Eisenstein series, which are holomorphic modular forms,

1991 *Mathematics Subject Classification.* Primary 11F33, 11F37; Secondary 11F85.
Key words and phrases. p-adic modular forms.
This paper is in final form and no version of it will be submitted for publication elsewhere.

and looking afterwards at the resulting constant terms of the q-expansions. This can be interpreted as evaluating at $q = 0$, the degenerate elliptic curve.

In [8], Katz interpolated the values of the Hecke L-series $L(\psi_k, r)$ by interpolating first "derivatives" of Eisenstein series, which are in general non-holomorphic, and then evaluating at elliptic curves.

In [18], we interpolated square roots of values of Hecke L-series by interpolating "derivatives" of theta functions, which are non-holomorphic half-integral weight modular forms, and evaluating at elliptic curves. This last step required appealing to the theory of integral weight p-adic modular forms, since there is not a complete theory for half-integral weight yet. There is ongoing research towards this (see the recent work of Jochnowitz [2], and Stevens [19]).

There is renewed interest in half-integral weight forms, after the work of Hida, Kohnen, Rodriguez-Villegas, Waldspurger, Zagier.

We here present a p-adic interpolation of the sequence $\left\{ \left(Nq\frac{d}{dq} \right)^k f \right\}_{k \geq 0}$ where f is a given p-adic modular form of half-integral weight and of level $N \geq 1$, $4|N$, $p \nmid N$. The integral weight version of this appears in [8, 1].

For integral weight, we use Serre's and Katz's theories of p-adic modular forms ([12, 8, 6, 5, 7, 1]). We use Koblitz's ([10]) and Jochnowitz's ([2, 3, 4]) definitions for half-integral weight p-adic modular forms.

I would like to thank N. Jochnowitz for very helpful conversations, and the referee for many suggestions and corrections.

2. Definitions and properties

2.1. p-adic modular forms. Let $p \geq 3$ be a rational prime. Fix $\mathbb{C}_p$ the completion of an algebraic closure of $\mathbb{Q}_p$. Let $B \subset \mathbb{C}_p$ be the ring of integers of a finite extension of $\mathbb{Q}_p$. We fix an isomorphism $\mathbb{C} \simeq \mathbb{C}_p$ and use it to consider B as included in $\mathbb{C}$.

The reader is referred to [16] or [9, Ch.4] for the following definitions.

Let j be the multiplier system on $\Gamma_0(N)$, $4|N$, corresponding to the standard theta series $\theta = 1 + 2\sum_{n \geq 1} e^{2\pi i n^2 z}$, and $\chi : (\mathbb{Z}/N\mathbb{Z})^* \to \mathbb{C}$ be a character acting on $\gamma = \begin{pmatrix} a & b \\ c & d \end{pmatrix} \in \Gamma_0(N)$ via $\chi(d)$. $\mathcal{H}$ is the upper-half plane in $\mathbb{C}$.

For $k \in \mathbb{Z}$, we denote by $M_k(\Gamma_0(N), \chi)$ the set of holomorphic $f : \mathcal{H} \to \mathbb{C}$ such that $f(\gamma z) = \chi(d)(cz + d)^k f(z)$ for every $\gamma = \begin{pmatrix} a & b \\ c & d \end{pmatrix} \in \Gamma_0(N)$ and f is holomorphic at all cusps.

For $k \in \mathbb{Z} + \frac{1}{2}$, we denote by $M_k(\Gamma_0(N), \chi)$ the set of holomorphic $f : \mathcal{H} \to \mathbb{C}$ such that $f(\gamma z) = \chi(d)j(\gamma)^{2k+1} f(z)$ for every $\gamma = \begin{pmatrix} a & b \\ c & d \end{pmatrix} \in \Gamma_0(N)$ and f is holomorphic at all cusps.

In both cases, and for every cusp t for $\Gamma_0(N)$, we associate to

$f \in M_k(\Gamma_0(N), \chi)$ a q-expansion $f_t \in \mathbb{C}[[q_N]]$, $(q_N = e^{2\pi i z/N})$ in a consistent manner. We choose $f_\infty(q_N) = f(z)$. Given a ring $A \subset \mathbb{C}$, we denote $M_k(\Gamma_0(N), \chi, A)$ the set of $f \in M_k(\Gamma_0(N), \chi)$ satisfying $f_t \in A[[q_N]]$ for every t. We extend these definitions to $M_k(\Gamma_1(N)) \simeq \bigoplus M_k(\Gamma_0(N), \chi)$.

Recall that θ does not vanish on $\mathcal{H}$, and that its q-expansions are:

$$(2.1) \qquad \begin{cases} \theta_\infty &= 1 + 2\sum_{n \geq 1} q^{n^2} \\ \theta_0 &= \frac{1}{\sqrt{2}}\left(1 + 2\sum_{n \geq 1} q_4^{n^2}\right) \\ \theta_{\frac{1}{2}} &= \theta(\frac{z}{4}) - \theta(z) \end{cases}$$

DEFINITION 2.1 (SERRE-KOBLITZ). We call a p-adic modular form over B for $\Gamma_1(N)$ a collection of q-expansions $\{f_t(q_N)\}_{\{t \text{ cusp}\}} \subset B[[q_N]]$ such that there is a sequence of complex modular forms $f_j \in M_{k_j}(\Gamma_1(N), B)$, $k_j \in \mathbb{Z}$ (or $k_j \in \mathbb{Z} + \frac{1}{2}$) with

$$(f_j)_t(q_N) \xrightarrow{p} f_t(q_N)$$

uniformly on the coefficients, for every cusp t.

DEFINITION 2.2 (SERRE-JOCHNOWITZ). We call a p-adic modular form over B for $\Gamma_1(N)$ a power series $f(q_N) \in B[[q_N]]$ such that there is a sequence of complex modular forms $f_j \in M_{k_j}(\Gamma_1(N), \overline{\mathbb{Q}})$ with p-adically integral coefficients, $k_j \in \mathbb{Z}$ (or $k_j \in \mathbb{Z} + \frac{1}{2}$) with

$$f_j(q_N) \xrightarrow{p} f(q_N)$$

uniformly on the coefficients.

The Corollary in the next section shows that these two definitions are equivalent. Until that is established, we will work with definition 2.1.

Any p-adic modular form has a weight $k \in X$ (or $k + \frac{1}{2} \in X + \frac{1}{2}$) (see [**2**, Theorem 1.2], [**12**, **10**]), where $X = \mathbb{Z}_p \times \mathbb{Z}/(p-1)\mathbb{Z}$. Moreover, if $p \nmid N$, k is the p-adic limit of $(k_j, k_j \bmod p - 1)$. We denote $\mathcal{M}_k^p(\Gamma_1(N), B)$ (or $\mathcal{M}_{k+\frac{1}{2}}^p(\Gamma_1(N), B)$) the set of such f's.

We immediately obtain that any classical holomorphic modular form with coefficients in B can be interpreted as a p-adic form.

PROPOSITION 2.1. *The Ramanujan function $P(z) = 1 - 24\sum_{n \geq 1} \sigma(n)e^{2\pi i n z}$ is not modular in the classical sense, but $P \in \mathcal{M}_2^p(SL_2(\mathbb{Z}), \mathbb{Z}_p)$.*

PROOF: [**12**, Example p.210].

Let "$q\frac{d}{dq}$" be the operator $D + \frac{1}{12}PH$, where D is the Halphen-Fricke operator, P is the Ramanujan function, and $H(g) = kg$ if $g \in M_k(\Gamma_0(N), \chi)$ (see [**8**, **18**]). We have $\left(Nq\frac{d}{dq}f\right)_\infty = \sum na_n q_N^n$ if $f_\infty = \sum a_n q_N^n$.

PROPOSITION 2.2. *Let $p \geq 5$, $k \in X \cup X + \frac{1}{2}$, and $f \in \mathcal{M}_k^p(\Gamma_1(N), B)$. Then $q\frac{d}{dq}f \in \mathcal{M}_{k+2}^p(\Gamma_1(N), B)$, where the weight $k + 2$ means $k + (2, 2)$.*

PROOF: $Nq\frac{d}{dq} = ND + \frac{N}{12}PH$. For $k \in \mathbb{Z} \cup \mathbb{Z} + \frac{1}{2}$, ND is a derivation $M_k(\Gamma_0(N), \chi, A) \to M_{k+2}(\Gamma_0(N), \chi, A[\frac{1}{12}])$ that acts continuously. Hence, for $k \in X \cup X + \frac{1}{2}$,

$$D : \mathcal{M}_k^p(\Gamma_1(N), B) \longrightarrow \mathcal{M}_{k+2}^p(\Gamma_1(N), B)$$
$$H : \mathcal{M}_k^p(\Gamma_1(N), B) \longrightarrow \mathcal{M}_k^p(\Gamma_1(N), B)$$
$$P \in \mathcal{M}_2^p(SL_2(\mathbb{Z}), \mathbb{Z}_p)$$

and we deduce the result.

NOTE: Also see an alternate proof of proposition 2.2 in [2].

2.2. The q-expansion principle. Recall the q-expansion principle for (integral weight) p-adic modular forms ([8, 5.2], [1, Theorem I.3.1, Prop. I.3.2]). We will prove a similar result for half-integral weight.

When $k \in \mathbb{Z}$, the weight k means (k, k), and $k + \frac{1}{2}$, $(k, k) + \frac{1}{2}$. Let us denote $\mathcal{M}_{\bullet + \frac{1}{2}}^p(\Gamma_1(N), B)$ (and $\mathcal{M}_\bullet^p(\Gamma_1(N), B)$) the set of $\mathbb{Z}$-linear combinations of forms in $\bigcup_{k \in \mathbb{Z} k \geq 0}(\mathcal{M}_{k+\frac{1}{2}}^p(\Gamma_1(N), B) \otimes \mathbb{Q}_p)$ (and, for integral weight, forms in $\bigcup_{k \in \mathbb{Z} k \geq 0}(\mathcal{M}_k^p(\Gamma_1(N), B) \otimes \mathbb{Q}_p)$) such that the q-expansions coefficients (of the $\mathbb{Z}$-linear combinations) at all cusps belong to B (the q-expansions are defined by linearity). Notice that any sequence approximating an element of $\mathcal{M}_{\bullet+\frac{1}{2}}^p(\Gamma_1(N), B)$ (or $\mathcal{M}_\bullet^p(\Gamma_1(N), B)$) will eventually have coefficients in B, rather than in $B \otimes \mathbb{Q}_p$, due to the uniform convergence. Also, for $k \in \mathbb{Z}$, $k \geq 0$,

$$(\mathcal{M}_{k+\frac{1}{2}}^p(\Gamma_1(N), B) \otimes \mathbb{Q}_p) \cap \mathcal{M}_{\bullet+\frac{1}{2}}^p(\Gamma_1(N), B) = \mathcal{M}_{k+\frac{1}{2}}^p(\Gamma_1(N), B),$$

and similarly for weight k. We define $q\frac{d}{dq}$ on $\mathcal{M}_{\bullet+\frac{1}{2}}^p(\Gamma_1(N), B)$ and on $\mathcal{M}_\bullet^p(\Gamma_1(N), B)$ by linearity.

Jochnowitz independently proved results similar to the next proposition.

PROPOSITION 2.3 (q-EXPANSION PRINCIPLE). *Let* $\mu_N \cup \{\sqrt{2}\} \subset B' \subset B \subset \mathbb{C}_p$, *$B$ and B' the ring of integers of $K, K' \subset \mathbb{C}_p$, finite extensions of $\mathbb{Q}_p$. Let* $p \geq 3$, $p \nmid N$, *and assume* $f \in \mathcal{M}_{\bullet+\frac{1}{2}}^p(\Gamma_1(N), B)$.

(i) *If* $f_\infty \in B'[[q_N]]$, *then* $f \in \mathcal{M}_{\bullet+\frac{1}{2}}^p(\Gamma_1(N), B')$.

(ii) *Let* π *be a prime element in* B. *The map*

$$\mathcal{M}_{\bullet+\frac{1}{2}}^p(\Gamma_1(N), B) \xrightarrow{\;\;q\text{-}exp \text{ at } \infty\;\;} B[[q_N]]$$

is injective, and the cokernel has trivial π-torsion.

PROOF:

1. Since $p \geq 3$, $\theta \in M_{\frac{1}{2}}(\Gamma_0(4), \mathbb{Z}[\frac{1}{\sqrt{2}}]) \subset \mathcal{M}^p(\Gamma_1(N), B')$. We know that $f\theta \in \mathcal{M}_\bullet^p(\Gamma_1(N), B)$, and $(f\theta)_\infty \in B'[[q_N]]$. By the integral-weight q-expansion principle, $f\theta \in \mathcal{M}_\bullet^p(\Gamma_1(N), B')$, i.e. $(f\theta)_t \in B'[[q_N]]$ for every cusp t. Since $\theta_t | (f\theta)_t$, and $\theta_t \in B'[[q_N]]$ with invertible first coefficient (see equations (2.1)), $(f)_t = \frac{(f\theta)_t}{(\theta)_t} \in B'[[q_N]]$ for every cusp t. Take a sequence of classical modular

forms, with coefficients in B approaching f in the following way: if $f = \sum_i h_i$, $h_i \in \mathcal{M}^p_{k_i+\frac{1}{2}}(\Gamma_1(N), B) \otimes \mathbb{Q}_p$, take sequences $f_{ij} \in M_{k_{i_j}}(\Gamma_1(N), B) \otimes \mathbb{Q}_p$ with

$$(f_{ji})_t \longrightarrow (h_i)_t.$$

We have

$$(f_j)_t = \sum_i (f_{ji})_t \longrightarrow f_t.$$

Since f has integral coefficients, the uniform convergence guarantees that f_j has coefficients in B, for every $j >> 0$.

Let $\bar{K}$ be the Galois closure of K in $\mathbb{C}_p$. Let $d = [\bar{K}, K']$. Define

$$g_j = \frac{1}{d} tr_{\bar{K}, K'} f_j = \sum_i \frac{1}{d} tr_{\bar{K}, K'} f_{ji},$$

a sequence such that dg_j has coefficients in $B' \cap \mathbb{C}$ for $j >> 0$. The forms $g_{ji} = \frac{1}{d} tr_{\bar{K}, K'} f_{ji}$ belong to $M_{k_{i_j}+\frac{1}{2}}(\Gamma_1(N), B') \otimes \mathbb{Q}_p$. Call $s_i \in \mathcal{M}^p_{k_i+\frac{1}{2}}(\Gamma_1(N), B') \otimes \mathbb{Q}_p$ the limit of g_{ji}. So

$$(g_j)_t = \sum_i (g_{ji})_t \longrightarrow (\sum_i s_i)_t.$$

Also

$$(g_j)_t \longrightarrow f_t.$$

We deduce that $f = \sum_i s_i \in \mathcal{M}^p_{\bullet+\frac{1}{2}}(\Gamma_1(N), B')$.

2. Assume $f, g \in \mathcal{M}^p_{\bullet+\frac{1}{2}}(\Gamma_1(N), B)$, $f_\infty = g_\infty$. Then $(f\theta)_\infty = (g\theta)_\infty$, and $f\theta, g\theta \in \mathcal{M}^p_\bullet(\Gamma_1(N), B)$. Again by the integral-weight q-expansion principle, $(f\theta)_t = (g\theta)_t$ for every t, hence $f_t = g_t$ for every t, and $f = g$.

Assume $f_\infty \in B[[q_N]]$ and $\pi^r f \in \mathcal{M}^p_{\bullet+\frac{1}{2}}(\Gamma_1(N), B)$ for some $r \geq 0$. Equivalently, $f \in \mathcal{M}^p_{\bullet+\frac{1}{2}}(\Gamma_1(N), B) \otimes \mathbb{Q}_p$. Again we multiply by θ and use the integral-weight q-expansion principle for $\pi^r f\theta \in \mathcal{M}^p_\bullet(\Gamma_1(N), B)$, and deduce that $f\theta \in \mathcal{M}^p_\bullet(\Gamma_1(N), B)$. Therefore, since θ_t is π-integral (see equations (2.1)), $\pi^r f_t$ is divisible by π^r for every t. Hence f_t has coefficients in B for every t. Take a sequence $(f_j)_j$ with coefficients in B, approaching $\pi^r f$. Then the sequence $(\pi^{-r} f_j)_j$ will approach f. We deduce that $f \in \mathcal{M}^p_{\bullet+\frac{1}{2}}(\Gamma_1(N), B)$. This ends the proof.

COROLLARY (JOCHNOWITZ). If $p \geq 3$, $p \nmid N$, $\mu_N \cup \{\sqrt{2}\} \subset B$, definitions 2.1 and 2.2 are equivalent.

PROOF: The reader is referred to [2, 3, 4].

Therefore, we identify f to f_∞ in what follows.

2.3. Measures. A measure on $\mathbb{Z}_p$ with values in a p-adic ring A (i.e. $A = \lim\limits_{\leftarrow} A/p^n A$) is an additive function

$$\mu : \{\text{compact open subsets of } \mathbb{Z}_p\} \longrightarrow A.$$

We call $Meas(\mathbb{Z}_p, A)$ the ring of measures (with convolution as product). The Fourier transform

$$\hat{\mu}(X) = \int_{\mathbb{Z}_p} (X+1)^x d\mu(x) = \sum_{n \geq 0} \left(\int_{\mathbb{Z}_p} \binom{x}{n} d\mu \right) X^n$$

gives a ring isomorphism between $Meas(\mathbb{Z}_p, A)$ and $A[[X]]$. For example, the Dirac measure at $a \in \mathbb{Z}_p$ has Fourier transform $(X+1)^a = \sum_{n \geq 0} \binom{a}{n} X^n$. By abuse of notation, we will often use $\hat{\mu}(T)$ instead of $\hat{\mu}(X)$, where $T = X+1$. Also, $Meas(\mathbb{Z}_p, A) = \lim\limits_{\leftarrow} Meas(\mathbb{Z}/p^n\mathbb{Z}, A)$, and

$$Meas(\mathbb{Z}/p^n\mathbb{Z}, A) \simeq A[\mathbb{Z}/p^n\mathbb{Z}] = \left\{ \sum_{a \bmod p^n} m_a T^a, m_a \in A \right\}.$$

If $\mu \in Meas(\mathbb{Z}_p, A)$ and C is a compact open subset of $\mathbb{Z}_p$, $\mu|_C$, the restriction of μ to C, extended by 0, is in $Meas(\mathbb{Z}_p, A)$ too.

The following properties can easily be checked by proving them first for the dense subring of finite linear combinations of Dirac measures.

- If $\hat{\mu}(T) = \sum_{n \geq 0} a_n T^n$, then $\hat{\mu}|_{p\mathbb{Z}_p}(T) = \sum_{n \geq 0} a_{pn} T^{pn}$.
- For any $a \in \mathbb{Z}_p$, and any measure $\mu \in Meas(\mathbb{Z}_p, A)$, define $\mu_a \in Meas(\mathbb{Z}_p, A)$ to be the unique measure with $\hat{\mu}_a(T) = \hat{\mu}(T^a)$. Then $\int h(x) d\mu_a = \int h(ax) d\mu$ for every continuous function $h : \mathbb{Z}_p \longrightarrow \mathbb{Z}_p$.

References for measures are [**8, 17, 20**].

3. Interpolation

3.1. The measure associated to a modular form. Let $p \geq 5$, $p \nmid N$. Define $\mathcal{M} = \mathcal{M}^p(\Gamma_1(N), B)$ to be the smallest p-adic ring containing $\mathcal{M}^p_{\bullet + \frac{1}{2}}(\Gamma_1(N), B)$. We define the q-expansions on $\mathcal{M}$ by continuity. Given $f \in \mathcal{M}^p_{\bullet + \frac{1}{2}}(\Gamma_1(N), B)$, we define μ_f by

$$\hat{\mu}_f(X) = \sum_{m \geq 0} \left(\sum_{r=0}^{m} c_{m,r} \left(Nq \frac{d}{dq} \right)^r f \right) X^m$$

where $c_{m,r} \in \mathbb{Q}$ are such that $\binom{x}{m} = \sum_{r=0}^{m} c_{m,r} x^r$. Notice that, if $f = \sum_{n \geq 0} a_n q_N^n$,

$$\hat{\mu}_f(T) = \sum_{n \geq 0} a_n q_N^n T^n.$$

PROPOSITION 3.1. $\mu_f \in Meas(\mathbb{Z}_p, \mathcal{M})$.

PROOF: To see that μ_f is a measure, it is enough to prove that $\hat{\mu}_f \in \mathcal{M}[[X]]$ (rather that in $\mathcal{M} \otimes \mathbb{Q}_p$ as the definition seems to suggest). An easy computation shows that

$$\hat{\mu}_f(X) = \sum_{m \geq 0} \left(\sum_{n \geq 0} \binom{n}{m} a_n q_N^n \right) X^m.$$

So the mth-coefficient is in $\mathcal{M}^p_{\bullet + \frac{1}{2}}(\Gamma_1(N), B) \otimes \mathbb{Q}_p$, and its q-expansion at ∞ is

$$\sum_{n \geq 0} \binom{n}{m} a_n q_N^n \in B[[q_N]].$$

Therefore, the mth-coefficient belongs to the p-torsion of

$$B[[q_N]] / \mathcal{M}^p_{\bullet + \frac{1}{2}}(\Gamma_1(N), B),$$

which is trivial by the q-expansion principle. We get that all coefficients are in $\mathcal{M}^p_{\bullet + \frac{1}{2}}(\Gamma_1(N), B)$.

By density, we extend the definition to $f \in \mathcal{M}$.

The expression $\int_{\mathbb{Z}_p} x^r d\mu_f$ is called the rth-moment of the measure ($r \in \mathbb{Z}$). The rth-moment of μ_f satisfies

$$\int_{\mathbb{Z}_p} x^r d\mu_f = \sum_{n \geq 0} a_n n^r q_N^n = \left(Nq \frac{d}{dq} \right)^r f.$$

Therefore, the rth-moment is $\left(Nq \frac{d}{dq} \right)^r f$.

Any measure $\mu \in Meas(\mathbb{Z}_p, A)$ "interpolates" its moments, via the Gamma transform

$$\begin{array}{ccc} \Gamma_\mu : \mathbb{Z}_p & \longrightarrow & A \\ s & \mapsto & \int_{\mathbb{Z}_p} <x>^s d\mu|_{\mathbb{Z}_p^*} \end{array}$$

This can be seen as follows: if $r \xrightarrow{p} s$, $|r| \to \infty$, $r \equiv 0 \bmod p-1$, then $\int_{\mathbb{Z}_p} x^r d\mu \to \Gamma_\mu(s)$. For $r \equiv 0 \bmod p-1$,

$$\Gamma_\mu(r) = \int_{\mathbb{Z}_p} x^r d\mu|_{\mathbb{Z}_p^*} = \int_{\mathbb{Z}_p} x^r d\mu - \int_{\mathbb{Z}_p} x^r d\mu|_{p\mathbb{Z}_p}.$$

This is the (r^{th}-moment) $-$ (something related to it). We next pay a closer look at $\Gamma_\mu(r)$.

3.2. Restriction of the measure to $p\mathbb{Z}_p$. If $\hat{\mu}_f(T) = \sum_{n \geq 0} a_n q_N^n T^n$, then

$$\hat{\mu}_f|_{p\mathbb{Z}_p}(T) = \sum_{n \geq 0} a_{pn} q_N^{pn} T^{pn} = \hat{\mu}_{Rf}(T)$$

where $Rf = \int_{\mathbb{Z}_p} d\mu|_{p\mathbb{Z}_p} \in \mathcal{M}$, with q-expansion at ∞: $\sum_{m \geq 0} a_{pm} q_N^{pm}$. The operator R equals VU, where $Vf = f(q^p)$, and $Uf = \sum a_{np} q_N^n$ if $f = \sum a_n q_N^n$. As in the p-adic integral-weight case, these operators do not increase the level; if f has level N, so do Vf and Uf. (See [4]).

THEOREM 3.2. *Let $p \geq 5$, $p \nmid N$, and $f \in \mathcal{M}^p_{\bullet + \frac{1}{2}}(\Gamma_1(N), B)$, and let μ_f be the measure on $\mathbb{Z}_p$ with values in $\mathcal{M}$ associated to it. Then its Gamma transform $\Gamma_f = \Gamma_{\mu_f}$ "interpolates" the sequence $\left(Nq\frac{d}{dq}\right)^r f$. More precisely, for $r \equiv 0 \bmod p - 1$,*

$$
\begin{aligned}
\Gamma_f(r) &= \left(Nq\frac{d}{dq}\right)^r f - R\left(\left(Nq\frac{d}{dq}\right)^r f\right) \\
&= \left(Nq\frac{d}{dq}\right)^r f - \left(\left(Nq\frac{d}{dq}\right)^r Rf\right) \\
&= \left(Nq\frac{d}{dq}\right)^r f - p^r V\left(\left(Nq\frac{d}{dq}\right)^r Uf\right)
\end{aligned}
$$

PROOF: We prove the first equality. The other two are left to the reader.

$$
\begin{aligned}
\Gamma_f(r) &= \int_{\mathbb{Z}_p} x^r d\mu_f - \int_{\mathbb{Z}_p} x^r d\mu_{Rf} \\
&= \left(Nq\frac{d}{dq}\right)^r f - \int_{\mathbb{Z}_p} x^r d(\sum_{n \geq 0} a_{pn} q_N^{pn} T^{pn}) \\
&= \left(Nq\frac{d}{dq}\right)^r f - \sum_{n \geq 0} a_{pn}(pn)^r q_N^{pn} \\
&= \left(Nq\frac{d}{dq}\right)^r f - R\left(\left(Nq\frac{d}{dq}\right)^r f\right)
\end{aligned}
$$

We include the following particular case.

COROLLARY.   *If $f \in \mathcal{M}^p_{\bullet + \frac{1}{2}}(\Gamma_1(N), B)$, $f_\infty(q_N) = \sum_{n \geq 0} a_n q_N^n$, satisfies $v_p(n) \neq 1$ when $a_n \neq 0$ (the standard theta function θ, and $\theta_{\psi,t}$, the generators of all forms of weight $\frac{1}{2}$ given in [13] are examples), then the interpolation can be further refined:*
For $r \equiv 0 \bmod p - 1$,

$$\Gamma_f(r) = \left(Nq\frac{d}{dq}\right)^r f - p^{2r}\left[\left(Nq\frac{d}{dq}\right)^r (U^2 f)\right](q^{p^2})$$

PROOF:

$$
\begin{aligned}
\Gamma_f(r) &= \left(Nq\frac{d}{dq}\right)^r f - \int_{\mathbb{Z}_p} x^r d\left(\sum_{n\geq 0} a_{np^2} q_N^{np^2} T^{np^2}\right) \\
&= \left(Nq\frac{d}{dq}\right)^r f - p^{2r}\int_{\mathbb{Z}_p} x^r d\left(\sum_{n\geq 0} a_{np^2} q_N^{np^2} T^n\right) \\
&= \left(Nq\frac{d}{dq}\right)^r f - p^{2r}\sum_{n\geq 0} a_{np^2} n^r q_N^{np^2} \\
&= \left(Nq\frac{d}{dq}\right)^r f - p^{2r}\left[\left(Nq\frac{d}{dq}\right)^r (U^2 f)\right](q^{p^2})
\end{aligned}
$$

4. Applications

As mentioned in the introduction, this kind of interpolation was used in several instances to obtain p-adic L-functions. For that, we start with a classical modular form f, we interpret it as a p-adic form, and we compose the interpolating function $\Gamma_f : \mathbb{Z}_p \longrightarrow \mathcal{M}$ from Proposition 3.2 with evaluation at an appropriate modular point. For half-integral weight, this can not be done directly, since we do not know how to evaluate half-integral weight p-adic forms yet. In [18], we evaluated, instead, the function Γ_f/θ, that takes values on Katz's generalized p-adic modular forms. This provides a function $\mathbb{Z}_p \longrightarrow B$, that interpolates the values obtained by evaluating the sequence $\left\{\frac{1}{\theta}\left(\left(Nq\frac{d}{dq}\right)^r f - R\left(\left(Nq\frac{d}{dq}\right)^r f\right)\right)\right\}$ at the modular point.

Via Shimura Reciprocity, these evaluations become an Euler factor times the evaluation of $\left\{\frac{1}{\theta}\left(Nq\frac{d}{dq}\right)^r f\right\}$.

Finally, the evaluation of $\frac{1}{\theta}\left(Nq\frac{d}{dq}\right)^r f$ is proved to equal the evaluation of $\frac{1}{\theta}W^r f$, where W is the Maass-Weil operator. This puts us in the realm of the classical theory, where the connection between $W^r f$ and special values of L-series is established. The extra θ in the denominator is finally absorbed by the power of the complex period that naturally appears in the resulting formula.

REFERENCES

1. F. Gouvêa, *Arithmetic of p-adic Modular Forms*, Springer-Verlag LNM **1304** (1988).
2. N. Jochnowitz, *Congruences between modular forms of half integral weights, and implications for class numbers and elliptic curves*, to appear in Invent. Math.
3. N. Jochnowitz, *An alternate approach to non-p divisibility of coefficients of half-integral forms with implications for Iwasawa λ-invariants and for Kolyvagin, Birch and Swinnerton-Dyer results*, to appear.
4. N. Jochnowitz, *A p-adic theory for half-integral weight forms*, to appear.
5. N. Katz, *p-adic properties of modular schemes and modular forms*, in Modular functions of one variable III, Springer-Verlag LNM **350** (1973), 69-190.
6. N. Katz, *p-adic L-functions via moduli of elliptic curves*, in Algebraic Geometry: Arcata 1974, Proceedings of Symposia in Pure Mathematics **29**, AMS (1975).

7. N. Katz, *Higher congruences between modular forms*, Ann. Math. **101** (1975), 332-367.
8. N. Katz, *p-adic interpolation of real analytic Eisenstein series*, Ann. of Math. **104** (1976), 459-571.
9. N. Koblitz, *Introduction to Elliptic Curves and Modular Forms*, Springer-Verlag (1984).
10. N. Koblitz, *p-adic Congruences and Modular Forms of Half Integral Weight*, Math. Ann. **274** (1986), 199-220.
11. S. Lang, *Elliptic Functions*, Springer-Verlag (1987).
12. J.P. Serre, *Formes Modulaires et Fonctions Zêta p-adiques*, in Modular functions of one variable III, Springer-Verlag LNM **350** (1973), 191-268.
13. J.P. Serre and H. Stark, *Modular Forms of Weight $\frac{1}{2}$*, in Modular functions of one variable VI, Springer-Verlag LNM **627** (1977), 28-67.
14. G. Shimura, *Introduction to the arithmetic theory of automorphic functions*, Princeton Univ. Press (1971).
15. G. Shimura, *Modular Forms of Half-Integral Weight*, in Modular Forms of One Variable I, Springer-Verlag LNM **320** (1972), 57-74.
16. G. Shimura, *On modular forms of half integral weight*, Ann. of Math. **97** (1973), 440-481.
17. W. Sinnott, *On the μ-invariant of the Γ-transform of a rational function*, Inv. Math **75** (1984), 273-282.
18. A. Sofer, *p-adic interpolation of square roots of central values of Hecke L-series*, Ph.D. thesis, The Ohio State University (1993).
19. G. Stevens, article in this volume.
20. R. Yager, *On two variable p-adic L-functions*, Annals of Math. **115** (1982), 411-449.

DEPARTMENT OF MATHEMATICS, PRINCETON UNIVERSITY, PRINCETON, NEW JERSEY 08544
E-mail address: asofer@math.princeton.edu

Contemporary Mathematics
Volume **174**, 1994

Λ-adic Modular Forms of Half-Integral Weight and a Λ-adic Shintani Lifting

GLENN STEVENS

§0. Introduction

The simplest example of a modular form of half-integral weight is the classical Riemann theta function

$$\theta(z) = \sum_{n \in \mathbf{Z}} e^{2\pi i n^2 z},$$

defined for z in the upper half plane $\mathbf{H}$. The well known transformation laws for $\theta(z)$ assert that for each $\gamma = \begin{pmatrix} a & b \\ c & d \end{pmatrix} \in \Gamma_0(4)$ (see (1.10) of [S3])

$$j(\gamma, z) := \frac{\theta(\gamma z)}{\theta(z)} = \epsilon_d^{-1} \cdot \left(\frac{c}{d}\right) \cdot (cz + d)^{1/2}$$

where $\epsilon_d = 1$ or i according as $d \equiv 1$ or 3 modulo 4, $\left(\frac{c}{d}\right)$ is the standard extension of the Legendre symbol (see p. 442 of [S3]), and the square root $(cz + d)^{1/2}$ is chosen so that $-\pi/2 < \arg(cz + d)^{1/2} \leq \pi/2$.

If M is a positive integer divisible by 4 and χ is a Dirichlet character defined modulo M, then for each $k \geq 0$ we let $\mathcal{M}_{k+\frac{1}{2}}(\Gamma_0(M), \chi)$ denote the space of cusp forms of level M, weight $k + 1/2$, and character χ. Hence an element of $\mathcal{M}_{k+\frac{1}{2}}(\Gamma_0(M), \chi)$ is a holomorphic function f on $\mathbf{H}$ that is finite at the cusps and transforms under $\Gamma_0(M)$ according to the rule

$$f(\gamma z) = \chi(d) \cdot j(\gamma, z)^{2k+1} \cdot f(z), \text{ for } \gamma \in \Gamma_0(M)$$

where d is the lower right hand entry of γ. If, moreover, f vanishes at the cusps, then f is called a cusp form, and we denote the space of all such cusp forms by $\mathcal{S}_{k+\frac{1}{2}}(\Gamma_0(M), \chi)$. For detailed definitions, see [S3].

The fourier coefficients of half-integral weight modular forms are well known to be rich in arithmetic content, and that is certainly one of the main reasons for

1991 *Mathematics Subject Classification.* Primary 11F37; Secondary 11F85.

This paper is in final form and no version of it will be submitted for publication elsewhere.

Supported in part by a grant from the National Science Foundation.

our interest in them. For example, the representation numbers of positive definite integral quadratic forms of odd dimension arise as fourier coefficients of theta functions of half-integral weight. Class numbers and "generalized class numbers" of imaginary quadratic fields arise as the fourier coefficients of Eisenstein series of half-integral weight (see [Coh, Zag] and also the remark after Proposition 2 in [S4]). Moreover, Waldspurger has shown [Wa2] that the coefficients of parabolic eigenforms are closely connected to the square roots of special values of L-functions of integral weight cusp forms.

It is natural to ask whether there might be a "mod p theory" of modular forms of half-integral weight. Such a theory might run parallel to the well-known integral weight theory initiated by Serre and extensively developed to great advantage by many other individuals. More generally, we might ask for a p-adic theory of modular forms of half-integral weight. Hida was apparently the first to ask whether congruences between integral weight modular forms are in some sense preserved under the Shimura correspondence (see [M]). The p-adic analytic properties of the Shimura correspondence were investigated by N. Koblitz in [Kob], who conjectured, among other things, that there are p-adic analytic families of half-integral weight cusp forms attached to powers of the Hecke character of an elliptic curve over $\mathbf{Q}$ with complex multiplication. This conjecture was proven shortly after by W. Kohnen in [Koh]. Recent work of N. Jochnowitz and S. Ren (see [J1, J2]) suggests that a p-adic theory of half-integral weight modular forms may hold some unexpected surprises. For example, it appears likely that the fourier coefficients of p-adic modular forms of half-integral weight may be connected not just to the special values of L-functions of integral weight modular forms (à la Waldspurger), but they may also be intimately connected with the *derivatives* at critical points of square roots of associated p-adic L-functions.

It is not difficult to construct examples of Λ-adic cusp forms of half-integral weight analogous to Hida's Λ-adic cusp forms of integral weight (see [Hi1, Hi2]). A construction was given by Hida in [Hi3] by multiplying a fixed half-integral weight cusp form by a Λ-adic Eisenstein series and normalizing appropriately. In [Hi3] Hida has also shown how the Shimura correspondence can be applied to "descend" to integral weight Λ-adic cusp forms. Our goal in this paper is to examine an inverse construction. Namely, we are going to construct a Λ-adic θ-lifting, whose specialization to arithmetic points is the classical Shintani lifting [Sh]. The simple observation underlying our construction is that the fourier coefficients of the Shintani lift $\theta(f)$ of an integral weight form f can be expressed as period integrals of f, hence can be described in terms of a certain cohomology class associated to f as in chapter 8 of [S2]. This leads to a purely cohomological construction of the Shintani lifting, which first appeared in the work of Shimura [S5]. This cohomological construction applied to the Λ-adic cohomology classes constructed in [G-S] gives rise to a Λ-adic version of Shintani's θ-lifting for Hida's universal ordinary Λ-adic modular forms.

The two basic ingredients needed for our paper are Hida's theory of Λ-adic modular forms and the theory of the Shimura correspondence, especially Shintani's θ-lifting. In section 1, we recall what is needed from Hida theory. Motivated by Hida, we introduce the notion of a Λ-adic analytic space $\mathcal{X}$ together with a certain subset of special points, $\mathcal{X}^{arith} \subseteq \mathcal{X}$, called the arithmetic points. We recall Hida's construction of the *universal ordinary Hecke algebra* $\mathcal{R}_N$ of tame conductor N. This is a finite Λ_N-algebra, where Λ_N is the completed group ring $\Lambda_N := \mathbf{Z}_p[[\mathbf{Z}_{p,N}^\times]]$. The space $\mathcal{X}_N := Hom_{cont}(\mathcal{R}_N, \overline{\mathbf{Q}}_p)$ is an example of a Λ-adic space, which we call the *universal ordinary modular Λ-adic space* of tame conductor N. At the end of section 1, we recall Hida's construction of a formal q-expansion $\mathbf{f}_N \in \mathcal{R}_N[[q]]$ having the remarkable property that specialization of the coefficients of $\mathbf{f}_N$ at arithmetic points induces a one-one correspondence between arithmetic points $\kappa \in \mathcal{X}_N^{arith}$ and normalized p-stabilized ordinary newforms $\mathbf{f}_N(\kappa)$ of tame conductor N. In section 2, we recall some basic facts about the Shimura correspondence. In particular we describe Shintani's θ-lifting from integral weight modular forms to half-integral weight modular forms and recall some of its basic properties. Most significantly, proposition 2.3.1 asserts that if f is an integral weight Hecke eigenform then $\theta(f)$ is essentially algebraic. More precisely, if Ω_f^- is an "imaginary" period of f, then $\theta(f)/\Omega_f^-$ has coefficients in the ring of algebraic integers generated by the Hecke eigenvalues of f. This is a version of a more general theorem of Shimura [S5], though Shimura only states his result up to multiplication by an algebraic number. In section 3 we state our main theorem as Theorem 3.3. We define the *universal ordinary metaplectic Hecke algebra* of tame conductor N to be

$$\tilde{\mathcal{R}}_N := \mathcal{R}_N \otimes_{\Lambda_N, \sigma} \Lambda_N$$

where $\sigma : \Lambda_N \longrightarrow \Lambda_N$ is the map induced by the continuous homomorphim $t \mapsto t^2$ on $\mathbf{Z}_{p,N}^\times \longrightarrow \mathbf{Z}_{p,N}^\times$. The main theorem asserts the existence of a formal q-expansion $\Theta \in \tilde{\mathcal{R}}_N[[q]]$ whose specializations to arithmetic points are, up to multiplication by scalars, the q-expansions obtained via Shintani's θ-lifting from the p-stabilized ordinary newforms of tame conductor N. In section 4, we give a purely cohomological description of the Shintani lifting and use this description to prove the algebraicity of $\theta(f)$ asserted in proposition 2.3.1. Our proof is basically the same as the proof given by Shimura in [S5]. In section 5, we define certain spaces of p-adic distributions and recall a result from [G-S] asserting the existence of a Λ-adic cohomology class associated to $\mathbf{f}_N$. Finally, in section 6 we define a Λ-adic analog of the Shintani lifting and prove our main theorem as a simple application of the tools developed in the paper.

In the interest of simplicity, we have applied our construction only to the original lifting defined by Shintani in [Sh]. It should be pointed out, though, that Shintani's construction has been generalized and strengthened considerably by Waldspurger in [Wa1]. Though we have gained simplicity by sticking to Shintani's original construction we have also paid a price, since we can not give a precise criterion for the non-vanishing of our Λ-adic lifting. On the other hand, it is clear that our construction can be recast in the adelic framework of Waldspurger and applied to his generalization of Shintani's lifting. It would certainly be worthwhile to work out the details of this construction.

The author thanks the referee for a number of thoughtful suggestions to improve this article.

§1. Λ-adic Modular Forms and a Theorem of Hida.

In this section we will recall some important results of Hida concerning the existence and nature of p-adic analytic families of ordinary cusp forms. For each prime number $p \geq 5$ and for each positive integer N prime to p, Hida has constructed a triple $(\mathcal{X}_N, \mathcal{X}_N^{arith}, \mathbf{f}_N)$ consisting of a "Λ-adic analytic space" $\mathcal{X}_N$, a subset $\mathcal{X}_N^{arith} \subseteq \mathcal{X}_N$ of certain special points called "arithmetic points", together with an "analytic" function $\mathbf{f}_N : \mathcal{X}_N \longrightarrow \overline{\mathbf{Q}}_p[[q]]$ of the form

$$\mathbf{f}_N(\kappa) = \sum_{n=1}^{\infty} \alpha_n(\kappa) q^n, \quad \kappa \in \mathcal{X}_N$$

with the property that for every arithmetic point $\kappa \in \mathcal{X}_N^{arith}$, $\mathbf{f}_N(\kappa)$ is the q-expansion of a classical Hecke eigenform that is ordinary at p.

1.1. Λ-adic analytic spaces. Let $\Lambda = \mathbf{Z}_p[[1+p\mathbf{Z}_p]]$ be the completed group ring on the principal unit group $1 + p\mathbf{Z}_p$. If $\mathcal{R}$ is a finite flat Λ-algebra then we set

$$\mathcal{X}(\mathcal{R}) := Hom_{cont}(\mathcal{R}, \overline{\mathbf{Q}}_p).$$

Following Hida's conventions, we will refer to the elements of $\mathcal{X}(\mathcal{R})$ as the "points" of $\mathcal{R}$. Restriction to Λ induces a surjective finite-to-one mapping

$$\pi : \mathcal{X}(\mathcal{R}) \longrightarrow \mathcal{X}(\Lambda).$$

We endow $\mathcal{X}(\Lambda)$ with the analytic structure inherited from the natural identification of $\mathcal{X}(\Lambda)$ with the group of continuous characters $\kappa : 1 + p\mathbf{Z}_p \longrightarrow \overline{\mathbf{Q}}_p^{\times}$. We then use the map π to define analytic charts around all points of $\mathcal{X}(\mathcal{R})$ that are unramified over Λ. Indeed, if $\kappa \in \mathcal{X}(\mathcal{R})$ is unramified over Λ and if $\kappa_0 = \pi(\kappa) \in \mathcal{X}(\Lambda)$, then there is a natural local section S_κ of π

$$S_\kappa : U_0 \subseteq \mathcal{X}(\Lambda) \longrightarrow \mathcal{X}(\mathcal{R})$$

defined on a neighborhood U_0 of $\kappa_0 \in \mathcal{X}(\Lambda)$ and sending κ_0 to κ. The definition of S_κ can be given as follows. The localizaton $\mathcal{R}_{(\kappa)}$ of $\mathcal{R}$ at κ is a discrete valuation ring finite and unramified over $\Lambda_{(\kappa_0)}$. Its residue field is a finite extension of $\mathbf{Q}_p$ and its maximal ideal is generated by a generator $T \in \Lambda$ of the kernel of κ_0. The completion $\mathcal{R}_\kappa$ of $\mathcal{R}_{(\kappa)}$ is isomorphic to the ring of formal power series in T over the residue class field. Moreover, from the finiteness of $\mathcal{R}$ over Λ it follows that there is a real number $r > 0$ such that the image of $\mathcal{R}$ in $\mathcal{R}_\kappa$ is contained in the subring of formal power series whose radius of convergence is $\geq r$. The desired section of π is defined over a neighborhood U_0 of κ_0 by sending a point $\tau \in U_0$ to the point $S_\kappa(\tau) \in \mathcal{X}(\mathcal{R})$ given by evaluation of these power series at $T(\tau)$. The maps S_κ give us analytic charts around each point $\kappa \in \mathcal{X}(\mathcal{R})$ that are unramified over $\mathcal{X}(\Lambda)$.

Definition. The Λ-*adic analytic space associated to a finite flat Λ-algebra* $\mathcal{R}$ is defined to be $\mathcal{X}(\mathcal{R})$ endowed with the above analytic charts S_κ around unramified points $\kappa \in \mathcal{X}(\mathcal{R})$.

The image of any local chart S_κ about an unramified point κ will be called an analytic neighborhood of κ. A function $f : U \subseteq \mathcal{X}(\mathcal{R}) \longrightarrow \overline{\mathbf{Q}}_p$ defined on an analytic neighborhood U of κ will be called analytic if $f \circ S_\kappa$ is analytic. For example, each element $\alpha \in \mathcal{R}$ gives rise to a function $\alpha : \mathcal{X}(\mathcal{R}) \longrightarrow \overline{\mathbf{Q}}_p$ defined by $\alpha(\kappa) = \kappa(\alpha)$ and this function is analytic at every unramified point of $\mathcal{X}(\mathcal{R})$. More generally, if κ is an unramified point, then every $\alpha \in \mathcal{R}_{(\kappa)}$ gives rise to an analytic function defined on some analytic neighborhood of κ.

For example, if N is a positive integer prime to p, we let

$$\Lambda_N := \mathbf{Z}_p[[\mathbf{Z}_{p,N}^\times]]$$

be the completed group ring on $\mathbf{Z}_{p,N}^\times := \varprojlim_m (\mathbf{Z}/Np^m\mathbf{Z})^\times$. The natural isomorphism $\Lambda_N \cong \Lambda[\Delta_{Np}]$, where $\Delta_{Np} = (\mathbf{Z}/Np\mathbf{Z})^\times$, shows that $\mathcal{X}(\Lambda_N)$ is just a product of $\varphi(Np)$ copies of $\mathcal{X}(\Lambda)$. The components are in natural one-one correspondence with the $\overline{\mathbf{Q}}_p$-valued characters of Δ_{Np}.

1.2. Arithmetic Points. We are interested in analytic functions on $\mathcal{X}(\mathcal{R})$ that interpolate global arithmetic data at certain special points called arithmetic points.

(1.2.1) Definition. A character $\kappa : 1 + p\mathbf{Z}_p \longrightarrow \overline{\mathbf{Q}}_p$ is said to be *arithmetic* if there is an integer $r \geq 0$ such that

$$\kappa(t) = t^r$$

for all t sufficiently close to 1 in $1+p\mathbf{Z}_p$. A point $\kappa \in \mathcal{X}(\Lambda)$ is said to be arithmetic if the associated character of $1 + p\mathbf{Z}_p$ is arithmetic. If $\mathcal{R}$ is a finite Λ-algebra, then a point $\kappa \in \mathcal{X}(\mathcal{R})$ is said to be arithmetic if it lies over an arithmetic point on $\mathcal{X}(\Lambda)$.

The set of arithmetic points of $\mathcal{R}$ will be denoted by

$$\mathcal{X}^{arith}(\mathcal{R}) := \left\{ \, \kappa \mid \kappa \text{ is arithmetic} \, \right\}.$$

For each positive integer N prime to p there is a canonical decomposition $\mathbf{Z}_{p,N}^\times = \mathbf{Z}_p^\times \times (\mathbf{Z}/N\mathbf{Z})^\times$. For $t \in \mathbf{Z}_{p,N}^\times$ we let t_p and t_N be the coordinates of t under this decomposition. In the applications, $\mathcal{R}$ will be a Λ_N-algebra. Since Λ_N has a natural Λ-algebra structure, such an $\mathcal{R}$ can be regarded as a Λ-algebra in a natural way.

(1.2.2) Definition. Let $r \geq 0$ be a rational integer and $\epsilon : \mathbf{Z}_{p,N}^{\times} \longrightarrow \overline{\mathbf{Q}}_p^{\times}$ be a finite character.

a. An arithmetic point $\kappa \in \mathcal{X}^{arith}(\Lambda_N)$ is said to have *signature* (r, ϵ) if the associated character of $\mathbf{Z}_{p,N}^{\times}$ is given by $\kappa(t) = \epsilon(t)t_p^r$, for $t \in \mathbf{Z}_{p,N}^{\times}$.

b. An arithmetic point $\kappa \in \mathcal{X}(\mathcal{R})$, for $\mathcal{R}$ a Λ_N-algebra finite and flat over Λ, is said to have signature (r, ϵ) if κ lies over an arithmetic point of Λ_N with signature (r, ϵ).

1.3. p-stabilized ordinary newforms. Let $M > 0$ be an integer and let ϵ be a Dirichlet character defined modulo M. For each $r \geq 0$ let $\mathcal{S}_{r+2}(\Gamma_0(M), \epsilon)$ denote the complex vector space of cusp forms of level M, weight $r + 2$, and nebentype character ϵ. Hence, if $f \in \mathcal{S}_{r+2}(\Gamma_0(M), \epsilon)$ then for each $\gamma = \begin{pmatrix} a & b \\ c & d \end{pmatrix}$ in $\Gamma_0(M)$ we have

$$(cz + d)^{-2-r} f\left(\frac{az + b}{cz + d}\right) = \epsilon(d)f(z), \quad z \in \mathbf{H}.$$

Now fix an embedding

$$(1.3.1) \qquad\qquad\qquad \overline{\mathbf{Q}} \subseteq \overline{\mathbf{Q}}_p.$$

We say that a Hecke eigenform $f \in \mathcal{S}_{r+2}(\Gamma_1(M))$ is *ordinary at p* if the eigenvalue a_p of the Hecke operator T_p is a p-adic unit with respect to this fixed embedding. In that case, the Euler factor at p of the L-function of f has a factorization $(1 - \alpha p^{-s})(1 - \beta p^{-s})$ where α is a p-adic unit and β is divisible by p. We will refer to α as the *unit root of Frobenius* and β as the *non-unit root of Frobenius*. Note that if $p|M$ then $\beta = 0$.

Definition. Suppose $M = Np^m$ for some $m > 0$ where $p \nmid N$. We say that a Hecke eigenform $f \in \mathcal{S}_{r+2}(\Gamma_0(Np^m), \epsilon)$ is a *p-stabilized ordinary newform* of tame conductor N if one of the following is true.

- f is a newform of conductor Np^m; or
- there is a newform g of conductor N such that $f(z) = g(z) - \beta g(pz)$ on the upper half-plane where β is the non-unit root of Frobenius attached to g.

Note that in the second case, we may take $m = 1$.

1.4. The universal p-stabilized ordinary newform of tame conductor N. For each integer $m > 0$, let $X_m = X_1(Np^m) = \Gamma_1(Np^m)\backslash \mathbf{H}^*$ be the complete modular curve obtained as the quotient of the extended upper half-plane $\mathbf{H}^*$ by the congruence group $\Gamma_1(Np^m)$. Let $V_m = H^1(X_m, \mathbf{Z}_p)$ be the simplicial cohomology of X_m with values in $\mathbf{Z}_p$. It is well-known (see e.g. chapter 8 of [S2]), that V_m is canonically isomorphic to the parabolic cohomology group $H_{par}^1(\Gamma_1(Np^m), \mathbf{Z}_p) \subseteq H^1(\Gamma_1(Np^m), \mathbf{Z}_p)$, defined as the image of the compactly supported cohomology under the natural map

$$H_c^1(\Gamma_1(Np^m), \mathbf{Z}_p) \longrightarrow H^1(\Gamma_1(Np^m), \mathbf{Z}_p).$$

We define the *abstract Λ-adic Hecke algebra* of tame conductor N to be the free polynomial algebra

$$\mathcal{H} = \Lambda_N[T_n \, (n \in \mathbf{Z}^+)]$$

generated over Λ_N by $T_n \, (n \in \mathbf{Z}^+)$. We let $\mathcal{H}$ act on V_m, by letting $\mathbf{Z}_{p,N}^\times$ act via the Nebentype and T_n act via the nth (covariant) Hecke correspondence. For each pair of integers $m_1 \geq m_2 > 0$, the natural inclusion $\Gamma_1(Np^{m_1}) \subseteq \Gamma_1(Np^{m_2})$ induces a corestriction map $V_{m_1} \longrightarrow V_{m_2}$ which commutes with the action of $\mathcal{H}$, hence we may form the projective limit and obtain an $\mathcal{H}$-module

$$V_\infty := \varprojlim_m V_m.$$

As Hida pointed out, we can use the operator T_p to decompose the space V_∞ into a natural direct sum

$$V_\infty = V_\infty^0 \oplus V_\infty^{nil}$$

of $\mathcal{H}[G_\mathbf{Q}]$-modules such that T_p acts invertibly on V_∞^0 and acts topologically nilpotently on V_∞^{nil}. Hida has proven that as a Λ-module, V_∞^0 is free of finite rank. Moreover, letting $\mathcal{L}$ denote the fraction field of Λ, Hida has constructed an idempotent e_{prim} in the image of $\mathcal{H} \otimes_\Lambda \mathcal{L}$ in $End_\mathcal{L}(V_\infty^0 \otimes_\Lambda \mathcal{L})$ analogous to projection to the space of N-primitive Hecke eigenforms in Atkin-Lehner theory. Let $\mathbf{V}_N$ be the $\mathcal{H}$-module obtained as the intersection of V_∞^0 and $e_{prim}V_\infty^0 \otimes_\Lambda \mathcal{L}$. Then $\mathbf{V}_N$ is a reflexive Λ-module of finite rank and is consequently free of finite rank over Λ. We will sometimes refer to $\mathbf{V}_N$ as the *universal ordinary parabolic cohomology group of tame conductor N*.

(1.4.1) Definition.

 a. The *universal ordinary p-adic Hecke algebra* of tame conductor N is defined to be the image $\mathcal{R}_N$ of $\mathcal{H}$ in $End_{\Lambda_N}(\mathbf{V}_N)$. We let $h : \mathcal{H} \longrightarrow \mathcal{R}_N$ be the associated homomorphism and endow $\mathcal{R}_N$ with the induced structure of Λ_N-algebra.

 b. The *universal p-stabilized ordinary newform* of tame conductor N is defined to be the formal q-expansion $\mathbf{f}_N := \sum_{n=1}^\infty \alpha_n q^n \in \mathcal{R}_N[[q]]$ where the coefficients are given by $\alpha_n = h(T_n)$.

Let $\mathcal{K}_N := \mathcal{R}_N \otimes_\Lambda \mathcal{L}$ where $\mathcal{L}$ is the fraction field of Λ. Clearly, $\mathcal{R}_N \subseteq \mathcal{K}_N$. We let $\mathcal{X}_N := \mathcal{X}(\mathcal{R}_N)$ be the points of $\mathcal{R}_N$ and let $\mathcal{X}_N^{arith} := \mathcal{X}^{arith}(\mathcal{R}_N)$ be the arithmetic points.

(1.5) HIDA'S THEOREM. *Let p be a prime ≥ 5 and let N be a positive integer with $p \nmid N$. Let $\mathcal{R}_N \subseteq \mathcal{K}_N$, $\mathbf{f}_N$, and $\mathcal{X}_N^{arith} \subseteq \mathcal{X}_N$ be defined as in 1.4. Then $\mathcal{K}_N$ is a product of finite field extensions of $\mathcal{L}$. Every arithmetic point in $\mathcal{X}_N^{arith}$ is unramified over $\mathcal{X}(\Lambda)$. Moreover, the map $\mathcal{X}_N^{arith} \longrightarrow \overline{\mathbf{Q}}_p[[q]]$ given by $\kappa \mapsto \mathbf{f}_N(\kappa) := \sum \alpha_n(\kappa)q^n$ induces a one-one correspondence*

$$\left\{ \begin{array}{c} \text{Arithmetic points} \\ \kappa \in \mathcal{X}_N^{arith} \\ \text{of signature } (r, \epsilon) \end{array} \right\} \longleftrightarrow \left\{ \begin{array}{c} \text{Ordinary p-stabilized newforms} \\ \text{of tame conductor } N \\ \text{with weight } r + 2 \text{ and character } \epsilon \end{array} \right\}$$

for every integer $r \geq 0$ and every finite character $\epsilon : \mathbf{Z}_{p,N}^{\times} \longrightarrow \overline{\mathbf{Q}}_p^{\times}$.

In the spirit of section 1.1, we will regard $\mathbf{f}_N$ as an analytic function on $\mathcal{X}_N$ interpolating the q-expansions of p-stabilized ordinary newforms at arithmetic points.

§2. Shintani's Lifting.

In this section we recall Shintani's theta correspondence [Sh] and a few of its properties.

2.1. Indefinite binary quadratic forms. Fix a positive integer M and consider the set $\mathcal{F}_M$ of all integral indefinite binary quadratic forms, $Q(X,Y) = aX^2 + bXY + cY^2$, whose elements are characterized by the conditions

$$(2.1.1) \qquad Q \in \mathcal{F}_M \Longleftrightarrow \begin{cases} b^2 - 4ac > 0; & \text{and} \\ (a, M) = 1, & b \equiv c \equiv 0 \bmod M. \end{cases}$$

We let Hecke's congruence group $\Gamma_0(M)$ act on $\mathcal{F}_M$ by linear substitutions: for $Q \in \mathcal{F}_M$ and $\gamma \in \Gamma_0(M)$ we define $(Q|\gamma)(X,Y) := Q((X,Y)\gamma^{-1})$. A straightforward calculation confirms that this action preserves $\mathcal{F}_M$.

(2.1.2) To each quadratic form $Q(X,Y) = aX^2 + bXY + cY^2$ in $\mathcal{F}_M$ we associate the following data (following Shintani [Sh]).
a. The discriminant $\delta_Q = b^2 - 4ac$.
b. The pair ω_Q, ω_Q' of points in $\mathbf{P}^1(\mathbf{R}) = \mathbf{R} \cup \{i\infty\}$ given by

$$(\omega_Q, \omega_Q') := \begin{cases} \left(\dfrac{b + \sqrt{\delta_Q}}{2c}, \dfrac{b - \sqrt{\delta_Q}}{2c} \right) & \text{if } c \neq 0; \\[2ex] \left(i\infty, \dfrac{a}{b} \right) & \text{if } c = 0 \text{ and } b > 0; \\[2ex] \left(\dfrac{a}{b}, i\infty \right) & \text{if } c = 0 \text{ and } b < 0. \end{cases}$$

c. The oriented geodesic path C_Q in the upper half-plane, defined as follows.
  (i) If δ_Q is a perfect square, then C_Q is the oriented geodesic path joining ω_Q to ω_Q'.
  (ii) If δ_Q is not a perfect square, then consider the stabilizer $\Gamma_0(M)_Q$ of Q in $\Gamma_0(M)$ and its index 2 subgroup $\Gamma_0(M)_Q^+ \subseteq \Gamma_0(M)_Q$ consisting of elements with positive trace. Let $\gamma_Q = \begin{pmatrix} r & s \\ t & u \end{pmatrix} \in \Gamma_0(M)_Q^+$ be the unique generator for which $r - t\omega_Q > 1$. We then define C_Q to be the oriented geodesic path from $i\infty$ to $\gamma_Q(i\infty)$.

Note that δ_Q is a positive integer divisible by M. The two lines in $\mathbf{R}^2$ determined by $\frac{y}{x} = \omega_Q$ and $\frac{y}{x} = \omega_Q'$ are the components of $Q(x, -y) = 0$. If δ_Q is not a perfect square, the group $\Gamma_0(M)_Q$ is equal to the stabilizer in $\Gamma_0(M)$ of the line spanned by $(1, -\omega_Q)$ in $\mathbf{R}^2$ (under matrix multiplication on the right) and $\Gamma_0(M)_Q^+$ is the subgroup of elements that act on this line in an orientation-preserving manner. From this description it follows that $\Gamma_0(M)_Q^+$ is infinite cyclic. The generator γ_Q is the one acting on $(1, -\omega_Q)$ via an eigenvalue > 1. More explicitly, if $(x, y) \in (\frac{1}{2}\mathbf{Z})^2$ is the smallest half-integral solution of Pell's equation $x^2 - \delta_Q y^2 = 1$ with $x + y\sqrt{\delta_Q} > 1$, then (compare [Sh])

$$(2.1.3) \qquad\qquad \gamma_Q = \begin{pmatrix} x - by & 2ay \\ -2cy & x + by \end{pmatrix}.$$

2.2. Shintani's Theorem. We now suppose for simplicity that M is odd. For a given Dirichlet character χ defined modulo M we define a new Dirichlet character χ' modulo $4M$ by

$$\chi'(d) := \chi(d) \cdot \left(\frac{(-1)^{k+1}M}{d} \right), \qquad d \in (\mathbf{Z}/4M\mathbf{Z})^\times.$$

For each quadratic form $Q(X, Y) = aX^2 + bXY + cY^2$ in $\mathcal{F}_M$ we set $\chi(Q) = \chi(a)$. Now fix an integer $k \geq 0$. A simple verification shows that for each $f \in \mathcal{S}_{2k+2}(\Gamma_0(M), \chi^2)$, the integral

$$(2.2.1) \qquad\qquad I_{k,\chi}(f, Q) := \chi(Q) \cdot \int_{C_Q} f(\tau) Q(1, -\tau)^k \, d\tau$$

converges and depends only on the $\Gamma_0(M)$-orbit of Q in $\mathcal{F}_M$.

For $z \in \mathcal{H}$, we define

$$(2.2.2) \qquad\qquad \theta_{k,\chi}(f, z) := \sum_{Q \in \mathcal{F}_M/\Gamma_0(M)} I_{k,\chi}(f, Q) q^{\delta_Q/M}, \qquad q := e^{2\pi i z}.$$

From the definitions we see that $C_{-Q} = -C_Q$ for each $Q \in \mathcal{F}_M$, and consequently that

$$I_{k,\chi}(f, -Q) = \chi(-1)(-1)^{k+1} I_{k,\chi}(f, Q).$$

Hence the sum defining $\theta_{k,\chi}(f)$ vanishes identically unless $\chi(-1) = (-1)^{k+1}$. The following theorem is due to Shintani [Sh].

$(2.2.3)$ THEOREM. *Let $k \geq 0$ and χ be a Dirichlet character defined modulo M. Then for each $f \in \mathcal{S}_{2k+2}(\Gamma_0(M), \chi^2)$, the series $\theta_{k,\chi}(f)$ is the q-expansion of a cusp form in $\mathcal{S}_{k+\frac{3}{2}}(\Gamma_0(4M), \chi')$. Moreover, the map*

$$\mathcal{S}_{2k+2}(\Gamma_0(M), \chi^2) \xrightarrow{\theta_{k,\chi}} \mathcal{S}_{k+\frac{3}{2}}(\Gamma_0(4M), \chi')$$

is a Hecke equivariant linear function: for ℓ an arbitrary prime $\theta_{k,\chi}(f|T_\ell) = \theta_{k,\chi}(f)|T_{\ell^2}$.

$(2.2.4)$ **Remark.** In fact, the definition $(2.2.2)$ of the theta lifting $\theta_{k,\chi}$ depends implicitly on the choice of level M. For each multiple M' of M we have a corresponding theta lifting which we denote by $\theta_{k,\chi}^{(M')}(f)$. If p is a prime divisor of M, then we have the following simple relation:

$$\theta_{k,\chi}(f)|T_p^m = \theta_{k,\chi}^{(Mp^m)}(f)$$

where T_p is the operator on power series defined by

$$(2.2.5) \qquad \left(\sum_{n=1}^{\infty} \beta_n \, q^n \right) \Big| \, T_p := \sum_{n=1}^{\infty} \beta_{np} q^n.$$

It follows from Proposition 1.5 of Shimura's paper [S3] that, assuming $p|M$, this map on q-expansions induces a map

$$T_p : \mathcal{S}_{k+3/2}(\Gamma_0(4M), \chi') \longrightarrow \mathcal{S}_{k+3/2}\left(\Gamma_0(4M), \ \chi' \cdot \left(\frac{p}{\cdot} \right) \right).$$

Hence, T_p preserves the level but multiplies the Nebentype character by the quadratic character $\left(\frac{p}{\cdot} \right)$.

2.3. Algebraicity and integrality. If R is a subring of $\mathbf{C}$ then we will denote by $\mathcal{S}_{k+\frac{3}{2}}(\Gamma_0(4M), \chi'; R)$ the R-submodule of $\mathcal{S}_{k+\frac{3}{2}}(\Gamma_0(4M), \chi')$ consisting of forms whose fourier coefficients lie in R. It is well known that $\mathcal{S}_{k+\frac{3}{2}}(\Gamma_0(4M), \chi')$ is spanned by $\mathcal{S}_{k+\frac{3}{2}}(\Gamma_0(4M), \chi'; \mathbf{Z}[\chi])$, consequently that

$$\mathcal{S}_{k+\frac{3}{2}}(\Gamma_0(4M), \chi'; R) \cong \mathcal{S}_{k+\frac{3}{2}}(\Gamma_0(4M), \chi'; \mathbf{Z}[\chi]) \otimes_{\mathbf{Z}[\chi]} R.$$

For an arbitrary $\mathbf{Z}[\chi]$-algebra R, we may therefore define

$$\mathcal{S}_{k+\frac{3}{2}}(\Gamma_0(4M), \chi'; R) := \mathcal{S}_{k+\frac{3}{2}}(\Gamma_0(4M), \chi'; \mathbf{Z}[\chi]) \otimes_{\mathbf{Z}[\chi]} R.$$

A glance at the integrals $I_{k,\chi}(f, Q)$ reveals that these integrals can be expressed as integral linear combinations of critical values of the L-function of f. In case f is a Hecke eigenform, it is well-known that there are two non-zero complex periods $\Omega_f^{\pm}$ of which the critical values are algebraic integral multiples. From this it follows easily that the fourier coefficients of $\theta_{k,\chi}(f)$ are algebraic integral linear combinations of Ω_f^+ and Ω_f^-. In section 4 we will prove the following result of Shimura [S5]. Though our statement of the theorem is slightly stronger than the (more general) statement in [S5], in that we have located the algebraicity in the Hecke ring, the proof we give in section 4 is essentially the same as the one given by Shimura.

(2.3.1) PROPOSITION. *Suppose $f \in \mathcal{S}_{2k+2}(\Gamma_0(M), \chi^2)$ is a Hecke eigenform and let $\mathcal{O}_f$ be the ring generated by the Hecke eigenvalues. Then there is a non-zero complex number $\Omega_f^- \in \mathbf{C}^{\times}$ such that*

$$\frac{1}{\Omega_f^-} \cdot \theta_{k,\chi}(f) \in \mathcal{S}_{k+\frac{3}{2}}(\Gamma_0(4M), \chi'; \mathcal{O}_f).$$

Now fix such a period Ω_f^- once and for all for each Hecke eigenform $f \in \mathcal{S}_{2k+2}(\Gamma_0(M), \chi^2)$ and define

$$(2.3.2) \qquad \theta_{k,\chi}^*(f) := \frac{1}{\Omega_f^-} \cdot \theta_{k,\chi}(f) \in \mathcal{S}_{k+\frac{3}{2}}(\Gamma_0(4M), \chi'; \mathcal{O}_f).$$

We may sometimes refer to $\theta_{k,\chi}^*(f)$ as the "algebraic part" of $\theta_{k,\chi}(f)$. We caution the reader that this definition depends on the choice of the period Ω_f^-.

§3. Statement of the Theorem

Fix a prime $p \geq 5$ and a positive integer N with $p \nmid N$. Let $\sigma : \Lambda_N \longrightarrow \Lambda_N$ be the ring homomorphism associated to the group homomorphism $t \mapsto t^2$ on $\mathbf{Z}_{p,N}^\times$. Define the *universal ordinary metaplectic p-adic Hecke algebra* of tame conductor N to be the Λ_N-algebra

$$(3.1) \qquad \tilde{\mathcal{R}}_N := \mathcal{R}_N \otimes_{\Lambda_N, \sigma} \Lambda_N$$

where $\mathcal{R}_N$ is the universal ordinary p-adic Hecke algebra of tame conductor N and the tensor product is taken with respect to σ. We regard $\tilde{\mathcal{R}}_N$ as a Λ_N-algebra by equipping it with the structure homomorphism $\Lambda_N \longrightarrow \tilde{\mathcal{R}}_N$, $\lambda \mapsto 1 \otimes \lambda$. Note that the ring homomorphism

$$\mathcal{R}_N \longrightarrow \tilde{\mathcal{R}}_N$$
$$\alpha \longmapsto \alpha \otimes 1$$

is *not a homomorphism of Λ_N-algebras*. This is reflected in the fact that the map induced by pullback on Λ-adic spaces

$$(3.2) \qquad \tilde{\mathcal{X}}_N \longrightarrow \mathcal{X}_N$$

does not preserve the signatures of arithmetic points. Indeed, if $\tilde{\kappa} \in \tilde{\mathcal{X}}_N^{arith}$ lies over $\kappa \in \mathcal{X}_N^{arith}$, then the signature of κ is twice the signature of $\tilde{\kappa}$. More precisely, if $\tilde{\kappa}$ has signature (k, χ) then κ has signature $(2k, \chi^2)$.

(3.3) THEOREM. *Let $\kappa_0 \in \mathcal{X}_N^{arith}$ be a fixed arithmetic point. Then there is a formal q-expansion $\Theta = \sum_{n=1}^\infty \beta_n q^n \in \tilde{\mathcal{R}}_N[[q]]$ and a choice of p-adic periods $\Omega_\kappa \in \overline{\mathbf{Q}}_p$ ($\kappa \in \mathcal{X}_N^{arith}$) with the following properties:*
a. $\Omega_{\kappa_0} \neq 0$.
b. *For every arithmetic point $\tilde{\kappa} \in \tilde{\mathcal{X}}_N^{arith}$ of signature (k, χ),*

$$\Theta(\tilde{\kappa}) := \sum_{n=1}^\infty \beta_n(\tilde{\kappa}) q^n \in \mathcal{S}_{k+\frac{3}{2}}(\Gamma_0(4Np^m), \chi^*; \overline{\mathbf{Q}}_p)$$

where m is the smallest positive integer for which χ is defined modulo Np^m and χ^* is defined by $\chi^*(d) = \chi(d)\left(\frac{(-1)^{k+1}Np}{d}\right)$. Moreover, if κ is the image of $\tilde{\kappa}$ in $\mathcal{X}_N^{arith}$, then

$$\Theta(\tilde{\kappa}) = \Omega_\kappa \cdot \theta_{k,\chi}^*\left(\mathbf{f}_N(\kappa)\right)\bigg| T_p^{1-m}.$$

(3.4) Remark. By Shintani's theorem we have

$$\theta_{k,\chi}^* \big(\mathbf{f}_N(\kappa)\big) \mid T_p^2 = \theta_{k,\chi}^* \big(\mathbf{f}_N(\kappa)|T_p\big)$$

and this in turn is given by $\alpha_p(\kappa) \cdot \theta_{k,\chi}^* \big(\mathbf{f}_N(\kappa)\big)$ where $\alpha_p \in \mathcal{R}_N^\times$ is the pth coefficient of $\mathbf{f}_N$. Hence

$$\theta_{k,\chi}^* \left(\mathbf{f}_N(\kappa) \right) \bigg| T_p^{1-m} := \alpha_p(\kappa)^{1-m} \theta_{k,\chi}^* \left(\mathbf{f}_N(\kappa) \right) \bigg| T_p^{m-1}.$$

We will give a proof of theorem 3.3 in section 6. Note that for given $\kappa \in \mathcal{X}_N^{arith}$, choosing a $\tilde{\kappa} \in \tilde{\mathcal{X}}_N^{arith}$ lying over κ is equivalent to choosing the lifting data that goes into Shintani's theta lifting. The theorem says that up to non-zero scalars $\Theta(\tilde{\kappa})|T_p^{m-1}$ is the theta lifting of $\mathbf{f}_N(\kappa)$ associated to that choice of lifting data.

(3.5) Remark. It is possible that Θ vanishes identically. However, it does follow from (3.3)**a** that our Λ-adic Shintani lifting is "as good as" the classical Shintani lifting. In other words, if (k_0, χ_0) is the signature of an arithmetic point $\tilde{\kappa}_0 \in \tilde{\mathcal{X}}_N^{arith}$ lying over $\kappa_0 \in \mathcal{X}_N^{arith}$, and if the classical Shintani lift $\theta_{k_0,\chi_0}(\mathbf{f}_N(\kappa_0))$ is non-zero, then the Λ-adic Shintani lift Θ does not vanish in a neighborhood of κ_0, hence gives rise to an analytic family of non-zero half-integral weight forms passing through $\theta_{k_0,\chi_0}(\mathbf{f}_N(\kappa_0))$. For a detailed analysis of the non-vanishing properties of the classical Shintani lifting and its generalizations, we refer the reader to the fundamental work of Waldspurger [Wa1].

§4. A Cohomological Shintani Lifting

In this section we will describe the Shintani lifting in terms of compactly supported one-dimensional cohomology. Recall from [A-S] that if A is a right $\mathbf{Z}[\frac{1}{6}][\Gamma_0(M)]$-module, then the group $H_c^1(\Gamma_0(M), A)$ of compactly supported cohomology classes can be described explicitly as a group of "modular symbols". Let $\mathcal{D}_0 = Div^0(\mathbf{P}^1(\mathbf{Q}))$ be the group of divisors of degree 0 supported on the rational cusps $\mathbf{P}^1(\mathbf{Q})$ of the upper half plane. Then $\Gamma_0(M)$ acts on $\mathcal{D}_0$ by fractional linear transformations. We define the group of A-*valued modular symbols over* $\Gamma_0(M)$ to be

$$(4.0.1) \qquad\qquad \mathrm{Symb}_{\Gamma_0(M)}(A) := \mathrm{Hom}_{\Gamma_0(M)}(\mathcal{D}_0, A).$$

Hence a modular symbol $\Phi \in \mathrm{Symb}_{\Gamma_0(M)}(A)$ is just an additive homomorphism $\Phi : \mathcal{D}_0 \longrightarrow A$ for which $\Phi|\gamma = \Phi$ for all $\gamma \in \Gamma_0(M)$ where we define the action of $\gamma \in \Gamma_0(M)$ on $\mathrm{Hom}(\mathcal{D}_0, A)$ by

$$(4.0.2) \qquad\qquad (\Phi|\gamma)(D) := \Phi(\gamma D)|\gamma.$$

In [A-S] it is proven that there is a canonical isomorphism

$$(4.0.3) \qquad\qquad H_c^1(\Gamma_0(M), A) = \mathrm{Symb}_{\Gamma_0(M)}(A).$$

Accordingly, we will speak of compactly supported cohomology and modular symbols interchangeably (assuming A is a $\mathbf{Z}[\frac{1}{6}][\Gamma_0(M)]$-module). For details, see [S2] or [A-S].

The natural map $H_c^1(\Gamma_0(M), A) \longrightarrow H^1(\Gamma_0(M), A)$ sends a modular symbol φ to the cohomology class represented by the 1-cocycle $\gamma \longmapsto \varphi(\{\gamma x_0\} - \{x_0\})$ for any fixed $x_0 \in \mathbf{P}^1(\mathbf{Q})$. Recall that the parabolic cohomology group $H_{par}^1(\Gamma_0(M), A)$ (defined for example as in chapter 8 of [S2]) is the image of $H_c^1(\Gamma_0(M), A)$ in $H^1(\Gamma_0(M), A)$, hence we have a natural surjective map

(4.0.4) $$H_c^1(\Gamma_0(M), A) \longrightarrow H_{par}^1(\Gamma_0(M), A).$$

If the action of $\Gamma_0(M)$ on A extends to an action of the semigroup $S_0(M)$ of 2×2 integral matrices $\gamma = \begin{pmatrix} a & b \\ c & d \end{pmatrix}$ with $(a, M) = 1$ and $c \equiv 0$ modulo M, then each of the cohomology groups described above inherits a natural action of the Hecke operators. Moreover the maps

$$H_c^1(\Gamma_0(M), A) \longrightarrow H^1(\Gamma_0(M), A) \longrightarrow H_{par}^1(\Gamma_0(M), A)$$

are Hecke equivariant. In a similar fashion, the matrix $\iota = \begin{pmatrix} 1 & 0 \\ 0 & -1 \end{pmatrix}$ induces natural involutions on the cohomology groups $H^1(\Gamma_0(M), A)$ and $H_c^1(\Gamma_0(M), A)$. On modular symbols this involution is given by $\varphi \mapsto \varphi|\iota$ where $\varphi|\iota : D \mapsto \varphi(\iota D)|\iota$ for $D \in \mathcal{D}_0$. Assuming A is a $\mathbf{Z}[\frac{1}{2}]$-module, the involution ι decomposes each of the cohomology groups into $\pm$ eigenspaces:

(4.0.5) $$H = H^+ \oplus H^-.$$

Indeed, each cohomology class φ decomposes as $\varphi = \varphi^+ + \varphi^-$ where $\varphi^\pm := \frac{1}{2}(\varphi \pm \varphi|\iota)$ satisfies $\varphi|\iota = \pm\varphi$.

4.1. Some $\Gamma_0(M)$-modules. Let R be a commutative ring. For each integer $r \geq 0$ we define $Sym^r(R^2)$ (respectively, $Sym^r(R^2)^*$) to be the R-module of homogeneous "divided powers polynomials" (repectively, polynomials) of degree r in two variables X, Y over R. Hence $Sym^r(R^2)$ (respectively, $Sym^r(R^2)^*$) is the free R-module generated by the divided powers monomials $\dfrac{X^n}{n!} \cdot \dfrac{Y^{r-n}}{(r-n)!}$ (respectively, the monomials $X^n \cdot Y^{r-n}$) for $0 \leq n \leq r$. The semigroup $M_2(R)$ of 2×2 matrices over R acts naturally on both $Sym^r(R^2)$ and $Sym^r(R^2)^*$ as follows. Define an anti-involution $* : M_2(R) \longrightarrow M_2(R)$ by $\gamma = \begin{pmatrix} a & b \\ c & d \end{pmatrix} \mapsto$ $\gamma^* := \begin{pmatrix} d & -b \\ -c & a \end{pmatrix}$. Then the action of $\gamma \in M_2(R)$ is defined by the formula

(4.1.1) $$(F|\gamma)(X, Y) = F\big((X, Y)\gamma^*\big)$$

for either $F \in Sym^r(R^2)$ or $F \in Sym^r(R^2)^*$.

There is a unique perfect pairing $\langle \cdot, \cdot \rangle_r : Sym^r(R^2) \times Sym^r(R^2)^* \longrightarrow R$ satisfying $\left\langle \dfrac{X^i Y^{r-i}}{i!(r-i)!}, X^{r-j}Y^j \right\rangle = (-1)^j \delta_{ij}$, where $\delta_{ij} = 1$ if $i = j$ and $= 0$ otherwise. In particular, this pairing has the property

$$(4.1.2) \qquad \left\langle \frac{(aY - bX)^r}{r!}, P(X,Y) \right\rangle = P(a,b)$$

for every $(a,b) \in R^2$ and every $P \in Sym^r(R^2)^*$. This pairing satisfies the relation

$$(4.1.3) \qquad \langle P_1 | \gamma, P_2 | \gamma \rangle = det(\gamma)^r \langle P_1, P_2 \rangle$$

for every pair $(P_1, P_2) \in Sym^r(R^2) \times Sym^r(R^2)^*$ and every $\gamma \in M_2(R)$.

Let ϵ be an R-valued Dirichlet character modulo a positive integer M. Define $L_{r,\epsilon}(R)$ (respectively, $L^*_{r,\epsilon}(R)$) to be the $R[\Gamma_0(M)]$-module whose underlying R-module is $Sym^r(R^2)$ (respectively, $Sym^r(R^2)^*$) equipped with the natural action (4.1.1) of $\Gamma_0(M)$ twisted by ϵ^{-1} (respectively, ϵ): for $\gamma = \begin{pmatrix} a & b \\ c & d \end{pmatrix} \in \Gamma_0(M)$:

$$(4.1.4) \qquad (F|\gamma)(X,Y) := \begin{cases} \epsilon(a) \cdot F((X,Y)\gamma^{-1}) & \text{for } F \in L_{r,\epsilon}(R); \\[2mm] \epsilon(d) \cdot F((X,Y)\gamma^{-1}) & \text{for } F \in L^*_{r,\epsilon}(R). \end{cases}$$

4.2. Parabolic cohomology and cusp forms. If $R = K$ is a field of characteristic zero, then by the "Manin-Drinfeld principle" there is a unique Hecke equivariant section

$$(4.2.1) \qquad s_{r,\epsilon} : H^1_{par}(\Gamma_0(M), L_{r,\epsilon}(K)) \longrightarrow H^1_c(\Gamma_0(M), L_{r,\epsilon}(K))$$

of the surjection (4.0.4). For example, to each cusp form $f \in \mathcal{S}_{r+2}(\Gamma_0(M), \epsilon)$ we may attach an $L_{r,\epsilon}(\mathbf{C})$-valued cohomology class Ψ_f over Γ as follows. Consider the $L_{r,\epsilon}(\mathbf{C})$-valued differential form $\omega_f := \dfrac{1}{r!} \cdot f(\tau)(\tau X + Y)^r \, d\tau$ on $\mathbf{H}$. A simple calculation confirms that ω_f satisfies the identity $\gamma^*(\omega_f)|\gamma = \omega_f$, for every $\gamma \in \Gamma_0(M)$. As in chapter 8 of [S2] we see that integration of ω_f gives rise to a parabolic cohomology class

$$(4.2.2) \qquad \Psi_f \in H^1_{par}(\Gamma_0(M), L_{r,\epsilon}(\mathbf{C})).$$

We have the following important result of Eichler and Shimura.

(4.2.3) THEOREM. (Eichler, Shimura) *For either choice of sign $\pm$, the map $f \mapsto \Psi_f^{\pm}$ is a Hecke equivariant isomorphism*

$$\mathcal{S}_{r+2}(\Gamma_0(M), \epsilon) \xrightarrow{\sim} H_{par}(\Gamma_0(M), L_{r,\epsilon}(\mathbf{C}))^{\pm}.$$

The additive map $\psi_f : \mathcal{D}_0 \longrightarrow L_{r,\epsilon}(\mathbf{C})$ given by

$$(4.2.4) \qquad \{c_2\} - \{c_1\} \longmapsto \int_{c_1}^{c_2} \omega_f$$

defines a modular symbol over $\Gamma_0(M)$ with values in $L_{r,\epsilon}(\mathbf{C})$. (The above integral is taken over the oriented geodesic path joining c_1 to c_2 in $\mathbf{H}$). Hence ψ_f determines a compactly supported cohomology class

$$(4.2.5) \qquad \psi_f \in H^1_c(\Gamma_0(M),\ L_{r,\epsilon}(\mathbf{C})).$$

One easily verifies that Ψ_f is the image of ψ_f in $H^1_{par}(\Gamma_0(M), L_{r,\epsilon}(\mathbf{C}))$. Moreover, the map $f \mapsto \psi_f$ is Hecke equivariant. Hence, by theorem 4.2.3 we see that for either choice of sign $\pm$,

$$s_{r,\epsilon}\left(\Psi_f^\pm\right) = \psi_f^\pm.$$

4.3. A cohomological Shintani lifting. The integrals (2.2.1) defining the fourier coefficients of the Shintani lift are taken over geodesic paths C_Q which join cusps to cusps in the upper half plane. For each $Q \in \mathcal{F}_M$, let $D_Q = \partial C_Q \in \mathcal{D}_0$ be the *boundary* of C_Q. Hence, using the notation of §2.1, $D_Q \in \mathcal{D}_0$ is given by

$$(4.3.1) \qquad D_Q := \begin{cases} \{\omega'_Q\} - \{\omega_Q\} & \text{if } \delta_Q \text{ is a perfect square;} \\[2ex] \{\gamma_Q(i\infty)\} - \{i\infty\} & otherwise. \end{cases}$$

Analogy with the Shintani correspondence (see (2.2.1) and (2.2.2)) motivates the following definition.

(4.3.2) Definition. Let R be a $\mathbf{Z}[\frac{1}{6}]$-algebra. Let $k \geq 0$ be an integer and χ be an R-valued Dirichlet character defined modulo an integer M.

a. For each cohomology class $\varphi \in H^1_c\big(\Gamma_0(M), L_{2k,\chi^2}(R)\big)$ and each $Q \in \mathcal{F}_M$ define

$$J_{k,\chi}(\varphi, Q) := \chi(Q)\left\langle \varphi(D_Q),\ Q^k \right\rangle \in R$$

where D_Q is defined as in (2.1.2.d) and $\varphi(D_Q)$ is the value at D_Q of any cocycle representing φ. (A simple calculation shows that our definition of $J_{k,\chi}(\varphi, Q)$ is independent of the choice of a representative cocycle and depends only on the $\Gamma_0(M)$-equivalence class of Q.)

b. Define an R-linear map $\Theta_{k,\chi} : H^1_c(\Gamma_0(M), L_{2k,\chi^2}(R)) \longrightarrow R[[q]]$ by

$$\Theta_{k,\chi}(\varphi) := \sum_{Q \in \mathcal{F}_M/\Gamma_0(M)} J_{k,\chi}(\varphi, Q)\, q^{\delta_Q/M} \in R[[q]]$$

for each $\varphi \in H^1_c\big(\Gamma_0(M), L_{2k,\chi^2}(R)\big)$.

(4.3.3) PROPOSITION. *Let R be a $\mathbf{Z}[\frac{1}{6}]$-algebra, $k \geq 0$ be an integer and χ be an R-valued Dirichlet character defined modulo M.*

a. *For an arbitrary commutative ring R and $\varphi \in H^1_c(\Gamma_0(M), L_{2k,\chi^2}(R))$ we have $\Theta_{k,\chi}(\varphi|\iota) = -\Theta_{k,\chi}(\varphi)$.*

b. *Let $f \in \mathcal{S}_{2k+2}(\Gamma_0(M), \chi^2)$ and define ψ_f be as in (4.2.2). Then*

$$\Theta_{k,\chi}(\psi_f) = \Theta_{k,\chi}(\psi_f^-) = \theta_{k,\chi}(f)$$

where $\theta_{k,\chi}(f)$ is as defined in (2.2.2).

c. *If $R = K$ is a field of characteristic zero and $\varphi \in H_c^1(\Gamma_0(M), L_{2k,\chi^2}(K))$ is in the image of s_{2k,χ^2} then*

$$\Theta_{k,\chi}(\varphi) \in \mathcal{S}_{k+\frac{3}{2}}(\Gamma_0(4M), \chi'; K)$$

with $\chi' = \chi \cdot \left(\frac{(-1)^{k+1} M}{-} \right)$.

Proof. From (2.1.2) we see that for any $Q \in \mathcal{F}_M$ we have $\iota D_Q = -D_{Q|\iota}$. It follows that $J_{k,\chi}(\varphi|\iota, Q)$ is equal to

$$\chi(Q) \left\langle \varphi(\iota D_Q)|\iota, \ Q^k \right\rangle = \chi(Q|\iota) \left\langle \varphi(-D_{Q|\iota}), \ (Q|\iota)^k \right\rangle$$
$$= -\chi(Q|\iota) \left\langle \varphi(D_{Q|\iota}), \ (Q|\iota)^k \right\rangle.$$

Hence $J_{k,\chi}(\varphi|\iota, Q) = -J_{k,\chi}(\varphi, Q|\iota)$. Assertion **a** is now an immediate consequence of the definition of $\Theta_{k,\chi}$.

To prove **b** we note that from (4.1.2) we have

$$\left\langle (\tau X + Y)^{2k}, \ Q^{k-1} \right\rangle = (2k)! \cdot Q(1, -\tau)^k.$$

Using the definition (2.2.1) of $I_{k,\chi}(f, Q)$ we calculate

$$I_{k,\chi}(f, Q) = \frac{\chi(Q)}{(2k)!} \cdot \int_{C_Q} f(\tau)\langle (\tau X + Y)^{2k}, \ Q^k \rangle d\tau = \chi(Q) \cdot \left\langle \psi_f(D_Q), \ Q^k \right\rangle.$$

Hence $I_{k,\chi}(f, Q) = J_{k,\chi}(\psi_f, Q)$ and the assertion $\Theta_{k,\chi}(\psi_f) = \theta_{k,\chi}(f)$ is immediate from the definitions. Now decompose $\psi_f = \psi_f^+ + \psi_f^-$ into its $\pm$ components with respect to ι. Then part **a** implies $\Theta_{k,\chi}(\psi_f^+) = 0$. Hence $\Theta_{k,\chi}(\psi_f) = \Theta_{k,\chi}(\psi_f^-)$.

To prove **c** it suffices to assume K is a subfield of $\mathbf{C}$, in which case the assertion follows at once from **b** and theorem 4.2.3. This completes the proof of the proposition.

We will say that a cohomology class $\varphi \in H_c^1(\Gamma_0(M), L_{r,\epsilon}(\mathbf{C}))$ is defined over a subring R of $\mathbf{C}$ if φ is in the image of the natural map

$$H_c^1(\Gamma_0(M), L_{r,\epsilon}(R)) \longrightarrow H_c^1(\Gamma_0(M), \ L_{r,\epsilon}(\mathbf{C})).$$

Equivalently, φ is defined over R if the corresponding modular symbol takes values in $L_{r,\epsilon}(R)$.

Now suppose $f \in \mathcal{S}_{r+2}(\Gamma_0(M), \epsilon)$ is a Hecke eigenform and let $\mathcal{O}_f$ be the ring generated by the Hecke eigenvalues. Then it is well known (see, for example, [Man]) that there are two complex numbers $\Omega_f^\pm \in \mathbf{C}^\times$ such that for either choice of sign

$$(4.3.4) \qquad\qquad \varphi_f^\pm := \frac{1}{\Omega_f^\pm} \cdot \psi_f^\pm \text{ is defined over } \mathcal{O}_f.$$

We fix, once and for all, periods $\Omega_f^\pm \in \mathbf{C}^\times$ for each Hecke eigenform f so that the classes $\varphi_f^\pm$ are defined over $\mathcal{O}_f$ as above. Then, as in 2.3.2, we define

$$(4.3.5) \qquad\qquad \theta_{k,\chi}^*(f) := \frac{1}{\Omega_f^-} \cdot \theta_{k,\chi}(f).$$

We summarize the results of this section in the following theorem.

(4.3.6) THEOREM. *Let $f \in \mathcal{S}_{2k+2}(\Gamma_0(M), \chi^2)$ be a Hecke eigenform. Let ψ_f be as in (4.2.2) and φ_f^- be the minus component with respect to ι. Then*

$$\theta_{k,\chi}^*(f) = \Theta_{k,\chi}(\varphi_f^-) \in \mathcal{S}_{k+\frac{3}{2}}(\Gamma_0(4M), \chi'; \mathcal{O}_f).$$

Proof. The assertion $\theta_{k,\chi}^*(f) = \Theta_{k,\chi}(\varphi_f^-)$ is just a restatement of proposition 4.3.3. Since φ_f^- is defined over $\mathcal{O}_f$, we have $\chi(Q) \cdot \langle \varphi_f^-(D_Q), Q^k \rangle \in \mathcal{O}_f$ for each $Q \in \mathcal{F}_M$. Hence $J_{k,\chi}(\varphi_f^-, Q) \in \mathcal{O}_f$ and consequently, the coefficients of $\Theta_{k,\chi}(\varphi_f^-)$ are in $\mathcal{O}_f$. This proves the theorem.

§5. p-Adic Measures and Λ-adic Cohomology Classes

As in section 1 we fix a prime $p \geq 5$ and a positive integer N prime to p. Let $\mathbf{f} = \mathbf{f}_N$ be the universal p-stabilized ordinary newform of tame conductor N as in 1.4.1(b). For each $\kappa \in \mathcal{X}_N^{arith}$ fix appropriate complex periods $\Omega_{\mathbf{f}(\kappa)}^\pm \in \mathbf{C}^\times$ so that the cohomology classes

$$(5.1) \qquad\qquad \varphi_\kappa^\pm := \frac{1}{\Omega_{\mathbf{f}(\kappa)}^\pm} \cdot \psi_{\mathbf{f}(\kappa)}$$

are defined over $\mathcal{O}_{\mathbf{f}(\kappa)}$ as in (4.3.4). In [G-S] a technique was described which unites the cohomology classes $\varphi_\kappa^\pm$ ($\kappa \in \mathcal{X}_N^{arith}$) into "Λ-adic cohomology classes" $\Phi^\pm$. The classes $\varphi_\kappa^\pm$ ($\kappa \in \mathcal{X}_N^{arith}$) are then retrieved (up to scalars) by specialization of $\Phi^\pm$ to $\kappa \in \mathcal{X}_N^{arith}$. Our goal in this section is to review these results.

Let $(\mathbf{Z}_p^2)'$ denote the set of *primitive* vectors in $\mathbf{Z}_p^2$ (i.e. vectors which are not divisible by p). Let $\mathbf{D} = \mathrm{Meas}((\mathbf{Z}_p^2)')$ be the group of $\mathbf{Z}_p$-valued measures on $(\mathbf{Z}_p^2)'$. Hence, if $\mathrm{Cont}((\mathbf{Z}_p^2)')$ is the space of continuous $\mathbf{Z}_p$-valued functions on $(\mathbf{Z}_p^2)'$ with the uniform norm topology, then $\mathbf{D}$ is the space of continuous $\mathbf{Z}_p$-valued $\mathbf{Z}_p$-linear functionals on $\mathrm{Cont}((\mathbf{Z}_p^2)')$. We will adopt the conventions of integration theory and write

$$\int_U f \, d\mu$$

for the value of a measure $\mu \in \mathbf{D}$ on the product of a continuous function f and the characteristic function of a compact open set $U \subseteq (\mathbf{Z}_p^2)'$. The scalar action of $\mathbf{Z}_p^\times$ on $(\mathbf{Z}_p^2)'$ induces a natural action of $\mathbf{Z}_p[[\mathbf{Z}_p^\times]]$ on $\mathbf{D}$. Viewing the elements of $(\mathbf{Z}_p^2)'$ as row vectors, we let $\Gamma_0(N)$ act by matrix multiplication on the right. This induces a natural right action of $\Gamma_0(N)$ on $\mathbf{D}$. This action is characterized by the following integration formulas, for $\mu \in \mathbf{D}$, $\gamma \in \Gamma_0(N)$, and $f \in \mathrm{Cont}((\mathbf{Z}_p^2)')$:

$$\int f(x, y) \, d(\mu|\gamma)(x, y) = \int f((x, y)\gamma) \, d\mu(x, y).$$

The action of $\Gamma_0(N)$ clearly commutes with the action of $\mathbf{Z}_p[[\mathbf{Z}_p^\times]]$. Hence we may view $\mathbf{D}$ as a $\mathbf{Z}_p[[\mathbf{Z}_p^\times]][\Gamma_0(N)]$-module.

Recall that $\Lambda_N = \mathbf{Z}_p[[\mathbf{Z}_{p,N}^\times]]$ and note that $\Lambda_N \cong \mathbf{Z}_p[[\mathbf{Z}_p^\times]][\Delta_N]$ where $\Delta_N = (\mathbf{Z}/N\mathbf{Z})^\times$. Define the Λ_N-module

$$\mathbf{D}_N := \mathbf{D} \otimes_{\mathbf{Z}_p[[\mathbf{Z}_p^\times]]} \Lambda_N$$

and let $\Gamma_0(N)$ act on $\mathbf{D}_N$ by the formula $(\mu \otimes \lambda)|\gamma := \mu|\gamma \otimes [a]_N \lambda \in \mathbf{D}_N$, where a is the upper left hand entry of γ and $[a]_N$ is its image in $\Delta_N \subseteq \Lambda_N^\times$. Finally, let $\mathcal{R}_N$ be the universal ordinary Hecke algebra of tame conductor N as in 1.4.1(a) and define

$$(5.2) \qquad \mathbf{D}_{\mathcal{R}_N} := \mathbf{D}_N \otimes_{\Lambda_N} \mathcal{R}_N.$$

We let $\Gamma_0(N)$ act on $\mathbf{D}_{\mathcal{R}_N}$ through the first factor and note that $\mathbf{D}_{\mathcal{R}_N}$ is a $\mathcal{R}_N[\Gamma_0(N)]$-module.

For each point $\kappa \in \mathcal{X}_N$, let $R_\kappa = \kappa(\mathcal{R}_N)$. If $\kappa \in \mathcal{X}_N^{arith}$ is an arithmetic point of signature (r, ϵ), we factor $\epsilon = \epsilon_p \epsilon_N$ where ϵ_N is defined modulo N and ϵ_p is defined modulo a power of p, and define the map $\phi_\kappa : \mathbf{D}_{\mathcal{R}_N} \longrightarrow L_{r,\epsilon}(R_\kappa)$ by

$$(5.3) \qquad \phi_\kappa(\mu \otimes \alpha) = \kappa(\alpha) \cdot \int_{\mathbf{Z}_p^\times \times \mathbf{Z}_p} \epsilon_p(x) \cdot \frac{(xY - yX)^r}{r!} \cdot d\mu(x, y)$$

for $\mu \in \mathbf{D}$ and $\alpha \in \mathcal{R}_N$. A simple calculation will confirm that if ϵ is defined modulo Np^m then ϕ_κ is a $\Gamma_0(Np^m)$-homomorphism, hence induces a map on cohomology:

$$(5.4) \qquad \phi_{\kappa,*} : H_c^1(\Gamma_0(N), \mathbf{D}_{\mathcal{R}_N}) \longrightarrow H_c^1(\Gamma_0(Np^m), L_{r,\epsilon}(R_\kappa)).$$

Moreover, the map $\phi_{\kappa,*}$ commutes with the natural action of the Hecke operators. For $\Phi \in H_c^1(\Gamma_0(N), \mathbf{D}_{\mathcal{R}_N})$ and $\kappa \in \mathcal{X}_N^{arith}$ we set

$$\Phi_\kappa := \phi_{\kappa,*}\Phi.$$

For more details, see [G-S].

(5.5) THEOREM. *Let $\kappa_0 \in \mathcal{X}_N^{arith}$ be a fixed arithmetic point. There is a cohomology class $\Phi \in H_c^1(\Gamma_0(N), \mathbf{D}_{\mathcal{R}_N})$ and a choice of p-adic periods $\Omega_\kappa \in R_\kappa$ ($\kappa \in \mathcal{X}_N^{arith}$) with the following properties:*

    a. *$\Omega_{\kappa_0} \neq 0$;*

    b. *For every arithmetic point $\kappa \in \mathcal{X}_N^{arith}$, we have $\Phi_\kappa = \Omega_\kappa \cdot \varphi_\kappa^-$.*

Proof Sketch. This is an easy consequence of theorem 5.13 of [G-S]. Suppose κ_0 has signature (r_0, ϵ_0) and let m_0 be the smallest positive integer for which ϵ_0 is defined modulo Np^{m_0}. Then $\varphi_{\kappa_0}^- \in H_c^1(\Gamma_0(Np^{m_0}), L_{r_0,\epsilon_0}(R_{\kappa_0}))$ is a Hecke eigenclass. Let $\mathcal{R}_{(\kappa_0)}$ be the localization of $\mathcal{R}_N$ at κ_0. Also, let $h : \mathcal{H} \longrightarrow \mathcal{R}_{(\kappa_0)}$ be the canonical map (see 1.4.1). According to Theorem 5.13 of [G-S], the $\mathcal{R}_{(\kappa_0)}$-module of h-eigenclasses in $H_c^1(\Gamma_0(N), \mathbf{D}_{\mathcal{R}_{(\kappa_0)}})^-$ is free of rank one and is generated by an element Ψ whose image in $H_c^1(\Gamma_0(Np^{m_0}), L_{r_0, \epsilon_0}(R_\kappa))$ is equal to $\varphi_{\kappa_0}^-$. Now Ψ has only a finite number of "zeroes and poles" (see the discussion on page 435 of [G-S]). We may therefore choose an element $\alpha \in \mathcal{R}_N$ such that $\alpha(\kappa_0) \neq 0$ and $\alpha\Psi$ is everywhere regular. Let $\Phi = \alpha\Psi$. By the weak multiplicity one theorem, we know that Φ_κ is a multiple of φ_κ^- for each arithmetic point κ, hence we may choose periods $\Omega_\kappa \in R_\kappa$ verifying the conditions of the theorem.

§6. A Λ-adic Shintani Lifting

As in the last section, p is a prime ≥ 5, N is a positive integer prime to p, and $\mathbf{D}_N = \mathbf{D} \otimes_{\mathbf{Z}_p[[\mathbf{Z}_p^\times]]} \Lambda_N$. Recall from section 3 that $\tilde{\mathcal{R}}_N$ is defined to be $\mathcal{R}_N \otimes_{\Lambda_N, \sigma} \Lambda_N$ where the tensor product is with respect to the homomorphism $\sigma : \Lambda_N \longrightarrow \Lambda$ induced by $t \mapsto t^2$ on $\mathbf{Z}_{p,N}^\times$. Also $\tilde{\mathcal{X}}_N$ is the Λ-adic analytic space attached to $\tilde{\mathcal{R}}_N$ and $\tilde{\mathcal{X}}_N^{arith}$ is the subset of arithmetic points. Recall that the homomorphism $\mathcal{R}_N \longrightarrow \tilde{\mathcal{R}}_N$ is not a Λ_N-morphism. Indeed, the induced map $\tilde{\mathcal{X}}_N \longrightarrow \mathcal{X}_N$ has the effect of *doubling* the signatures of arithmetic points.

(6.1) LEMMA. *For each $Q \in \mathcal{F}_{Np}$ there is a unique $\mathcal{R}_N$-homomorphism*

$$J_Q : \mathbf{D}_{\mathcal{R}_N} \longrightarrow \tilde{\mathcal{R}}_N$$

such that for all $\tilde{\kappa} \in \tilde{\mathcal{X}}_N^{arith}$ of signature (k, χ) lying over $\kappa \in \mathcal{X}_N^{arith}$, we have

$$\tilde{\kappa}(J_Q(\mu)) = \chi_N(Q) \cdot \langle \phi_\kappa(\mu), \, Q^k \rangle$$

for all $\mu \in \mathbf{D}_{\mathcal{R}_N}$. Here χ_N is the Dirichlet character associated to the composition $(\mathbf{Z}/N\mathbf{Z})^\times = \Delta_N \subseteq \mathbf{Z}_{p,N}^\times \xrightarrow{\chi} R_\kappa$; ϕ_κ is the map defined by (5.3); and $\langle \, , \, \rangle$ is the pairing defined by (4.1).

Proof. The uniqueness assertion is clear. To prove existence, we define J_Q in stages. Recall that $\mathbf{Z}_p[[\mathbf{Z}_p^\times]]$ is canonically isomorphic to the group of $\mathbf{Z}_p$-valued measures $\mathrm{Meas}(\mathbf{Z}_p^\times)$. For each $\nu \in \mathbf{D}$ define $j_Q(\nu) \in \mathbf{Z}_p[[\mathbf{Z}_p^\times]]$ by the integration formulas

$$\int_{\mathbf{Z}_p^\times} f \, dj_Q(\nu) = \int_{\mathbf{Z}_p^\times \times \mathbf{Z}_p} f(Q(x,y)) \, d\nu(x,y)$$

for continuous functions $f : \mathbf{Z}_p^\times \longrightarrow \mathbf{Z}_p$. For each quadratic form $Q(X,Y) = aX^2 + bXY + cY^2$ in $\mathcal{F}_{Np}$ let $[Q]_N := [a]_N \in \Delta_N$. For $t \in \mathbf{Z}_p^\times$, we have $j_Q([t]\nu) = [t^2] \cdot j_Q(\nu)$. Hence there is a unique $\mathbf{Z}_p$-linear map

$$J_Q : \mathbf{D}_N \longrightarrow \Lambda_N \quad \text{such that} \quad J_Q(\nu \otimes \lambda) = [Q]_N \cdot \sigma(\lambda) \cdot j_Q(\nu)$$

for all $\lambda \in \Lambda_N$ and $\nu \in \mathbf{D}$. Here $\sigma : \Lambda_N \longrightarrow \Lambda_N$ is the map induced by $t \mapsto t^2$ on $\mathbf{Z}_{p,N}^\times$ as in section 3. Finally, we extend J_Q by $\mathcal{R}_N$-linearity to obtain a map

$$J_Q : \mathbf{D}_{\mathcal{R}_N} \longrightarrow \tilde{\mathcal{R}}_N$$

$$\mu \otimes \alpha \longmapsto \alpha \otimes_\sigma J_Q(\mu).$$

Now let $\mu = \nu \otimes \alpha \in \mathbf{D} \otimes_\sigma \tilde{\mathcal{R}}_N$, and let $\tilde{\kappa} \in \tilde{\mathcal{X}}_N^{arith}$ be an arithmetic point on $\tilde{\mathcal{R}}_N$ of signature (k, χ) lying over $\kappa \in \mathcal{X}_N^{arith}$. We then have

$$
\tilde{\kappa}(J_Q(\mu)) = \kappa([Q]_N \alpha) \cdot \int_{\mathbf{Z}_p^\times} \kappa \, dj_Q(\nu)
$$

$$
= \chi_N(Q)\kappa(\alpha) \cdot \int_{\mathbf{Z}_p^\times \times \mathbf{Z}_p} \kappa(Q(x,y)) \, d\nu(x,y)
$$

$$
= \chi_N(Q)\kappa(\alpha) \cdot \int_{\mathbf{Z}_p^\times \times \mathbf{Z}_p} \chi_p(Q(x,y)) \, Q(x,y)^k \, d\nu(x,y)
$$

$$
= \chi(Q)\kappa(\alpha) \int_{\mathbf{Z}_p^\times \times \mathbf{Z}_p} \chi_p(x^2) \left\langle \frac{(xY - yX)^{2k}}{(2k)!}, \ Q^k \right\rangle d\nu(x,y)
$$

$$
= \chi(Q) \left\langle \kappa(\alpha) \cdot \int_{\mathbf{Z}_p^\times \times \mathbf{Z}_p} \chi_p^2(x) \cdot \frac{(xY - yX)^{2k}}{(2k)!} \cdot d\nu(x,y), \ Q^k \right\rangle
$$

$$
= \chi(Q)\langle \phi_\kappa(\mu), \ Q^k \rangle.
$$

This completes the proof of lemma 6.1.

We are now ready to define a Λ-adic Shintani lifting.

(6.2) Definition.

a. For each $\Phi \in H_c^1(\Gamma_0(N), \mathbf{D}_{\mathcal{R}_N})$ and each $Q \in \mathcal{F}_{Np}$ define $J(\Phi, Q) \in \tilde{\mathcal{R}}_N$ by

$$
J(\Phi, Q) := J_Q(\Phi(D_Q)) \in \tilde{\mathcal{R}}_N
$$

where $\Phi(D_Q) \in \mathbf{D}_{\mathcal{R}_N}$ is the value of any cocycle representing Φ on D_Q. (As before, a simple calculation confirms that this definition of $J(\Phi, Q)$ does not depend on the representative cocycle, and depends only on the $\Gamma_0(Np)$-equivalence class of Q.)

b. Define $\Theta : H_c^1(\Gamma_0(N), \mathbf{D}_{\mathcal{R}_N}) \longrightarrow \Lambda_N$ by

$$
\Theta(\Phi) := \sum_{Q \in \mathcal{F}_{Np}/\Gamma_0(Np)} J(\Phi, Q) \, q^{\delta_Q/Np}.
$$

We regard $\Theta(\Phi)$ as the Λ-adic Shintani lift of Φ. The interpolation properties of $\Theta(\Phi)$ are expressed in the following theorem.

(6.3) THEOREM. *For each cohomology class $\Phi \in H_c^1(\Gamma_0(N), \mathbf{D}_{\mathcal{R}_N})$ and for each arithmetic point $\tilde{\kappa} \in \tilde{\mathcal{X}}_N^{arith}$ of signature (k, χ) lying over $\kappa \in \mathcal{X}_N^{arith}$ we have*

$$
\Theta(\Phi)(\tilde{\kappa})|T_p^{m-1} = \Theta_{k,\chi}(\Phi_\kappa)
$$

where m is the smallest positive integer for which χ is defined modulo Np^m.

Proof. This is a consequence of the following simple calculation:

$$
\begin{aligned}
\Theta(\Phi)(\tilde{\kappa})|T_p^{m-1} &= \left(\sum_{Q \in \mathcal{F}_{Np}/\Gamma_0(Np)} \tilde{\kappa}(J(\Phi,\,Q))\, q^{\delta_Q/Np} \right) \Bigg| T_p^{m-1} \\
&= \sum_{\substack{Q \in \mathcal{F}_{Np}/\Gamma_0(Np) \\ p^m | \delta_Q}} \chi(Q)\langle \Phi_\kappa, Q^k \rangle \, q^{\delta_Q/Np^m} \\
&= \sum_{Q \in \mathcal{F}_{Np^m}/\Gamma_0(Np^m)} \chi(Q)\langle \Phi_\kappa, Q^k \rangle \, q^{\delta_Q/Np^m} \\
&= \sum_{Q \in \mathcal{F}_{Np^m}/\Gamma_0(Np^m)} J_{k,\chi}(\Phi_\kappa,\,Q)\, q^{\delta_Q/Np^m} \\
&= \Theta_{k,\chi}(\Phi_\kappa).
\end{aligned}
$$

With theorem 6.3 in hand, we can now conclude the proof of our main theorem.

Proof of Theorem 3.3. Choose $\Phi \in H_c^1(\Gamma_0(N), \mathbf{D}_{\mathcal{R}_N})$ and periods $\Omega_\kappa \in R_\kappa$ for each $\kappa \in \mathcal{X}_N^{arith}$ as in theorem 5.5. Then 3.3.a is just 5.5.a. To prove 3.3.b we set

$$
\Theta := \Theta(\Phi) = \sum_{n=1}^{\infty} \beta_n\, q^n.
$$

Let $\tilde{\kappa} \in \tilde{\mathcal{X}}_N^{arith}$ have signature (k, χ) and let κ be its image in $\mathcal{X}_N^{arith}$. Letting m be the smallest positive integer for which χ is defined modulo Np^m, theorem 6.3 implies $\Theta(\tilde{\kappa})|T_p^{m-1} = \Theta_{k,\chi}(\Phi_\kappa)$. Since $\Phi_\kappa = \Omega_\kappa \cdot \varphi_\kappa$, we conclude

$$
(6.4) \qquad \Theta(\tilde{\kappa})|T_p^{m-1} = \Omega_\kappa \theta_{k,\chi}^*(\mathbf{f}_N(\kappa)).
$$

If $m = 1$, this is what we want to prove. Moreover, for $m = 1$, it follows that $\Theta(\tilde{\kappa})|T_p^2 = \alpha_p(\kappa) \cdot \Theta(\tilde{\kappa})$ and therefore $(\beta_{np^2})(\tilde{\kappa}) = \alpha_p(\kappa)\beta_n(\tilde{\kappa})$ for every n and every $\tilde{\kappa} \in \tilde{\mathcal{X}}_N^{arith}$ of signature (k, χ) with χ defined modulo Np. From this we conclude that $\beta_{np^2} = \alpha_p \cdot \beta_n$ for all $n \geq 1$ and consequently that

$$
(6.5) \qquad \Theta|T_p^2 = \alpha_p \cdot \Theta
$$

Now apply T_p^{m-1} to both sides of (6.4), use (6.5), and multiply by $\alpha_p(\kappa)^{1-m}$ to obtain

$$
(6.6) \qquad \Theta(\tilde{\kappa}) = \alpha_p(\kappa)^{1-m} \cdot \Omega_\kappa \theta_{k,\chi}^*\left(\mathbf{f}_N(\kappa) \right) \Bigg| T_p^{m-1}
$$

Theorem 2.2.3 tells us $\theta_{k,\chi}^*(\mathbf{f}_N(\kappa)) \in \mathcal{S}_{k+\frac{3}{2}}(\Gamma_0(Np^m), \chi')$ where χ' is the character of $(\mathbf{Z}/4Np^m\mathbf{Z})^\times$ defined by $\chi'(d) := \chi(d) \left(\frac{(-1)^{k+1}Np^m}{d} \right)$. By proposition 1.5 of [S3], each application of T_p multiplies the Nebentype by $\left(\frac{p}{\cdot} \right)$ (also, see 2.2.4). Hence

$$
\theta_{k,\chi}^*(\mathbf{f}_N(\kappa)) \in \mathcal{S}_{k+\frac{3}{2}}(\Gamma_0(Np^m), \chi^*)
$$

where χ^* is defined by $\chi^*(d) := \chi(d) \left(\frac{(-1)^{k+1} Np}{d} \right)$ for $d \in (\mathbf{Z}/4Np^m\mathbf{Z})^\times$. Hence, by (6.6) we have $\Theta(\tilde{\kappa}) \in \mathcal{S}_{k+\frac{3}{2}}(\Gamma_0(Np^m), \chi^*)$ and

$$\Theta(\tilde{\kappa}) = \Omega_\kappa \theta_{k,\chi}^* \left(\mathbf{f}_N(\kappa) \right) \bigg| \, T_p^{1-m}.$$

This completes the proof of Theorem 3.3.

REFERENCES

[A-S] Ash, A., Stevens, G.: Modular forms in characteristic ℓ and special values of their L-functions. *Duke Math. J.* **53** (1986), No.3, 849-868.

[Coh] Cohen, H.: Sums involving the values at negative integers of L-functions of quadratic characters. *Math. Ann.* **217** (1975), 271-285.

[G-S] Greenberg, R., Stevens, G.: p-adic L-functions and p-adic periods of modular forms. *Invent. Math.* **111** (1993), 407-447.

[H1] Hida, H.: Galois representations into $GL_2(\mathbf{Z}_p[[X]])$ attached to ordinary cusp forms. *Invent. Math.* **85** (1986), 545-613.

[H2] Hida, H.: Iwasawa modules attached to congruences of cusp forms. *Ann. Sci. École Norm. Sup.* **19** (1986), 231-273.

[H3] Hida, H.: On Λ-adic forms of half integral weight for $SL(2)_{/\mathbf{Q}}$. Preprint, June 7, 1993.

[Kob] Koblitz, N.: p-Adic congruences and modular forms of half integer weight. *Math. Ann.* **274** (1986) 199-220.

[J1] Jochnowitz, N.: A p-adic conjecture about derivatives of L-series attached to modular forms. To appear in: p-Adic Monodromy and the Birch-Swinnerton-Dyer Conjecture. *AMS Contemporary Math.*

[J2] Jochnowitz, N.: Congruences between modular forms of half integral weight and implications for class numbers and elliptic curves. To appear in: *Invent. Math.*

[Koh] Kohnen, W.: p-Adic congruences for cycle integrals associated to elliptic curves with complex multiplication. *Math. Ann.* **280** (1988), 267-283.

[M] Maeda, Y.: A congruence between modular forms of half-integral weight. *Hokkaido Math. J.* **12** (1983), 64-73.

[Man] Manin, J.: Periods of parabolic forms and p-adic Hecke series. *Math. Sbornik* (1973), AMS translation 371-393.

[O] Ohta, M.: On ℓ-adic representations attached to automorphic forms. *Japan J. Math.* **8**, No. 1 (1982), 1-47.

[S1] Shimura, G.: An ℓ-adic method in the theory of automorphic forms. (1968), unpublished.

[S2] Shimura, G.: *Introduction to the Arithmetic Theory of Automorphic Functions.* Princeton University Press, 1971.

[S3] Shimura, G.: On modular forms of half integral weight. *Ann. of Math.* **97** (1973), 440-481.

[S4] Shimura, G.: On the holomorphy of certain Dirichlet series. *Proc. London Math. Soc.* (3) **31**, no. 1 (1975), 79-98.

[S5] Shimura, G.: The periods of certain automorphic forms of arithmetic type. *J. Fac. Sci. Univ. Tokyo* Sect. IA Math. **28** (1981), no. 3, 605-632 (1982).

[Sh] Shintani, T.: On construction of holomorphic cusp forms of half integral weight. *Nagoya Math. J.* **58** (1975), 83-126.

[Wa1] Waldspurger, J.-L.: Correspondence de Shimura. *J. Math. pures et appl.* **59** (1980), 1-133.

[Wa2] Waldspurger, J.-L.: Sur les coefficients de Fourier des formes modulaires de poids demi-entier. *J. Math. pures et appl.* **60** (1981), 375-484.

[Zag] Zagier, D.: Modular forms whose Fourier coefficientws involve zeta-functi0ns of quadratic fields. Modular functions of one variable. VI. *Lect. Notes Math.* **627**, 105-169. Berlin, Heidelberg, New York: Springer 1977.

MATHEMATICS DEPARTMENT, BOSTON UNIVERSITY, BOSTON, MA 02215

E-mail address: ghs@math.bu.edu

Contemporary Mathematics
Volume **174**, 1994

The Non-Existence of Certain Galois Extensions of $\mathbb{Q}$ Unramified Outside 2

JOHN TATE

On May 1, 1973, Serre wrote me about his idea that perhaps every two-dimensional semisimple representation in characteristic ℓ of $\mathrm{Gal}(\overline{\mathbb{Q}}/\mathbb{Q})$ which is unramified outside ℓ and has odd determinant, is "modular." (For a precise statement, see [**1**, p. 229]. This idea has been developed much further by Serre in recent years; see [**2**].) For $\ell = 2$ this means simply that such a representation is trivial [**1**, p. 232], and it occurred to me that it might be possible to prove this by a discriminant estimate. After finding that it was, I sent Serre the following proof and remarks, in a letter dated July 2, 1973. He has made a sketch of the method available [**1**, p. 710] but the details have never appeared. I am happy to have the opportunity to publish them here, in response to recent interest.

THEOREM. *Let G be the Galois group of a finite extension $K/\mathbb{Q}$ which is unramified at every odd prime. Suppose there is an embedding $\rho : G \hookrightarrow SL_2(k)$, where k is a finite field of characteristic 2. Then $K \subset \mathbb{Q}(\sqrt{-1}, \sqrt{2})$ and $\mathrm{Tr}\,\rho(\sigma) = 0$ for each $\sigma \in G$.*

Let S be a 2-Sylow subgroup of G. Then S is elementary abelian, because the 2-Sylow subgroup $T = \{ \begin{pmatrix} 1 & x \\ 0 & 1 \end{pmatrix} \mid x \in k \}$ of $SL_2(k)$ is isomorphic to k, so is elementary abelian. Note that elements of T have trace 0, so the statement about traces will follow if we prove G is a 2-group.

The case in which G is solvable is a simple exercise in class field theory. Let G' be the commutator subgroup of G and let F' be its fixed field. By class field theory (even Kronecker-Weber), $F' \subset \mathbb{Q}(\mu_{2^r})$ for some r, and G/G' is a 2-group. Hence G/G' is a homomorphic image of S, so is elementary abelian, and $F' \subset \mathbb{Q}(\mu_8)$.

Each subfield of $\mathbb{Q}(\mu_8)$ has class number 1 (even in the strong sense if it is real), and has only one prime above 2, of absolute degree 1. By class field

---

1991 *Mathematics Subject Classification.* Primary 11S15, 11S31, 11S37.

theory, an abelian extension of such a field, if it is unramified outside 2 and ∞, has 2-power degree. Hence G'/G'' is a 2-group, G/G'' is a 2-group, and G/G'' is therefore abelian, being a homomorphic image of S. Hence $G' = G''$. Thus if G is solvable, we have $G' = (1)$, $K = F' \subset \mathbb{Q}(\mu_8)$, and the theorem is true.

Suppose now G is not solvable. We will get a contradiction via Minkowski's estimate for the discriminant d of $K/\mathbb{Q}$. If $n = [K : \mathbb{Q}]$, then the estimate we use is

$$|d| \geq \left(\frac{\pi}{4}\right)^n \frac{n^{2n}}{(n!)^2} \geq \left(\frac{\pi e^2}{4}\right)^n \frac{1}{2\pi n e^{1/(6n)}}.$$

The proof works mainly because of the following lucky fact

$$2^{5/2} = 5.656\ldots < \frac{\pi e^2}{4} = 5.80\ldots \qquad (!)$$

Namely, we shall see below that

$(*)$
$$\begin{cases} |d| \leq 2^n, & \text{if } K/\mathbb{Q} \text{ is only tamely ramified,} \\ |d| \leq 2^{n(5/2-2^{1-m})}, & \text{if } K/\mathbb{Q} \text{ has wild ramification} \\ & \text{of degree } 2^m, \ m \geq 1. \end{cases}$$

Granting this, we get in the tame case $(2\pi n)e^{1/(6n)} \geq (\pi e^2/8)^n > (2.9)^n$, which is probably a contradiction for $n \geq 3$, and certainly a contradiction for our n, which is ≥ 60, since it is the order of a non-solvable group.

In the wild case we get

$$n\left(\frac{5}{2} - 2^{1-m}\right)\log 2 \geq \log|d| \geq n\log(\pi e^2/4) - \log(2\pi n) - \frac{1}{6n},$$

or

$$\log 2\pi n + \frac{1}{6n} \geq n\left(\log(\pi e^2/4) - \frac{5}{2}\log 2\right) + \frac{n}{2^{m-1}}\log 2.$$

Now $n/2^{m-1} \geq 30$ because 2^m divides n, and n, the order of a non-solvable group, is divisible by at least three distinct primes. Thus:

$$\log n + \frac{1}{6n} \geq n(0.0255\ldots) + 30\log 2 - \log 2\pi \geq \frac{n}{40} + 18,$$

which is clearly a contradiction.

To complete the proof of the theorem we must prove $(*)$. Let $D \supset I \supset R$ be the Decomposition, Inertia, and Ramification group for a place v of K above 2, and let $|R| = 2^m$. Let $\mathcal{D}$ be the different of $K/\mathbb{Q}$. Then $(*)$ is trivially the same as

$(**)$
$$\begin{cases} v(\mathcal{D}) \leq v(2) & \text{in tame case } (m=0), \\ v(\mathcal{D}) \leq (5/2 - 2^{1-m})v(2) & \text{if there is wild ramification } (m \geq 1). \end{cases}$$

In the tame case it is well-known that $v(\mathcal{D}) = \frac{e-1}{e}v(2)$, where $e = v(2)$ is the ramification index. Hence we need only discuss the case of wild ramification. Let S_D be a 2-Sylow subgroup of D and choose the embedding $\rho : G \hookrightarrow SL_2(k)$ so that $\rho(S_D) \subset T = \{(\begin{smallmatrix} 1 & x \\ 0 & 1 \end{smallmatrix})\}$. Then $\rho(R)$ is a non-zero subgroup of T, and hence, as is easy to see, its normalizer is contained in the Borel subgroup $B = \{\begin{pmatrix} a & x \\ 0 & a^{-1} \end{pmatrix}\}$. Since D normalizes R, $\rho(D) \subset B$. Hence $\rho(S_D) = \rho(D) \cap T$ is normal in $\rho(D)$, and $D/S_D \approx \rho(D)T/T \subset B/T \approx k^*$ is cyclic. Thus, D/S_D is the Galois group of an abelian tamely ramified extension F of $\mathbb{Q}_2$. Such an extension is unramified! Hence $S_D \supset I = R$, and we are reduced to the following lemma.

LEMMA. *Let F be a finite unramified extension of $\mathbb{Q}_2$. Let E/F be a totally ramified extension whose Galois group, R, is an elementary abelian 2-group of order 2^m, $m \geq 1$. Then the different of E/F divides 2^c, where $c = 5/2 - 2^{1-m}$.*

This is easy to show via the "Führerdiskriminantenproduktformel." If A is the ring of integers in F, then $(1 + 4A)^2 = 1 + 8A$ and $(1 + 2A)^2$ is a subgroup of index 2 in $1 + 4A$. Hence, if X is a group of quadratic characters of $1 + 2A$, then either all of the non-trivial characters in X have conductor 4, or half of them have conductor 8, and the other half (except for the trivial character) have conductor 4. Hence the discriminant divides

$$8^{2^{m-1}} 4^{2^{m-1}-1} = 2^{3 \cdot 2^{m-1} + 2^m - 2}$$

and accordingly, the different divides $2^{\frac{3}{2}+1-2^{1-m}}$, as claimed.

REMARKS.

1. If we replace 2 by an arbitrary prime ℓ, then the local arguments can still be carried out, and if I've done it right, we get instead of $(*)$:

$$(*)_\ell \begin{cases} |d| \leq \ell^n, & \text{if } K/\mathbb{Q} \text{ tame,} \\ |d| \leq \ell^{cn}, \ c = 2 + \dfrac{1}{\ell} - \dfrac{1}{(\ell-1)\ell^{m-1}}, & \text{if } K/\mathbb{Q} \text{ has wild index } \ell^m, \\ & m \geq 1. \end{cases}$$

Since $\ell^{2+(1/\ell)} > \pi e^2/4$ for $\ell \geq 3$, this is not good enough. To show that there is only a finite set of fields $K/\mathbb{Q}$ for a given ℓ by this method, we would need the type of improvement in Minkowski which Golod-Shafaryevitch have shown does not exist in general.

2. For a few low values of ℓ, e.g., 3, 5, 7, one might be able to handle the solvable cases by class field theory in a manner analogous to the discussion for $\ell = 2$ above, but I have no idea how to take care of the non-solvable ones.

As Serre points out in comment 229.2 on p. 710 of [1], the bounds of Odlyzko-Poitou are good enough to allow the case $\ell = 3$ of Serre's 1973 conjecture to be treated. He found they give $n \leq 38$, hence G is solvable, and then by class field theory he concludes $n = 2$, and the representation is $1 \oplus \chi_3$.

JOHN TATE

REFERENCES

1. Jean-Pierre Serre, *Œuvres Volume III*, Springer-Verlag, Berlin, Heidelberg, and New York, 1986.
2. _____, *Sur les représentations modulaires de degré 2 de* $\mathrm{Gal}(\overline{\mathbb{Q}}/\mathbb{Q})$, Duke Math. J. **54** (1987), 179–230.

DEPARTMENT OF MATHEMATICS, UNIVERSITY OF TEXAS, AUSTIN, TEXAS 78712

E-mail address: tate@math.utexas.edu

Contemporary Mathematics
Volume **174**, 1994

Iwasawa theory and Cyclotomic Function Fields

DINESH S. THAKUR

ABSTRACT. We will describe and put in the perspective of Drinfeld's theory, some therorems and conjectures relating class numbers and zeta values at positive and negative integers (as we will see, these are two distinct theories in contrast to the classical case), analogues of results and conjectures of Kummer and Vandiver, growth rates of class numbers, zeta measures and other aspects of Iwasawa theory.

Iwasawa theory started as an attempt by Iwasawa to carry out analogue for number fields of a well-developed theory for function fields, due to Andre Weil and others. This theory related the zeta function of the function field to the characteristic polynomial of Frobenius acting on p-power order torsion in the Jacobian of the corresponding curve. Here p can be any prime unequal to the characteristic. In this theory, to get good structural results, one needs to take algebraic closure of the finite field of constants of the function field and hence consideration of the tower of constant field extensions enters naturally. Since the constant field extensions are just the extensions obtained by adjoining roots of unity, to get a better analogy between p-power order torsion in the Jacobian on the function field side and the p-Sylow subgroup of the class group on the number field side, Iwasawa considered the tower of number fields obtained by adjoining p-power roots of unity or more generally $\mathbf{Z}_p$-extensions for some fixed prime p.

Over $\mathbf{Q}$, the cyclotomic extensions (i.e. (subfields of) the extensions obtained by adjoining the roots of unity) coincide with the abelian extensions by the Kronecker-Weber theorem. Over a function field this is far from the case. Indeed, the cyclotomic extensions are then just the constant field extensions and there are, of course, many more abelian extensions, for example, various Kummer and

1991 *Mathematics Subject Classification.* 11G09, 11R58; Secondary 11R23, 11T22, 11M41.
Key words and phrases. Drinfeld modules, cyclotomic, Iwasawa, zeta, Bernoulli.
Supported in part by NSF grants # DMS 9207706 and # DMS 9314059
This paper is in final form and no version of it will be submitted for publication elsewhere

Artin-Schreier extensions. Carlitz, Drinfeld and Hayes [C2, D, H1] developed other families of abelian extensions of function fields, which also can be thought of as 'cyclotomic' because of the strong analogies with the classical case. We will now briefly describe this 'cyclotomic' theory. For more details see [H1,H2]. For the corresponding classical cyclotomic theory, see the books by S. Lang and L. Washington.

At the most basic level we have, as analogues of $\mathbf{Q}$, the archimedean place ∞, $\mathbf{Z}$, $\mathbf{R}$, $\mathbf{C}$ respectively, their counterparts: a function field K with a finite field $\mathbf{F}_q$ (of characteristic p) of constants, any place ∞ of K (we will assume it to be rational for simplicity), the ring A of integers outside ∞, K_∞ and the completion Ω of an algebraic closure of K_∞ respectively. The simplest example, where the analogies also turn out to be the strongest, is when $K = \mathbf{F}_q(T)$ and $A = \mathbf{F}_q[T]$.

Now the roots of unity can be interpreted as the torsion of the rank one object '$\mathbf{Z}$ inside the endomorphism ring of the multiplicative group' (where n is viewed as the n-th power map). In finite characteristic, we have a shorter supply of multiplicative functions, (i.e., there is no nontrivial 'exponential' from additive to multiplicative groups in characteristic p: we would have $e(0) = e(px) = e(x)^p$, for all x, for such an exponential e) but have a larger supply of additive functions. The endomorphism ring of the additive group in characteristic p is a huge non-commutative ring of polynomials in the Frobenius endomorphism. Hence Drinfeld considered, as an analogue of roots of unity, the torsion of a rank one object 'A inside this ring' (where $a \in A$ is viewed as the map $u \mapsto \rho_a(u) = \sum_{i=0}^{\deg a} a_i u^{q^i}$, with the normalizations $a_0 = a$ and $a_{\deg a} = \operatorname{sgn}(a)$). Here deg and sgn are the degree and a fixed sign function on A. The conditions on $\rho_a(u)$ assure that the a-torsion $\Lambda_a := \{u \in \Omega : \rho_a(u) = 0\}$ is an $A/(a)$ module of rank one. The simplest example is the 'Carlitz module' : $A = \mathbf{F}_q[T]$, $\rho_T(u) = Tu + u^q$. (Exercise: The T^2-th cyclotomic equation is $u^{q^2} + (T + T^q)u^q + T^2 u = 0$).

To the 'Drinfeld module' (*we have restricted the terminology more than in the original definition in* [D]) ρ one associates the exponential function $e: =$ $e_\rho : \Omega \to \Omega$ defined to be an entire additive function $e(z) = \sum_{i=0}^\infty e_i z^{q^i}$ satisfying $e(az) = \rho_a(e(z))$ (in analogy with $e^{nz} = (e^z)^n$) for all $a \in A$ and normalized by $e_0 = 1$. The kernel of e, which can be thought of as an analogue of $2\pi i \mathbf{Z}$, can be written as $\tilde{\pi} A$ (if the class number of A is greater than one, it can be $\tilde{\pi} I$ for some ideal I of A and in fact there are class number many Drinfeld modules corresponding to these rank one A-lattices). In terms of these analogues of the exponential and $2\pi i$, the a-torsion can be written as $e(\tilde{\pi}a'/a)$ for some $a' \in A$ and can be thought of as analogue of n-th root of unity ζ_n or $1 - \zeta_n$. (See the next paragraph).

To illustrate these close analogies a little further, we note that if K is of class number one, and $\wp$ is a prime of A, then $K(\Lambda_\wp)$ is an extension of K with Galois group $(A/\wp)^*$, in which $\wp$ is totally ramified (so that the extension is geometric in contrast to the constant field extension), all other finite primes are unramified, and for $\lambda \in \Lambda_\wp - \{0\}$, (λ) is a prime above $\wp$. (Comparing this with the fact that

$1 - \zeta_p$ is a prime above p in $\mathbf{Q}(\zeta_p)$, we see that λ is an analogue of $1 - \zeta_p$. This reflects the fact that we are now dealing with the additive group rather than the multiplicative group.) The Galois action is given by $\mathrm{Frob}_v(\lambda) = \rho_v(\lambda)$. (For I an ideal of A, ρ_I is defined to be the monic generator of the ideal generated by ρ_i, $i \in I$). If we denote the 'maximal totally real (i.e., ∞ splits completely) subfield' by $K(\Lambda_\wp)^+$, the Galois group over it is $A^* = \mathbf{F}_q^*$. Compare to $\mathbf{Z}^* = \{\pm 1\}$ of the classical case. The respective cardinalities, namely $q - 1$ and 2 play an analogous role and we call multiples of $q - 1$ 'even'.

We remark that the arbitrary choice of the infinite place ∞, which made possible this strong analogy also makes all the abelian extensions obtained by adjoining the a-torsion tamely ramified at ∞ by the above. To get the maximal abelian extension of K one needs to play the whole game again by switching to a different infinite place and by taking the compositum of all such extensions.

We now turn to some deeper aspects of the cyclotomic theory. Let $L = K(\Lambda_a)$ and $F = K(\Lambda_a)^+$, where $a \in A$ is nonconstant. The intersection of the subgroup of L^* generated by the elements of $\Lambda_a - \{0\}$ with $\mathcal{O}_L^*$ is the group E of cyclotomic units. We denote by $h(R)$ the class number of R, for R a Dedekind domain or a field. We then have the following analogue of Kummer's theorem.

THEOREM 1. (GALOVICH-ROSEN [G-R]): *Let $A = \mathbf{F}_q[T]$ and $a = \wp^n$, where $\wp$ is a prime of A. Then $h(\mathcal{O}_F) = [\mathcal{O}_L^* : E]$.*

Remark: This was later generalized to the case of any a (i.e., an analogue of Sinnott's result). Shu has announced [S1] generalization to any A.

SKETCH OF THE PROOF. We have $\mathcal{O}_L^* = \mathcal{O}_F^*$ just as in the classical case. Let $S = \{\infty_i\}$ denote the set of the infinite primes of F. Let $\mathrm{Div}^0(S) \supset P(S) \supset \mathcal{E}$ be the groups of divisors of degree zero supported on S, divisors of elements of $\mathcal{O}_F^*$ and the divisors of cyclotomic units respectively. It is elementary to see that the index in the theorem is equal to that of $\mathcal{E}$ in $\mathrm{Div}^0(S)$. The calculation of the divisors of cyclotomic units using the basis $\infty_i - \infty_0$ of $\mathrm{Div}^0(S)$ allows us to express this index as determinant, which by the Dedekind determinant formula, can be expressed as a product of certain character sums. Finally, the analytic class number formula for the Artin-Weil zeta and L-functions for the function fields identifies this product as the class number $h(F)$ of F. The theorem follows by noticing that $[\mathrm{Div}^0(S) : P(S)] = h(F)/h(\mathcal{O}_F)$. $\square$

Next we discuss Tate's proof of the analogue of the Stickelberger theorem. Let K' be a geometric (i.e., with the same field of constants) extension over K, with an abelian Galois group G. Let $Cl(K')$ be the class group of K' and let $\theta(T) = \sum_{\sigma \in G} Z(\sigma^{-1}, T)\sigma$ where Z is the partial zeta function for σ. Then the Stickelberger element is $\theta = \theta(1)$.

THEOREM 2. (TATE [TA]): *$(q - 1)\theta \in \mathbf{Z}[G]$ kills $Cl(K')$.*

SKETCH OF THE PROOF. Class group is just the group of $\mathbf{F}_q$- rational points of the Jacobian (of the corresponding curve), i.e., the part of the $\overline{\mathbf{F}_q}$- points of the Jacobian where the Frobenius acts as the identity. Now by fundamental results of Weil, the L-function of a character of G is the characteristic function of the Frobenius on the corresponding component of the Jacobian or rather the Tate module, and hence it kills the component when $T = F$ by the Cayley-Hamilton theorem. Hence $\theta(T)$, which is just a linear combination of L-functions with projection to the components operators, kills the class group when $T = 1$. (One needs $(q-1)$ factor to clear out the denominators to get polynomials, when one makes this sketch precise.) $\square$

The interpretation of the L-function as the characteristic polynomial in the above is precisely the result on which Iwasawa's main conjecture is based. Since this is already known, the Gras conjecture giving the componentwise version of Theorem 1, which follows classically from the main conjecture, is known here. This was recognised in [G-S].

In contrast to the number field case, we have class groups of fields as well as the ring of integers outside the infinite places above the chosen ∞. Let $a = \wp$ be a prime of A of degree d, so that $L = K(\Lambda_\wp)$. Let C, $\tilde{C}$ be the p-primary components of the class groups of L and $\mathcal{O}_L$ respectively. Let W be the Witt ring of $A/\wp$. Then if w denotes the Teichmuller character, we have the decomposition $C \otimes_{\mathbf{Z}_p} W = \bigoplus C(w^k)$ into isotypical components according to the characters of $(A/\wp)^*$.

THEOREM 3. (GOSS- SINNOTT [G-S]): *For $0 < k < q^d - 1$, $C(w^{-k}) \neq 0$ if and only if p divides $L(w^k, 1)$*

SKETCH OF THE PROOF. The duality between the Jacobian and the p-adic Tate module T_p transforms the connection between the Jacobian and the class group in the proof of Theorem II to $T_p(w^{-k})/(1 - F)T_p(w^{-k}) \cong C(w^{-k})$. On the other hand, we have a Weil type result: $\det(1 - F : T_p(w^{-k})) = L_u(w^{-k}, 1)$. Here L_u is the unit root part of the L-function and hence has the same p-power divisibility as the complete L-function. Hence $\mathrm{ord}_p(L(w^k, 1))$ is the length of $C(w^{-k})$ as a $\mathbf{Z}_p[G]$-module and the theorem follows. $\square$

Comparison with the corresponding classical result shows that we are looking at divisibilty by p, the characteristic, rather than the prime $\wp$ relevant to the cyclotomic field. To bring $\wp$ in, we need to look at another zeta function introduced by Carlitz and Goss:

Let $\zeta(s) := \sum n^{-s} \in K_\infty$, where the sum is over monic polynomials of A and s is a positive integer. If s is any integer,

$$\zeta(s) := \sum_{i=0}^{\infty} \sum_{\substack{\deg n = i \\ n \text{ monic}}} n^{-s}$$

makes sense and in fact belongs to A for s a negative integer, since in that case the second sum vanishes for large i.

The identification $W/pW \cong A/\wp$ provides us with the Teichmuller character $w : (A/\wp)^* \to W^*$ satisfying $w^k(n \mod \wp) = (n^k \mod \wp) \mod p$. Hence the reduction of L value in the theorem 3 modulo p is $\zeta(-k) \mod \wp$. (This works for k 'odd' (i.e. not a multiple of $q-1$), for 'even' k, we need to use 'the leading term' when there are 'trivial zeros'; but we will ignore this aspect below). Hence we get

THEOREM 4. (GOSS-SINNOTT [G-S]): *For k 'odd', $0 < k < q^d - 1$, $C(w^k) \neq 0$ if and only if $\wp$ divides $\zeta(-k)$.*

For simplicity, we will now restrict to the case $A = \mathbf{F}_q[T]$. Classically, Bernoulli numbers occur in the special values of the Riemann zeta function at both positive and negative integers and these values are connected by the functional equation for the Riemann zeta function. In our case, there is no simple functional equation known and in fact we get two distinct analogues of Bernoulli numbers B_k (or rather the more fundamental B_k/k) both connecting to class groups: those coming from the positive values relate to class groups of rings of integers (see Theorem 6 below) in contrast to class groups of fields (as in Theorem 4).

Let us define the factorial function $\Pi(m)$ and Bernoulli numbers B_m by analogy with the classical case: For a positive integer m, define $\Pi(m) := \prod_\wp \wp^{m_\wp} \in \mathbf{F}_q[T]$, where $m_\wp := \sum_{e \geq 1}[m/\mathrm{Norm}(\wp)^e]$. Define $B_m \in \mathbf{F}_q(T)$ by the formula $z/e(z) = \sum B_m/\Pi(m)z^m$ (compare with the classical generating function $z/(e^z - 1)$). The connection with the special zeta values at the positive integers is through the following analogue of the Euler's result: $\zeta(m) = -B_m(2\pi i)^m/2(m!)$ for even m.

THEOREM 5. (CARLITZ [C1]): *For 'even' m, $\zeta(m) = -B_m\tilde{\pi}^m/(q-1)\Pi(m)$.*

SKETCH OF THE PROOF. First note that $q - 1 = -1$ in the formula. The proof follows by taking the logarithmic derivative of $e(z)$ (to get the Bernoulli numbers through the generating function on one hand and to get zeta values using the geometric sum expansion through the product formula for $e(z)$ on the other hand) and by comparing coefficients. $\square$

Since $\zeta(-k)$ turns out to be a finite sum of n^k's, by Fermat's little theorem, the $\zeta(-k)$'s satisfy Kummer congruences enabling us to define a $\wp$-adic interpolation $\zeta_\wp$. On the other hand, the B_m satisfy analogues of the von-Staudt congruences and the Sylvester-Lipschitz theorem. We have now two distinct analogues of B_k/k: $-\zeta(-k+1)$ for $k-1$ 'odd' on one hand and $\Pi(k-1)\zeta(k)/\tilde{\pi}^k$, with k 'even' on the other. It should be noticed that the shift by one does not transform 'odd' to 'even' unless $q = 3$, and we do not know any reasonable functional equation linking the two.

THEOREM 6. (OKADA, GOSS [O]): *Let $A = \mathbf{F}_q[T]$. Then for $0 < k < q^d - 1$, k 'even', if $\tilde{C}(w^k) \neq 0$, then $\wp$ divides B_k.*

SKETCH OF THE PROOF. We define analogues of Kummer homomorphisms $\psi_i \colon \mathcal{O}_F^* \to A/\wp$ $(0 < i < q^d - 1)$ by $\psi_i(u) = u_{i-1}$, where u_i is defined as follows. Let $u(t) \in A[[t]]$ be such that $u = u(\lambda)$ and define u_i to be $\Pi(i)$ times the coefficient of z^i in the logarithmic derivative of $u(e(z))$. Using the definition of the Bernoulli numbers, we calculate that the i-th Kummer homomorphism takes the basic cyclotomic unit $\lambda^{\sigma_a - 1}$ to $(a^i - 1)B_i/\Pi(i)$. If $\tilde{C}(w^k) \neq 0$, then by the componentwise version of Theorem 1 ('Gras conjecture'), $\psi_k(\lambda^{\sum w^{-k}(\sigma)\sigma^{-1}}) = 0$ and hence the calculation above implies that $\wp$ divides B_k. $\square$

Using her generalization of Theorem 1, Shu has announced [S2] a generalization of Theorem 6 to any A.

Now we describe some results by Anderson and myself [A-T] about the zeta function for $A = \mathbf{F}_q[T]$, defined above. Classical counterparts of these results are not known. To avoid defining a lot more terminology, we describe the result roughly as an expression of $\zeta(n)$ ($\zeta_\wp(n)$ resp.) (where n can be 'odd' as well as 'even') as a logarithm ($\wp$-adic logarithm resp.) of an explicit algebraic point on the n-th tensor power of the Carlitz module. Using this result together with his theorems on the transcendence properties of the exponential (analogues of Hermite- Lindemann, Gelfond-Schneider and Mahler theorems), Jing Yu has proved

THEOREM 7. (YU [Y]): *For a positive integer n, $\zeta(n)$ is transcendental over K and if further n is 'odd', then $\zeta(n)/\tilde{\pi}^n$ and $\zeta_\wp(n)$ are transcendental.*

Now we mention some curious consequence of these results. K. Kato has raised the question of whether, for a given n, $\wp$ divides $\zeta_\wp(n)$ for infinitely many $\wp$'s or not. The expressions in [A-T] mentioned above show that whether $\wp$ divides $\zeta_\wp(1)$ is equivalent to whether $\wp^2$ divides $\rho_{\wp-1}(1)$. This last statement is clearly an analogue of the well-known Wieferich criterion in classical cyclotomic theory: whether $(1+1)^{p-1} - 1 \equiv 0 \mod p^2$. We can then also write down 'higher Wieferich criteria' using higher zeta values. It might be interesting to understand their classical counterparts and their significance. It should be stressed though that classically the zeta function has a pole at $n = 1$, in contrast to the case here.

Two important statements in the cyclotomic theory concerning the class numbers are Kummer's result that p divides $h(\mathbf{Q}(\zeta_p)^+)$ implies that p divides $h(\mathbf{Q}(\zeta_p)^-)$ and the conjecture of Kummer and Vandiver that p does not divide $h(\mathbf{Q}(\zeta_p)^+)$. Ireland and Small [I-S] exhibited a simple example showing that an analogue of both these statements is false. Namely, if $A = \mathbf{F}_q[T]$, with $q = p = 3$ and $\wp = 2 + T^2 + T^4$, then p divides $h(\mathcal{O}_L^+)$ but does not divide $h(\mathcal{O}_L^-)$. We would like to point out that there are various possible analogues that might be explored. First note that we have p and $\wp$ instead of just p as in the classical case. How about checking the divisibilty by the norm of $\wp$? Some examples show that the analogue of the Kummer result still fails if we use the class numbers of fields, but it is not known with the rings of integers. Another naive analogy would be to note that p divides the order of a group if and only if $\mathbf{Z}/p$ sits in the group and we may want to ask whether $A/\wp$ sits in the group: a question a priori different from whether the norm divides the order. Instead of the class group, which is an abelian group or a $\mathbf{Z}$- module we may need to get some A-module to formulate an analogue. (See also [G3] pa. 391).

Let us now look at the class number growth in towers of fields. We can use the class number formula for the zeta function which gives $h = \prod_1^{2g}(1-\alpha_i)$, where α_i are the eigenvalues of Frobenius of absolute value $\sqrt{q}$, by Weil's result. (Analogue of the Riemann hypothesis). This implies that $(\sqrt{q}+1)^{2g} \geq h \geq (\sqrt{q}-1)^{2g}$, where g is the genus of the field. For the constant field extensions, on which the Iwasawa theory was based, we get for the n-th layer of a $\mathbf{Z}_l$ -tower the asymptotics $h_n \sim q^{gl^n}$, because $\prod \alpha_i = q^g$. Note that by class field theory, for $l \neq p$, all $\mathbf{Z}_l$-extensions are essentially constant field extensions, whereas for $l = p$, the characteristic, there are many more $\mathbf{Z}_p$-extensions. Gold and Kisilevsky [G-K] have shown that for geometric $\mathbf{Z}_p$-extensions, $\log h_n \geq p^{2(n-n_1)-1}/3$ and in fact they could construct such towers with arbitrarily large growth.

Analogies we have been discussing suggest that we might look at $A_\wp$-towers rather than $\mathbf{Z}_p$-towers. But note that $A_\wp$'s are much too wild to arise as Galois groups in a similar fashion. The cyclotomic tower $K(\Lambda_{\wp^n})$ corresponds to an $A_\wp^*$-extension (a general theory of such extensions or of extensions over 'the first level' is not much developed yet; it is interesting to note that the dependence of the group on $\wp$ is just through its degree) and putting the ramification data mentioned above in the Riemann-Hurwitz formula gives $g_n \sim d(q^d-1)nq^{d(n-1)}$. This implies that $\log h_n \asymp n(\mathrm{Norm}(\wp))^n$. A similar asymptotic is not known classically. It is known for the minus part in the classical as well as our case. How about the class numbers of the rings of integers?

We now come to the zeta measure associated to $\mathbf{F}_q[T]$. (See [T1], [G3] and references there). Under the Iwasawa isomorphism, $\mathbf{Z}_p$-valued measures on $\mathbf{Z}_p$ can be identified with power series in such a way that the convolution of measures corresponds to the multiplication of the corresponding power series. In fact, the binomial coefficients $\binom{x}{k}$ give basis of polynomials in x over $\mathbf{Q}$, which map $\mathbf{Z}$ to $\mathbf{Z}$ and the power series associated to the measure μ is just $\sum \mu_k X^k$ where $\mu_k := \int_{\mathbf{Z}_p} \binom{x}{k} d\mu$. The analogue for $A = \mathbf{F}_q[T]$ of $(1+t)^x - 1 = \sum \binom{x}{k}t^k$ is $\rho_x(t) = \sum \{\begin{smallmatrix} x \\ q^k \end{smallmatrix}\} t^{q^k}$ and in fact $\{\begin{smallmatrix} x \\ q^k \end{smallmatrix}\}$ gives a basis of 'additive' polynomials over K which map A into A. There is a way to extend this definition of binomial coefficients to any $\{\begin{smallmatrix} x \\ k \end{smallmatrix}\}$ and if we associate to an $A_\wp$-valued measure μ on $A_\wp$ a divided power series $\sum \mu_k(X^k/k!)$, with μ_k defined analogously, then the convolution corresponds to the multiplication of the divided power series.

Classically, the measure μ whose moments $\int_{\mathbf{Z}_p} x^k d\mu$ are $(1 - a^{k+1})\zeta(-k)$ for some $a \geq 2$, $(a, p) = 1$ has the associated power series $(1 + X)/(1 - (1 + X)) - a(1 + X)^a/(1 - (1 + X)^a)$. We need a twisting factor in front of the zeta values to compensate for the fact that the zeta values are rational rather than integral, in contrast to our case. Comparison of this result with the following is not well-understood.

THEOREM 8. (See[T1]): *For $A = \mathbf{F}_q[T]$, the divided power series corresponding to the measure μ whose i-th moment is $\zeta(-i)$ is given by $\sum \mu_k(X^k/k!)$ with μ_k being $(-1)^n$ when $k = cq^n + (q^n - 1)$, $0 < c < q - 1$, and $\mu_k = 0$ otherwise.*

The k's for which $\mu_k \neq 0$ can be characterised as those 'odd' k's for which any smaller positive integer has each base q digit no larger than the corresponding base q digit of k. For such k's, all the binomial coeffecients $\binom{k}{i}$ are nonzero modulo q. When q is a prime, this last property characterises such k's among the 'odd' numbers. Some other properties of k's are described in [T1, G3]. More results on the interesting influence of the base q digits on the zeta values and their orders of vanishing will be described elsewhere.

Let us now look at an analogue of Fermat equation, which was, of course, a motivation for the study of cyclotomic fields. The usual Fermat equation is well understood. Writing the Fermat equation as $z^p = y^p((x/y)^p - 1)$ to bring in the analogies with the cyclotomic theory, Goss [G2] looked at the equation in the cyclotomic theory for $A = \mathbf{F}_q[T]$ analogous to the one here, namely $z^{q^d} = y^{q^d}\rho_\wp(x/y)$. He proved various analogous features of the theory and made conjectures about its non-trivial solutions. (There are no non-trivial solutions except for some small exceptions, just as in the classical case). These have been recently settled in a very nice work [De] by Laurent Denis. The idea is to rewrite the equation as $(z/y)^{q^d} = \sum a_i(x/y)^{q^i}$, differentiate with respect to T (note $q = 0$ in characteristic p) and then clear out the denominators by multiplying by y^{q^d} to see that y divides a power of x. But essentially by the usual reductions, y could have been taken prime to x and hence apart from some low cases and trivial solutions, there are no more!

We end this paper by just remarking that there are analogues of Gauss sums, Gamma functions, Gross-Koblitz formula, etc. In fact there is a family of such objects associated to each 'cyclotomic family': one for the constant field extensions and one for the geometric extensions of Drinfeld. There is even more (see [T2] and references there) to the story when K is not $\mathbf{F}_q(T)$.

Acknowledgements: This paper is a written version of the talk given at the conference on Arithmetic Geometry with an emphasis on Iwasawa theory, held at Tempe in March 93. It is a pleasure to thank the organisers Nancy Childress and John Jones for arranging a nice conference.

REFERENCES

[A-T] Anderson G. and Thakur D. - *Tensor powers of the Carlitz module and zeta values*, Annals of Math. 132 (1990), 159-191.

[C1]   Carlitz L. - *An analogue of the von-Staudt-Clausen theorem*, Duke Math. J. 3 (1937), 503-517.

[C2]   Carlitz L. - *A class of polynomials*, Trans A. M. S. 43 (1938), 167-182.

[De]   Denis L. - *Le théorème de Fermat-Goss*, To Appear in Trans. A. M.S.

[D]   Drinfeld V. - *Elliptic modules*, (Translation), Math. Sbornik 23 (1974), 561-592.

[G-R] Galovich S. and Rosen M. - *The class number of cyclotomic function fields*, J. Number theory 13 (1981), 363-375.

[G-K] Gold R. and Kisilevsky H. - *On geometric $\mathbf{Z}_p$ extensions of function fields*, Manuscripta Math. 62 (1988), 145-161.

[G1]   Goss D. - *The arithmetic of function fields 2: The cyclotomic theory*, J. Algebra 81 (1983), 107-149.

[G2]   Goss D. - *On a Fermat equation arising in the arithmetic theory of function fields*, Math. Annalen, 261 (1982), 269-286.

[G3]   Goss D. - *L-series of t-motives and Drinfeld modules*, The Arithmetic of Function Fields, Ed. David Goss et al, Walter de Gruyter, NY (1992), 313-402.

[G-S] Goss D. and Sinnott W. - *Class groups of function fields*, Duke Math. J. 52 (1985), 507-516.

[H1]   Hayes D. - *Explicit class field theory in global function fields*, Studies in Algebra and Number theory - Ed. G. C. Rota, Academic press (1979), 173-217.

[H2]   Hayes D. - *A brief introduction to Drinfeld modules*, The Arithmetic of Function Fields, Ed. David Goss et al, Walter de Gruyter, NY (1992), 1-32.

[I-S] Ireland K. and Small R. - *Class numbers of cyclotomic function fields*, Math. of Computation, 46 (1986), 337-340.

[O]   Okada S. - *Kummer's theory for function fields*, J. Number theory, 38 (1991), 212-215.

[S1]   Shu L. - *Cyclotomic units and class number formulas over global function fields*, To appear in J. Number theory.

[S2]   Shu L. - *Kummer's criterion over global function fields*, To appear in J. Number theory.

[Ta]   Tate J. - *Les conjectures de Stark sur les fonctions L d'Artin en $s = 0$*, Progress in Math. 47, Birkhauser, Boston (1984).

[T1]   Thakur D. - *Zeta measure associated to $\mathbf{F}_q[T]$*, J. Number theory, 35 (1990), 1-17.

[T2]   Thakur D. - *On Gamma functions for function fields*, The Arithmetic of Function Fields, Ed. David Goss et al, Walter de Gruyter, NY (1992), 75-86.

[Y]   Yu J. - *Transcendence and special zeta values in characteristic p*, Ann. Math. 134 (1991), 1-23.

DEPARTMENT OF MATHEMATICS, UNIVERSITY OF ARIZONA, TUCSON, AZ 85721
E-mail address: thakur@math.arizona.edu

Contemporary Mathematics
Volume **174**, 1994

Slopes of Modular Forms

DOUGLAS L. ULMER

ABSTRACT. We formulate a strategy for proving that the number of modular eigenforms of weight w and level pN whose eigenvalue for the Hecke operator U_p has a fixed p-adic valuation is bounded independently of w. We also carry out the first step in this program.

1. Introduction

In this note, we continue our study of the p-adic valuations of eigenvalues of Hecke operators on modular forms on $\Gamma_1(pN)$. Our aim is to discuss the possibility of proving a "control theorem" in the style of Hida. Such a theorem would assert roughly that the number of normalized eigenforms of weight w and level pN whose eigenvalue for U_p has some fixed valuation λ is bounded independently of w. This assertion is known when $\lambda = 0$ and it is the starting point for one possible approach to Hida's theory of p-adic analytic families of *ordinary* modular eigenforms (cf. [**H**, Ch. 7]). Our motivation for studying this question comes from the possibility of generalizing Hida's theory to the non-ordinary case, i.e., to $\lambda > 0$.

Here we take the first step toward a control theorem by extending the results of [**U1**], which were valid only for small weights, to all weights. As we will explain below, the truth of a control theorem is then equivalent to a suitable extension of the results of [**U2**] to all weights. We also discuss the relationship between our results and recent conjectures of Gouvêa and Mazur. It is interesting to note that their conjectures (which predict a very simple periodicity property, as a function of the weight, for the number of eigenforms with a given λ) imply the existence of many cohomology classes of a type usually considered to be pathological.

This paper is an outgrowth of the talk I gave at the conference on Arithmetic Geometry at the Arizona State University in March, 1993. It is a pleasure to

1991 *Mathematics Subject Classification*. Primary 11G18; Secondary 11F30 11F85.

This research was partially supported by NSF grant DMS 9302976.

This paper is in final form and no version of it will be submitted for publication elsewhere.

thank the organizers of this conference, Nancy Childress and John Jones, for their efforts in putting together a conference with both interesting mathematics and a truly pleasant atmosphere. Thanks are due also to Gouvêa and Mazur for supplying me with early versions of some of their works.

2. Slopes of modular forms

We begin by introducing spaces of modular forms with p-adic coefficients and their slope decompositions. Our main objects of study are the dimensions of the factors in this decomposition.

Fix a positive integer M and a non-negative integer k. Then we have the complex vector space $S_{k+2}(\Gamma_1(M))$ of cusp forms of weight $k+2$ for the congruence subgroup $\Gamma_1(M)$ of $\mathrm{SL}_2(\mathbf{Z})$. This space carries an action of Hecke operators T_ℓ for all primes $\ell \nmid M$, U_ℓ for $\ell | M$, and $\langle d \rangle_M$ for $d \in (\mathbf{Z}/M\mathbf{Z})^\times$ ([Sh], Ch. 3 or [Mi], 4.5). If $M = M_1 M_2$ with $(M_1, M_2) = 1$, then the operators $\langle d \rangle_M$ can be factored $\langle d \rangle_M = \langle d \rangle_{M_1} \langle d \rangle_{M_2}$ and $\langle d \rangle_{M_i}$ depends only on the image of d in $(\mathbf{Z}/M_i\mathbf{Z})^\times$. We have a direct sum decomposition

$$S_{k+2}(\Gamma_1(M)) \cong \bigoplus_{\psi:(\mathbf{Z}/M\mathbf{Z})^\times \to \mathbf{C}} S_{k+2}(\Gamma_0(M), \psi)$$

where $S_{k+2}(\Gamma_0(M), \psi)$ is the subspace on which the $\langle d \rangle_M$ act via the character ψ. This decomposition is preserved by the Hecke operators.

We need a rational version of these decompositions. Let

$$S_{k+2}(\Gamma_1(M); \mathbf{Q}) \subseteq S_{k+2}(\Gamma_1(M))$$

be the subspace of modular forms all of whose Fourier coefficients at the standard cusp ∞ are rational numbers. A fundamental theorem asserts that this space is a $\mathbf{Q}$-structure on $S_{k+2}(\Gamma_1(M))$, i.e.,

$$S_{k+2}(\Gamma_1(M); \mathbf{Q}) \otimes_{\mathbf{Q}} \mathbf{C} \cong S_{k+2}(\Gamma_1(M)).$$

(This even holds with $\mathbf{Q}$ replaced with $\mathbf{Z}$; cf. [Sh] 3.52.) Moreover, the Hecke operators preserve this structure. For any commutative $\mathbf{Q}$-algebra R, define

$$S_{k+2}(\Gamma_1(M); R) = S_{k+2}(\Gamma_1(M); \mathbf{Q}) \otimes_{\mathbf{Q}} R;$$

we will be interested mostly in the case where R is a field of p-adic numbers.

If $M = M_1 M_2$ is a factorization with $(M_1, M_2) = 1$ and R contains the $\phi(M_1)$-th roots of unity $\mu_{\phi(M_1)}$, then we have a direct sum decomposition

$$S_{k+2}(\Gamma_1(M); R) \cong \bigoplus_{\psi:(\mathbf{Z}/M_1\mathbf{Z})^\times \to R} S_{k+2}(\Gamma_1(M); R)(\psi)$$

where $f \in S_{k+2}(\Gamma_1(M); R)$ lies in $S_{k+2}(\Gamma_1(M); R)(\psi)$ if and only if $\langle d \rangle_{M_1} f = \psi(d) f$ for all $d \in (\mathbf{Z}/M_1\mathbf{Z})^\times$.

Now fix a prime number p, a non-negative integer k, and an integer N prime to p. Let R be $\mathbf{Q}_p$, the p-adic numbers. We can apply the above constructions with $M_1 = p$, $M_2 = N$, and we get a direct sum decomposition:

$$S_{k+2}(\Gamma_1(pN); \mathbf{Q}_p) \cong \bigoplus_{a=0}^{p-2} S_{k+2}(\Gamma_1(pN); \mathbf{Q}_p)(\chi^a)$$

where $\chi : (\mathbf{Z}/p\mathbf{Z})^\times \to \mathbf{Z}_p$ is the Teichmüller character (characterized by $\chi(d) \equiv d \pmod{p}$). (Note that we are not decomposing for characters modulo N.) For any integer a, define

$$S(k, N, a) = S_{k+2}(\Gamma_1(pN); \mathbf{Q}_p)(\chi^a).$$

We will sometimes abbreviate this space to just S when k, N, and a are fixed.

We want to consider the action of U_p on $S = S(k, N, a)$ when $0 \le a < p - 1$. This action is semi-simple and when $a \ne 0$, its eigenvalues α are algebraic integers all of whose complex embeddings satisfy $\alpha\overline{\alpha} = p^{(k+1)}$. When $a = 0$, the eigenvalues of U_p coming from forms new at p satisfy $\alpha^2 = \zeta p^k$ (where ζ is a root of unity) and the other eigenvalues satisfy $\alpha\overline{\alpha} = p^{(k+1)}$ ([Mi], 4.6.17). Let v be the valuation of $\overline{\mathbf{Q}_p}$ normalized so that $v(p) = 1$. Then if α is an eigenvalue of U_p on S, we have $0 \le v(\alpha) \le k + 1$. We define $v(\alpha)$ to be the *slope* of the eigenvalue, and if $f \in S \otimes \overline{\mathbf{Q}_p}$ is an eigenvector for U_p with eigenvalue α, we will also call $v(\alpha)$ the slope of f.

The characteristic polynomial of U_p is a polynomial over $\mathbf{Q}_p$ and all of the roots of each of its irreducible factors are conjugate over $\mathbf{Q}_p$ and thus have the same valuation. Taking the generalized eigenspaces corresponding to each irreducible factor and collecting together those with a fixed slope gives a unique $\mathbf{Q}_p$-rational decomposition

$$S \cong \bigoplus_\lambda S_\lambda$$

(compatible with the Hecke operators) such that all of the eigenvalues of U_p on S_λ have slope λ.

We want to study the numbers

$$d_\lambda = d(k, N, a)_\lambda = \dim S(k, N, a)_\lambda$$

as functions of k and a. Gouvêa and Mazur have conjectured the following periodicity property: if $p > 3$, $k > 2\lambda$, and $n \ge \lambda$, then

$$d(k, N, 0)_\lambda \overset{?}{=} d(k + p^n(p-1), N, 0)$$

[**G-M**, Conjecture 1]. It is reasonable to suppose that more generally,

$$d(k, N, a)_\lambda \overset{?}{=} d(k + p^n(p-1), N, a)$$

for all a; in private communication they have in fact made even stronger variations of their basic conjecture. The displayed assertions are theorems when $\lambda = 0$, by work of Hida. Note also that the second assertion certainly imples that for a fixed N and λ, $d(k, N, a)_\lambda$ is bounded independently of k and a. It is this weaker assertion (refered to as a "control theorem") that we wish to discuss for the moment.

It will be convenient to package the numbers d_λ into Newton polygons. Recall that if

$$H(T) = 1 + a_1 T + \cdots + a_d T^d = \prod_{i=1}^{d}(1 - \alpha_i T)$$

is a polynomial with $\overline{\mathbf{Q}_p}$ coefficients and with the factors α_i ordered so that $v(\alpha_i) \leq v(\alpha_{i+1})$, then the *Newton polygon* of H with respect to v is defined to be the graph of the continuous, piecewise linear, convex function f on $[0, d]$ with $f(0) = 0$ and $f'(x) = v(\alpha_i)$ for all $x \in (i-1, i)$. This polygon can be seen to be part of the boundary of the convex hull in the plane of the points $(0, 0)$ and $(i, v(a_i))$ for $i = 1, \ldots, d$. In particular, if the $v(a_i) \in \mathbf{Z}$, the break points of the Newton polygon (i.e., the points on the polygon where f changes slope) have integer coordinates.

For fixed k, N, and a, knowledge of the d_λ is clearly equivalent to knowledge of the Newton polygon of the characteristic polynomial

$$H(T) = \det(1 - U_p T | S(k, N, a)).$$

Given non-negative real numbers $l_0, \ldots, l_n$, define the associated *Hodge polygon* to be the graph of the continuous, piecewise linear, convex function f on $[0, l_0 + \cdots + l_n]$ with $f(0) = 0$ and $f'(x) = i$ for all $x \in (l_0 + \cdots + l_{i-1}, l_0 + \cdots + l_i)$. We want to consider the highest Hodge polygon (i.e., polygon with integral slopes) lying on or below the Newton polygon of H. Generally, given any Newton polygon, define its associated *contact polygon* as the highest Hodge polygon lying on or below it and having the same endpoints. From its definition, the contact polygon has the following properties: the Newton and contact polygons must meet at some point along each edge of the contact polygon; and, if $A_1, \ldots, A_n$ are the break points of the Newton polygon which lie on the contact polygon (in order), then for each j, either a) the Newton and contact polygons coincide along the segment $\overline{A_j A_{j+1}}$ (which thus has integral slope) or b) the contact polygon has two edges between A_j and A_{j+1}, with slopes i and $i+1$ for some integer i and the slopes of the Newton polygon between A_j and A_{j+1} all lie in the interval $(i, i+1)$. (Both properties hold since if they failed, we could raise some edge of the contact polygon.) Figure 1 illustrates the two possibilities.

Note that in the second case, if l_i and l_{i+1} are the lengths of the two edges of the contact polygon between A_j and A_{j+1} and if $v(\alpha_1), \ldots, v(\alpha_k)$ are the slopes of the Newton polygon between A_j and A_{j+1} (with multiplicities) we have $l_i + l_{i+1} = k$ and $i l_i + (i+1) l_{i+1} = \sum v(\alpha_j)$. Thus the following formula gives the lengths m_i of the sides of the contact polygon in terms of the slopes of the Newton polygon:

$$m_i = \sum_{v(\alpha_j) \in (i-1,i)} (v(\alpha_j) - (i-1)) + \sum_{v(\alpha_j)=i} 1 + \sum_{v(\alpha_j) \in (i,i+1)} (i+1-v(\alpha_j)).$$

Since all the $v(a_i)$ are integers, the break points of the Newton polygon occur at points with integer coordinates and so the m_i are also integers.

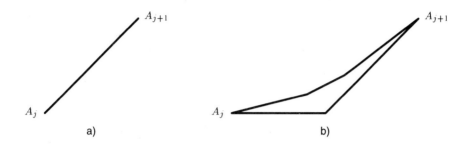

FIGURE 1

Knowledge of the contact polygon attached to $H(T)$ gives some control on the dimensions d_λ. Indeed, if $m_0, \ldots, m_{k+1}$ are the lengths of the contact polygon, we have

$$d_i \le m_i$$

and

$$\sum_{i \le \lambda \le i+1} d_\lambda \le m_i + m_{i+1}.$$

Also,

$$\sum_{j \le i} m_j \le \sum_{\lambda < i+1} d_\lambda \le \sum_{\lambda \le i+1} d_\lambda \le \sum_{j \le i+1} m_j$$

so a control theorem for a given p and N is equivalent to the assertion that the corresponding m_i are bounded independently of k and a.

In the next section, we will review some work of Crew and Ekedahl which allows one to relate the m_i to cohomological invariants and in Sections 4 and 5, we compute some of these invariants. This amounts to extending in some form the Newton-Hodge inequalities of [**U1**] to all weights k. (The results of [**U1**] were proven only for $k < p$ and the extension to all k gives a slightly less precise result.) The remaining invariant was studied in [**U2**], again for $k < p$, and we hope to extend those computations to all k.

3. Contact polygons and cohomology

Let $\mathbf{F}$ be a perfect field of characteristic p and X a smooth projective variety over $\mathbf{F}$. The action of Frobenius on the (crystalline or ℓ-adic) cohomology of X gives rise to a characteristic polynomial $H(T)$ with $\overline{\mathbf{Q}}$-coefficients and thus to a

Newton polygon which encodes the p-adic valuations of the roots of H. This is in general of course a rather delicate invariant of X and the associated contact polygon is also subtle. However, in some cases good control on the contact polygon can be obtained. We want to use this to study the polygons attached to modular forms.

We need to work in a more general setting, so let $\mathbf{F}$ and X be as above and assume further that $\Pi \in \mathbf{Q}_p[\mathrm{Aut}_\mathbf{F} X]$ is a projector, i.e., $\Pi^2 = \Pi$. The pair (X, Π) is then a (Chow) motive with $\mathbf{Q}_p$-coefficients defined over $\mathbf{F}$, which we denote by M. For each n, the group algebra element Π defines a projector on $H^n_{cris}(X/W) \otimes_{\mathbf{Z}_p} \mathbf{Q}_p$ (where $W = W(\mathbf{F})$ is the Witt ring of $\mathbf{F}$) and we set

$$H^n_{cris}(M/W) \otimes_{\mathbf{Z}_p} \mathbf{Q}_p = \Pi \left(H^n_{cris}(X/W) \otimes_{\mathbf{Z}_p} \mathbf{Q}_p \right).$$

The absolute Frobenius of $\mathbf{F}$ acts on $H^n_{cris}(X/W)$ and commutes with Π; extending scalars to K, the fraction field of the Witt ring of the algebraic closure of $\mathbf{F}$, we get a finite dimensional K vector space with a σ-linear automorphism (where σ is the automorphism of K induced by the absolute Frobenius of $\mathbf{F}$), i.e., an F-isocrystal.

Now finite dimensional K vector spaces with σ-linear automorphisms F are completely classified by their slopes, which are defined as follows. There exist non-negative rational numbers $r_1/s_1, \ldots, r_k/s_k$ (written in lowest terms and with 0 written as $0/1$) such that

$$V \cong \bigoplus_i K[T]/(T^{s_i} - p^{r_i})$$

with $F(T^j) = T^{j+1}$ and extended by semi-linearity. The r_i/s_i are unique up to permutation and by definition they are the *slopes* of V, with each r_i/s_i counted with multiplicity s_i. If V happens to come by extension of scalars from a $\mathrm{Frac}(W(\mathbf{F}_{p^f}))$ vector space V_0, so that F^f is a linear automorphism of V_0, then the slopes and multiplicities of V as defined here coincide with the valuations of the inverse roots of the characteristic polynomial $\det(1 - F^f T|V_0)$, computed with respect to the valuation with $v(p) = 1/f$.

In the next section we will find a pair (X, Π) over $\mathbf{F}_p$ so that the slopes of $H^n_{cris}(M/W) \otimes_{\mathbf{Z}_p} \mathbf{Q}_p$ for some fixed n are the slopes of U_p on a suitable combination of the $S(k, N, a)$. Thus we can apply all the techniques of p-adic cohomology to study these slopes. In fact we need to do something slightly more, namely to find a projector $\Pi \in \mathbf{Z}_p[\mathrm{Aut}_\mathbf{F}]$ (i.e., Π should have p-integral coefficients). Then Π acts on $H^n_{cris}(X/W)$ itself and we define $H^n_{cris}(M/W) = \Pi H^n_{cris}(X/W)$.

Return now to a general (X, Π), but assume that Π has $\mathbf{Z}_p$ coefficients. Let m_{ij} be the i-th contact number associated with the slopes of $H^{i+j}_{cris}(M/W)$. We want to review some work of Crew and Ekedahl which allows one to compute the m_{ij} in terms of two other (hopefully more tractable) cohomological invariants. More precisely, there exist non-negative integers h_W^{ij} and T^{ij} to be discussed below such that

$$(3.1) \qquad m_{ij} = h_W^{ij} - T^{ij} + 2T^{i-1,j+1} - T^{i-2,j+2}.$$

The h_W^{ij} are called Hodge-Witt numbers and are closely related to the usual Hodge numbers $h^{ij} = \dim_{\mathbf{F}} \Pi H^j(X, \Omega^i)$ (which make sense because Π has p-integral coefficients). The T^{ij} are a measure of the non-finitely generated torsion in the deRham-Witt cohomology of (X, Π).

The precise definitions of the h_W^{ij} and T^{ij} are rather technical, so for the moment we will just give some of the properties these numbers enjoy and which make it possible to compute them in some cases. We will give a precise definition of T^{ij} in Section 5 and then formula 3.1 can be taken as the definition of the h_W^{ij}.

First of all, $T^{ij} \geq 0$ and $T^{ij} = 0$ if $i < 0$ or $i \geq \dim X - 1$ or $j \leq 1$ or $j > \dim X$. Since $\sum_{i+j=n} m_{ij} = b_n = \dim_K H^n_{cris}(M/W) \otimes K$ by definition, the formula 3.1 gives that

$$\sum_{i+j=n} h_W^{i,j} = b_n.$$

If we define the Hodge-Witt polygon to be the Hodge polygon associated to the $h_W^{i,j}$ (where $n = i + j$ is fixed), then 3.1 can be interpreted as saying that $T^{i,j}$ is the number of units the slope $i + 1$ edge of the Hodge-Witt polygon should be raised in order that it coincide with the slope $i + 1$ edge of the contact polygon along some interval. In particular, the vanishing of the T^{ij} for (i, j) outside the range $0 \leq i \leq \dim X - 1$, $1 \leq j \leq \dim X$ and the formula for the sum of the h_W^{ij} imply that the Hodge-Witt and contact (and Newton) polygons have the same endpoints.

Secondly, there are two relations between the h_W^{ij} and the h^{ij}, namely *Crew's formula:*

$$\sum_j (-1)^j h_W^{ij} = \sum_j (-1)^j h^{ij} = \chi(M, \Omega^i)$$

(which we will use below) and *Ekedahl's inequality:*

$$h_W^{ij} \leq h^{ij}.$$

The latter, together with 3.1, implies that when $\sum_{i+j=n} h^{ij} = b_n$ we have $h_W^{ij} = h^{ij}$ for $i + j = n$.

The interested reader can find more details about the slopes associated to an F-crystal in [**Ma**, Ch. 2]. The work of Crew and Ekedahl, as well as results of Illusie and Raynaud upon which it is based, is beautifully surveyed in Illusie's article [**I**]; we will use it as a general reference for these ideas. See also the related work of Milne [**M**] which among other things gives another proof of Crew's formula.

4. Slopes of a modular variety

In this section we are going to define a modular variety $\tilde{X}$ and a projector Π such that the slopes of Frobenius on its cohomology are the slopes of U_p on certain spaces of modular forms. The arguments are essentially a simple modification of those of Section 2 of [**U1**], so we will be brief.

Fix a prime $p > 2$, an integer $N \geq 5$ and prime to p, and a positive integer k. Let $I = Ig_1(pN)$ be the Igusa curve of level pN, i.e., the complete modular curve parameterizing elliptic curves with a $\Gamma_1(N)$ structure and an Igusa structure of level p. Forgetting the Igusa structure induces a map $I \to X_1(N)$ of degree $p-1$ which is unramified off the points representing supersingular elliptic curves and is totally ramified over the supersingular points. Let $C \subseteq I$ be the reduced subscheme of cusps, i.e., the points representing singular elliptic curves and let $S \subseteq I$ be the reduced subscheme of supersingular points.

Let $\pi_N : \mathcal{E}_N \to X_1(N)$ be the universal generalized elliptic curve over $X_1(N)$ and $\pi : \mathcal{E} \to I$ its pull back to I. Define X to be the k-fold fiber product

$$\mathcal{E} \times_I \cdots \times_I \mathcal{E}.$$

When $k > 1$, this variety has singularities coming from the product of double points in the fibers of π over C. There is a canonical disingularization $\tilde{X} \to X$ defined by Deligne which is described in some detail in [**U1**], Section 4; it is obtained from X by blowing up points in the fibers over C. In particular, away from C, $f : \tilde{X} \to I$ is still the k-fold fiber product of a smooth elliptic curve.

The group $H = (\mathbf{Z}/N\mathbf{Z} \rtimes \mu_2)^k$ acts on X, with the factors $\mathbf{Z}/N\mathbf{Z}$ acting by translation by the canonical point of order N and the factors μ_2 acting by inversion; this action lifts to $\tilde{X}$. Also, $(\mathbf{Z}/p\mathbf{Z})^\times$ acts on I via its natural action on Igusa structures and on $\tilde{X}$; the map f is equivariant for these actions. Define a character $\epsilon : H \to \{\pm 1\}$ by requiring that ϵ be 1 on the factors $\mathbf{Z}/N\mathbf{Z}$ and the identity $\mu_2 \xrightarrow{\sim} \{\pm 1\}$ on the factors μ_2, and let Π be the corresponding projector in $\mathbf{Z}[1/2N][\mathrm{Aut}\,\tilde{X}]$. For any $\mathbf{Z}[1/2N][H]$-module V, we write $V(\epsilon)$ for ΠV.

Now define polynomials encoding the eigenvalues of U_p as follows:

$$H_k(T) = \prod_{\psi:(\mathbf{Z}/N\mathbf{Z})^\times \to \mathbf{C}} \det\left(1 - T_p T + \psi(p)p^{k+1}T^2 | S_{k+2}(\Gamma_1(N), \psi)\right)$$

$$\times \prod_{\substack{\psi:(\mathbf{Z}/pN\mathbf{Z})^\times \to \mathbf{C} \\ p | cond(\psi)}} \det\left(1 - U_p T | S_{k+2}(\Gamma_1(pN), \psi)\right).$$

(This is essentially the characteristic polynomial of U_p on $S_{k+2}(\Gamma_1(pN))$, except that we have removed the factors coming from eigenforms which are new at p but whose character has conductor prime to p; these forms all have slope $k/2$.)

To state the main result of this section, let c be the number of cusps on the modular curve $X_1(N)$, and set $c_1 = (p-1)c$; then c_1 is the number of cusps on

I, i.e., the degree of the divisor C. We also introduce multiplicities

$$e_{kj} = \binom{k}{j} - \binom{k}{j-1}.$$

THEOREM 4.1. *Let p be a prime number, $N \geq 5$ an integer relatively prime to p, and k a non-negative integer. Define the variety $\tilde{X}$, the character ϵ and the polynomials H_j as above. Then*

a)

$$\det(1 - \operatorname{Fr} T|(H_{cris}^{k+1}(\tilde{X}/W) \otimes \mathbf{Q}_p)(\epsilon)) = H_k(T) \prod_{1 \leq j \leq k/2} \left(H_{k-2j}(p^j T)P_{k,j}(T)\right)^{e_{kj}}$$

where $P_{kj}(T)$ is a product $\prod(1 - \alpha T)$ of degree c_1 if $k - 2j \neq 0$ or degree $c_1 - 1$ if $k - 2j = 0$ all of whose inverse roots α have the form ζp^{k+1-j} where ζ is a root of unity.

b)

$$\det(1 - \operatorname{Fr} T|(H_{cris}^{k}(\tilde{X}/W) \otimes \mathbf{Q}_p)(\epsilon)) = \begin{cases} (1 - p^{k/2}T)^{e_{k,k/2}} & \text{if k is even} \\ 0 & \text{otherwise.} \end{cases}$$

c)

$$H_{cris}^{i}(\tilde{X}/W)(\epsilon) = 0 \quad \text{if $i \neq k, k+1$.}$$

The proof of this theorem is a minor modification of that of Corollary 2.2 of [**U1**] and will be left to the reader. The main difference is that since we are not using the symmetric group in our projector, the computation using the Künneth formula, the coherent cohomology of abelian varieties, and linear algebra which takes place in Proposition 2.1b of [**U1**] has a different outcome. The new computation is, however, every bit as standard as the old. Not using the symmetric group has the advantage that our projector has p-integral coefficients for all k. For simplicity, we have not broken down the result according to characters of $(\mathbf{Z}/p\mathbf{Z})^{\times}$.

REMARKS. 1) Note that since $\tilde{X}$ is defined over $\mathbf{F}_p$, the valuations of the inverse roots of the polynomials of the Theorem are the slopes of $(\tilde{X}, \Pi)$ as defined in the last section.
2) If G_k denotes the polynomial on the right hand side of part a) of the Theorem, then knowledge of the valuations of the inverse roots of G_j for $j \leq k$ is clearly equivalent to knowledge of the valuations of the inverse roots of the H_j for $j \leq k$. In particular, the mixing together of different weights caused by not using the symmetric group does not prevent us from obtaining results on the slopes for individual weights.

COROLLARY 4.2. *We have $m_{ij} = 0$ if $i + j \neq k$ or $k + 1$, or if $i + j = k$ and $i \neq k/2$. Also, $m_{k/2,k/2} = e_{k/2,k/2} = \binom{k}{k/2} - \binom{k}{k/2-1}$.*

5. Hodge-Witt numbers of a modular variety

In this section we will compute the Hodge-Witt numbers h_W^{ij} of the pair $(\tilde{X}, \Pi)$ defined in the previous section.

To state the result, let g be the genus of $X_1(N)$, c the number of cusps on this curve and set $w = (g - 1) + c/2$; then w is the degree of the invertible sheaf $\omega = (R^1\pi_{N*}\mathcal{O}_{\mathcal{E}_N})^{-1}$ on $X_1(N)$. Also set $w_1 = (p - 1)w$ and $c_1 = (p - 1)c$. These are the degree of the pull back of ω to I and the number of cusps on I respectively. The genus g_1 of I satisfies $2g_1 - 2 = pw_1 - c_1$. Define integers $l_{k,i}$ by the formulae

$$l_{k,0} = l_{k,k+1} = g_1 - 1 + kw_1 + \begin{cases} 1 & \text{if } k = 0 \\ 0 & \text{otherwise} \end{cases}$$

$$l_{k,i} = (p - 2)w_1 \qquad 1 \leq i \leq k$$

THEOREM 5.1. *Let p be an odd prime number, $N \geq 5$ an integer prime to p, and k a non-negative integer. The Hodge-Witt numbers of the pair $(\tilde{X}, \Pi)$ defined above are as follows: for $0 \leq i \leq k + 1$*

$$h_W^{i,k+1-i} = \sum_{j \leq \min(i, k+1-i, k/2)} l_{k-2j,i-j} e_{kj} + \begin{cases} c_1 e_{k,k+1-i} & \text{if } i \geq \frac{k+3}{2} \\ (c_1 - 1) e_{k,k+1-i} & \text{if } i = \frac{k+2}{2} \\ 0 & \text{if } i \leq \frac{k+1}{2}, \end{cases}$$

$$h_W^{k/2,k/2} = \begin{cases} e_{k/2,k/2} & \text{if } k \text{ is even} \\ 0 & \text{otherwise,} \end{cases}$$

and all other $h_W^{ij} = 0$.

REMARKS. 1) If M_k denotes the pair $(\tilde{X}_k, \Pi_k)$ attached to modular forms of weight $k + 2$ and level pN studied in [**U1**] (so $k < p$), then we have that the non-zero Hodge-Witt (and Hodge) numbers of M_k are $h_W^{i,k+1-i} = l_{k,i}$ if $k > 0$; when $k = 0$, $h_W^{01} = h_W^{10} = g_1$ and $h_W^{00} = 1$. So, ignoring the terms involving c_1, the theorem says that the motive $M = (\tilde{X}, \Pi)$ "looks like" the direct sum of the M_{k-2j}, each with multiplicity e_{kj}. This of course should not be taken too seriously, since M_{k-2j} does not exist as a motive defined by a p-integral projector as soon as $k - 2j \geq p$.

2) The variety $\tilde{X}$ and the projector Π can be lifted to $\mathbf{Z}_p$ and the Hodge-Witt numbers of M are equal to the Hodge numbers of the generic fiber of this lifting. This equality of of Hodge-Witt and Hodge numbers is known to fail for a general smooth and proper variety over $\mathbf{Z}_p$, so it would be interesting to have some *a priori* criterion that could be applied to the case at hand.

3) On the other hand, the Hodge-Witt numbers of $(\tilde{X}, \Pi)$ are definitely not equal to the Hodge numbers of this pair over $\mathbf{F}_p$. This fails as soon as $k \geq 2p - 1$.

COROLLARY 5.2. *The Newton polygon of the polynomial*

$$\prod_{0 \leq j \leq k} H_{k-2j}(p^j T)^{e_{kj}}$$

lies on or above the Hodge polygon associated to the integers

$$l_i = \sum_{j \leq \min(i,k+1-i,k/2)} l_{k-2j,i-j} e_{kj} \qquad 0 \leq i \leq k+1$$

and these two polygons have the same endpoints.

PROOF. By the definition (3.1) of the Hodge-Witt numbers, the Newton polygon of M lies on or above the Hodge polygon associated to the Hodge-Witt numbers. By 4.1, the Newton polygon has an edge of slope i and length at least $c_1 e_{k,k+1-i}$ for each $i \geq (k+3)/2$, and an edge of slope $(k+2)/2$ and length at least $(c_1 - 1)e_{k,k/2}$ when k is even. Removing these edges and reducing the Hodge-Witt numbers by a corresponding amount gives the result.

REMARK. Corollary 5.2 would follow from an extension of the Newton-Hodge inequalities of [**U1**] (i.e., Theorem 1.2) to all weights, using the evident extension of the formulae for the Hodge numbers. Using other methods, Arthur Ogus has sketched a proof of such an extension [**O**, 8.4].

PROOF OF 5.1. The strategy will be first to show that the $T^{ij} = 0$ for $i+j \neq k+1$, then to use the definition of the h_W^{ij} (3.1) and the computation of the m_{ij} (4.2) to get the h_W^{ij} for all $i + j \neq k + 1$. Then Crew's formula reduces the problem to computing the Euler characteristics $\chi(M, \Omega^i)$, which is easy.

At this point, we must give the definition of the T^{ij}! The usual definition is in terms of certain non-finitely generated torsion in deRham-Witt cohomology (hence the name torsion numbers). By a Theorem of Illusie and Raynaud, the T^{ij} are also equal to another cohomological invariant which is much easier to compute. We will take this latter invariant as our definition. For more details on what follows, see [**I**] and the references there.

Let $W_n \Omega_{\tilde{X}}$ be the deRham-Witt complex of length n on $\tilde{X}$. This is a complex of étale sheaves of $\mathbf{Z}/p^n\mathbf{Z}$-modules, we have $W_1 \Omega_{\tilde{X}} = \Omega_{\tilde{X}}$, and $W_n \Omega_{\tilde{X}}^0 = W_n \mathcal{O}$ is the sheaf of Witt vectors introduced by Serre. Let $W_n \Omega_{log}^i$ be the subsheaf of $W_n \Omega_{\tilde{X}}^i$ generated additively by differentials

$$d\underline{x}_1/\underline{x}_1 \wedge \cdots \wedge d\underline{x}_i/\underline{x}_i$$

where $x_j \in \mathcal{O}^\times$ and $\underline{x}_j \in W_n \mathcal{O}$ is its multiplicative representative. There are exact sequences

(5.3) $$0 \to W_m \Omega_{log}^i \to W_{m+n} \Omega_{log}^i \to W_n \Omega_{log}^i \to 0$$

and we have $W_1 \Omega_{log}^i = \Omega_{log}^i$, the usual sheaf of logarithmic differentials. (So if $Z_{\tilde{X}}^i$ denotes the sheaf of closed differentials, then

$$\Omega_{log}^i = \ker\left(1 - \mathcal{C} : Z_{\tilde{X}}^i \to \Omega_{\tilde{X}}^i\right).$$

Here of course 1 denotes the natural inclusion $Z_{\tilde{X}}^i \to \Omega_{\tilde{X}}^i$ and $\mathcal{C}$ is the Cartier operator.)

Now consider, on the category of perfect $\mathbf{F}_p$-algebras equipped with the étale topology, the sheaf associated to the presheaf

$$A \longmapsto \amalg H^j_{\acute{e}t}(\tilde{X} \times \mathrm{Spec}\, A, W_n \Omega^i_{log}).$$

This sheaf turns out to be represented by a commutative algebraic perfect group scheme of finite dimension and this dimension is by definition $T^{i-1,j}$. Moreover, the value of the sheaf on an algebraically closed field is equal to the value of the presheaf, so using the exact sequence 5.3, to show $T^{ij} = 0$ it suffices to show that $\amalg H^j(\tilde{X} \times \mathrm{Spec}\, \mathbf{F}, \Omega^{i-1}_{log}) = 0$ for all algebraically closed fields $\mathbf{F}$ of characteristic p.

PROPOSITION 5.4. $T^{ij} = 0$ whenever $i + j \neq k + 1$.

PROOF. First some reductions: since $T^{ij} = T^{k-1-i,k+3-j}$ ([$\mathbf{I}$, 4.4.5]), it suffices to prove that $T^{ij} = 0$ for $i + j > k + 1$ and to prove this, it suffices to prove that $H^j(M \otimes \overline{\mathbf{F}_p}, \Omega^i_{log}) = 0$ for $i + j > k + 2$. For the rest of the proof, we replace M with $M \otimes \overline{\mathbf{F}_p}$, i.e., with $(\tilde{X} \otimes \overline{\mathbf{F}_p}, \Pi)$.

The next point is that $H^j(M, \Omega^i_{\tilde{X}}) = 0$ unless $i + j = k$, $k + 1$, or $k + 2$. The detailed proof of this requires a detour through the cohomology of the log schemes $X^\times$ and $I^\times$, as in [$\mathbf{U1}$], Section 4, which we will omit. The main point is to use the exact sequence

$$0 \to f^* \Omega^1_{I^\times} \otimes \Omega^{i-1}_{X^\times / I^\times} \to \Omega^i_{X^\times} \to \Omega^i_{X^\times / I^\times} \to 0$$

together with the Leray spectral sequence and the calculation

$$\amalg R^j f_* \Omega^i_{X^\times / I^\times} = \begin{cases} \left(\omega^{2i-k}\right)^{\oplus \binom{k}{i}} & \text{if } i + j = k \\ 0 & \text{otherwise} \end{cases}$$

and then compare the Hodge cohomology of $X^\times$ and $\tilde{X}$. This last comparison requires checking that Proposition 4.1 of [$\mathbf{U1}$] still holds using the projector of this paper in place of that of [$\mathbf{U1}$]; the crucial point here is the vanishing of the Hodge cohomology groups of the toric varieties $\tilde{Q}_r$ and $\tilde{P}_r$ after applying a suitable projector [$\mathbf{U1}$, 4.5]. Essentially the same proof works with the projector considered here.

Now let $Z^i_{\tilde{X}}$ and $B^i_{\tilde{X}}$ denote the sheaves of closed and exact differentials respectively. Taking cohomology of the sequences

$$0 \to Z^i_{\tilde{X}} \to \Omega^i_{\tilde{X}} \xrightarrow{d} B^{i+1}_{\tilde{X}} \to 0$$

and

$$0 \to B^i_{\tilde{X}} \to Z^i_{\tilde{X}} \xrightarrow{c} \Omega^i_{\tilde{X}} \to 0$$

and using that $Z^{k+1}_{\tilde{X}} = \Omega^{k+1}_{\tilde{X}}$, we have also that $H^j(M, Z^i_{\tilde{X}}) = H^j(M, B^i_{\tilde{X}}) = 0$ if $i + j > k + 3$. Using the sequence

$$0 \to \Omega^i_{log} \to Z^i_{\tilde{X}} \xrightarrow{1-c} \Omega^i_{\tilde{X}} \to 0$$

we already have that $H^j(M, \Omega^i_{log}) = 0$ for $i + j > k + 3$. To finish, it suffices to prove that

a) $1 - \mathcal{C} : H^{k+2-i}(M, Z^i_{\tilde{X}}) \to H^{k+2-i}(M, \Omega^i_{\tilde{X}})$ is onto, and

b) $H^{k+3-i}(M, Z^i_{\tilde{X}}) = 0$

for all i.

I claim that both a) and b) follow from the assertion that the natural map

$$(5.5) \qquad\qquad H^{k+2-i}(M, Z^i_{\tilde{X}}) \xrightarrow{n} H^{k+2-i}(M, \Omega^i_{\tilde{X}})$$

is onto for all i. Indeed, assume by induction that b) holds for some i. Then 5.5 implies that $H^{k+2-i}(M, B^{i+1}_{\tilde{X}}) = 0$ which implies that

$$H^{k+1-i}(M, Z^{i+1}_{\tilde{X}}) \xrightarrow{\mathcal{C}} H^{k+1-i}(M, \Omega^{i+1}_{\tilde{X}})$$

is onto. By Lemma 5.7 below, this implies a) for $i + 1$. It also implies the vanishing of $H^{k+2-i}(M, Z^{i+1}_{\tilde{X}})$, i.e., b) for $i + 1$. To start the induction, note that both a) and b) hold for $i = 0$.

Now we have a commutative diagram

$$
\begin{array}{ccc}
H^1(I, \amalg R^{k+1-i} f_* Z^i_{\tilde{X}}) & \xrightarrow{n} & H^1(I, \amalg R^{k+1-i} f_* \Omega^i_{\tilde{X}}) \\
\downarrow & & \downarrow \\
H^{k+2-i}(M, Z^i_{\tilde{X}}) & \xrightarrow{n} & H^{k+2-i}(M, \Omega^i_{\tilde{X}})
\end{array}
$$

and the right hand vertical map is an isomorphism, so it will suffice to prove that the top map is surjective. But this cohomology is on a curve I, so it suffices to show that the sheaf

$$(5.6) \qquad\qquad \mathrm{Coker}\left(\amalg R^{k+1-i} f_* Z^i_{\tilde{X}} \xrightarrow{n} \amalg R^{k+1-i} f_* \Omega^i_{\tilde{X}}\right)$$

is supported on points.

To see this, we use deRham cohomology; although we will not explicitly use the results of [**U2**] here, Sections 3 and 5 of that paper may help to unwind what follows. Recall the relative deRham cohomology sheaves $\underline{H}^j_{dR}(M/I) = \amalg R^j f_*(\Omega^{\cdot}_{\tilde{X}/I})$ and the deRham cohomology sheaves $\underline{H}^j_{dR}(M) = \amalg R^j f_*(\Omega^{\cdot}_{\tilde{X}})$, with their Hodge filtrations $F^{\cdot}$ and their conjugate filtrations $F_{\cdot}$. The relative Hodge to deRham spectral sequence $\amalg R^j f_* \Omega^i_{\tilde{X}/I} \Rightarrow \underline{H}^{i+j}_{dR}(M/I)$ degenerates and we have $\underline{H}^j_{dR}(M/I) = 0$ unless $j = k$. Moreover, over $U = I - \{C \cup S\}$, we have canonical isomorphisms

$$
\begin{aligned}
\underline{H}^k_{dR}(M/I) &\cong H^1_{dR}(\mathcal{E}/I)^{\otimes k} \\
&\cong \left(F^1 H^1_{dR}(\mathcal{E}/I) \oplus F_0 H^1_{dR}(\mathcal{E}/I)\right)^{\otimes k} \\
&\cong \left(\omega \oplus \omega^{-1}\right)^{\otimes k}
\end{aligned}
$$

Over U, the invertible sheaf ω is trivial with a canonical nowhere vanishing section ω_c; also there is a distinguished nowhere vanishing 1-form dq/q on U. In terms of ω_c and dq/q, the Gauss-Manin connection ∇ on $\underline{H}^k_{dR}(M/I)$ is the k-th tensor power of the connection on $\omega \oplus \omega^{-1}$ defined by $\nabla(\omega_c) = (dq/q)\omega_c^{-1}$ and $\nabla(\omega_c^{-1}) = 0$. Using the exact sequence

$$0 \to \underline{H}^k_{dR}(M) \to \underline{H}^k_{dR}(M/I) \xrightarrow{\nabla} \Omega^1_I \otimes \underline{H}^k_{dR}(M/I) \to \underline{H}^{k+1}_{dR}(M) \to 0$$

we can compute $\underline{H}^{k+1}_{dR}(M)$ and its Hodge and conjugate filtrations completely explicitly and see that the Hodge to deRham spectral sequence $\amalg R^j f_* \Omega^i_{\tilde{X}} \Rightarrow \underline{H}^{i+j}_{dR}(M)$ degenerates and that $\amalg R^j f_* \Omega^i_{\tilde{X}} = (F^i \cap F_i)(\underline{H}^{i+j}_{dR}(M))$ (again, all over U). But we have another commutative diagram over U

$$\begin{array}{ccc}
\amalg R^{k+1-i} f_* Z^i_{\tilde{X}} & \xrightarrow{n} & \amalg R^{k+1-i} f_* \Omega^i_{\tilde{X}} \\
\downarrow & & \downarrow \\
(F^i \cap F_i)(\underline{H}^{i+j}_{dR}(M)) & \longrightarrow & \mathrm{gr}^i \underline{H}^{k+1}_{dR}(M)
\end{array}$$

with the bottom and right arrows isomorphisms and the left arrow a surjection. This shows that the top arrow is surjective, so the sheaf 5.6 is supported on (at worst) the cusps and supersingular points.

To finish the proof of Proposition 5.4, we need to prove the lemma alluded to above.

LEMMA 5.7. *Suppose* **F** *is an algebraically closed field of characteristic p, V and W are finite dimensional* **F**-*vector spaces, and f and g are maps $V \to W$ with f linear and g p^{-1}-linear (i.e., $g(a^p v) = ag(v)$ for $a \in$ **F** and $v \in V$). Then*

$$V \xrightarrow{f-g} W$$

is surjective if g is.

PROOF. Choose a splitting h of g, i.e., a p-linear map $h : W \to V$ such that $g \circ h = \mathrm{id}_W$. Then the composition

$$W \xrightarrow{h} V \xrightarrow{f-g} W$$

is the difference of a p-linear endomorphism of W and the identity of W and as such is well-known to be surjective.

This completes the proof of Proposition 5.4.

Using the Proposition, we can compute the h^{ij}_W for all $i + j \neq k + 1$. Indeed, by Corollary 4.2, we know the m_{ij} when $i + j \neq k + 1$, and using formula 3.1 we find that when $i + j \neq k + 1$, $h^{ij}_W = 0$ unless k is even and $i = j = k/2$, in which case $h^{k/2, k/2}_W = e_{k,k/2}$. Applying Crew's formula, we have

$$h^{i, k+1-i}_W = \begin{cases} (-1)^{k+1-i}\chi(M, \Omega^i_{\tilde{X}}) & \text{if } i \neq k/2 \\ (-1)^{k+1-i}\chi(M, \Omega^i_{\tilde{X}}) + e_{k,k/2} & \text{if } i = k/2. \end{cases}$$

On the other hand, using the exact sequence of relative differentials and the computation alluded to in the proof of Proposition 5.4, we have

$$\chi(M, \Omega^i_{\tilde{X}}) = (-1)^{k+1-i} \binom{k}{i-1} \chi(I, \Omega^1_I \otimes \omega^{2i-2-k}) + (-1)^{k-i} \binom{k}{i} \chi(I, \omega^{2i-k}).$$

Now using that the degree of ω is w_1 and doing a bit of rearranging, we get Theorem 5.1.

6. A remark on other cases of the main theorems

Our main results, namely the calculation of the slopes and the Hodge numbers associated to the pair $(\tilde{X}, \mathrm{II})$, can be carried out in many other cases. Indeed, we can consider modular forms on $\Gamma_1(p^n N)$ with p odd, N prime to p, and $n \geq 0$. We need only assume that if $p^n \geq 3$, and if $p^n = 3$ then $N \geq 5$. Then Theorems 4.1 and 5.1 are true as stated, provided we make suitable modifications to the definitions of the Hecke polynomials $H_k(T)$ and the integers l_{kj}. Also, for any direct factor of $(\mathbf{Z}/p^n N\mathbf{Z})^\times$ of order prime to p, we can decompose 4.1 and 5.1 for characters of this factor. This is most useful for the factor $(\mathbf{Z}/p\mathbf{Z})^\times$.

7. What remains to be done

As we saw in Section 2, in order to prove a control theorem, i.e., that the number of eigenforms of weight k, level pN and slope λ is bounded independently of k, we need to show that certain contact numbers are bounded independently of k. In this section, we boil this down to an assertion about the torsion numbers T^{ij}.

Changing notation, let $m_{ki} \ 0 \leq i \leq k+1$ be the contact numbers associated to the prime p and the polynomial H_k defined in Section 3. Thus the Hodge polygon associated to $m_{k0}, \ldots, m_{k,k+1}$ is the highest polygon with integral slopes below the Newton polygon of the action of U_p on $S_{k+2}(\Gamma_1(pN))$ modulo the subspace of forms which are new at p but whose character is trivial at p.

A control theorem is then equivalent to the assertion that for fixed i, m_{ki} is bounded independently of k. Clearly we can assume that $k \geq 2i$. Let M_{ki} $0 \leq i \leq k+1$ be the contact numbers associated to $\left(H^{k+1}_{cris}(\tilde{X}_k/W) \otimes \mathbf{Q}_p \right)(\epsilon)$, as in Section 4 (where the subscript on $\tilde{X}_k$ is to emphasize the dependence on k). Then using Theorem 4.1, as soon as $k \geq 2i$, we have

$$M_{ki} = \sum_{0 \leq j \leq i} m_{k-2j,i-j} e_{kj}.$$

Inverting this formula, we have that

$$(7.1) \qquad m_{ki} = \sum_{0 \leq j \leq i} (-1)^j \binom{k-j}{j} M_{k-2j,i-j}.$$

Now for each k, let T_k^i be the torsion number $T^{i,k+1-i}$ associated to the pair $(\tilde{X}_k, \Pi)$. Also recall the integers l_{ki} introduced above Theorem 5.1. Then 7.1 becomes

$$m_{ki} = l_{ki} - \sum_{0 \le j \le i} (-1)^j \binom{k-j}{j} \left(T_k^{i-j} - 2T_k^{i-j-1} + T_k^{i-j-2} \right).$$

Summing up:

THEOREM 7.2. *For a fixed integer λ, if the integers*

$$l_{ki} - \sum_{0 \le j \le i} (-1)^j \binom{k-j}{j} \left(T_k^{i-j} - 2T_k^{i-j-1} + T_k^{i-j-2} \right)$$

are bounded independently of k for all $i \le \lambda$, then the number of eigenforms of weight $k+2$, level pN, and slope $\le \lambda$ is bounded independently of k. Conversely, if the number of such forms is bounded, then the integers above are also bounded for all $i \le \lambda - 1$.

In fact, slightly more is true, namely that a control theorem slopes $\le \lambda'$ where λ' is a real number $< \lambda$ follows from the boundedness of the integers of the theorem for $i \le \lambda - 1$. For example, to obtain a control theorem for slopes $\le \lambda' < 1$, it suffices to prove that

$$T_k^0 \ge kw_1 - b$$

for all k and some constant b. Using the techniques of [U2], it is not hard to show that at least

$$T_k^0 \ge k(1 - 1/p)w_1 - b.$$

In a sense we do not want to make precise, this inequality comes already from working modulo p, i.e., with the Ω_{log}^i; working modulo p^n, i.e., with the $W_n\Omega_{log}^i$ should allow one to prove that

$$T_k^0 \ge k(1 - 1/p^n)w_1 - b$$

and so to obtain the desired result, one should work with all of the $W_n\Omega_{log}^i$.

REFERENCES

[G-M] F. Gouvêa and B. Mazur, *Families of modular eigenforms*, Math. Comp. **58** (1992), 793-805.

[H]   H. Hida, *Elementary theory of L-functions and Eisenstein series*, Cambridge University Press, Cambridge, 1993.

[I]   L. Illusie, *Finiteness, duality, and Künneth theorems in the cohomology of the deRham Witt complex*, Algebraic Geometry (Tokyo/Kyoto) Lecture Notes in Math. vol. 1016, (M. Raynaud and T. Shioda, eds.), Springer, Berlin, 1982, pp. 20-72.

[Ma]   Y. Manin, *The theory of commutative formal groups over fields of finite characteristic*, Russian Math. Surveys **18** (1963), 1-83.

[M]   J. S. Milne, *Values of zeta functions of varieties over finite fields*, Am. J. Math. **108** (1986), 297-360.

[Mi]   T. Miyake, *Modular Forms*, Springer, New York, 1989.

[O]   A. Ogus, *F-crystals, Griffiths transversality, and the Hodge decomposition*, Preprint
         (To appear in Asterisque) (1993).
[Sh]   G. Shimura, *Introduction to the Arithmetic Theory of Automorphic Functions*, Prince-
         ton University Press, Princeton, 1971.
[U1]   D. Ulmer, *On the Fourier coefficients of modular forms*, Preprint (To appear in Ann.
         Sc. Éc. Norm. Sup.) (1992).
[U2]   _____ , *On the Fourier coefficients of modular forms II*, Preprint (1993).

DEPARTMENT OF MATHEMATICS, UNIVERSITY OF ARIZONA, TUCSON, AZ 85721

E-mail address: ulmer@math.arizona.edu

Contemporary Mathematics
Volume **174**, 1994

On the Taylor Coefficients of
Theta Functions of CM Elliptic Curves

FERNANDO RODRIGUEZ VILLEGAS

Dedicated to the memory of Enzo R. Gentile

ABSTRACT. Let θ be the standard theta function associated to a CM elliptic curve E and a nowhere vanishing differential ω on E, both defined over a number field. What is the significance, if any, of the Taylor coefficients of θ at 0? In this paper we do two things. First, we reformulate joint results with D. Zagier showing a relation between these coefficients and square roots of central values of Hecke L-series for certain curves. Second, we give two new proofs of a p-adic interpolation (due to A. Sofer and originally conjectured by N. Koblitz) of those square roots (suitably modified) in the case of good ordinary reduction; one of the proofs uses the relation just mentioned. For simplicity, details are only given for five curves defined over $\mathbf{Q}$; what to expect for arbitrary CM elliptic curves remains unclear, though we suspect there is a more general phenomenon.

CONTENTS

1991 *Mathematics Subject Classification*. Primary 11G40, 11F33; Secondary 11F27, 11G05.
Key words and phrases. theta functions, elliptic curves, p-adic L-functions.
Research of the author was supported in part by an NSF grant
This paper is in final form and no version of it will be submitted for publication elsewhere.

1. Introduction

1) Let (E, ω) be an elliptic curve with complex multiplication together with a nowhere vanishing differential on it, both defined over a number field F. Let $L \subset \mathbf{C}$ be the lattice of periods of ω and $\theta(u, L)$ for $u \in \mathbf{C}$ the standard theta function associated to L (see (2.9) for the exact definition). A fundamental result of Damerell implies that the Taylor coefficients of θ at $u = 0$ are in F. We may ask:

Question: *What is the significance, if any, of the Taylor coefficients of θ at the origin?*

The relation between theta functions and Eisenstein series via logarithmic differentiation has proved to be crucial in the study of the arithmetic of the curve. However, the relation between the Taylor coefficients of a function and those of its logarithmic derivative may be very intricate (consider $1 - e^u$ for example), and it is not a priori clear to us what the Taylor coefficients of θ itself might mean.

Our intention in this paper is to present some evidence for the relevance of the question we have raised. Indeed, in §3 we show how the main results of [15] can be reformulated in this setting, obtaining an answer to the question for the curves $A(l)$ studied by Gross [3], for a prime $l \equiv 3 \bmod 8$. For clarity of exposition we restrict ourselves to the simplest possible cases (five in all) where the curve $A(l)$ is defined over $\mathbf{Q}$. We obtain:

Partial answer: *For the curves $A(l)$ with $l = 11, 19, 43, 67$ or 163 the square of the $(k-1)$-st Taylor coefficient of θ at the origin is essentially the central value $L(\psi^{2k-1}, k)$, where ψ is its associated Hecke character.*

We refer the reader to §3.1 for a description of the five curves in question and to the theorem in §3.2 for the precise statement.

For other curves $A(l)$ with $l \equiv 3 \bmod 8$ prime (no longer defined over $\mathbf{Q}$) the central value is the square of *linear combinations* of Taylor coefficients of theta series associated to Galois conjugates of the curve. At present, is not clear to us what exactly to expect for arbitrary CM elliptic curves, though we suspect there is a more general phenomenon.

2) Koblitz conjectured in [10] that there exists a consistent choice of square roots of suitably modified central values $L^*(\psi^{2k-1}, k)$ having p-adic interpolation properties for all primes $p > 2$ of good ordinary reduction (we emphasize that the choice must be made independently of p). This has interesting implications in the Iwasawa theory of CM elliptic curves (cf. the work of Li Guo [4]).

The conjecture was first proved by A. Sofer in her thesis for all curves $A(l)$ with $l \equiv 3 \mod 4, l > 3$ prime. We present here two new proofs of her results (again we give details for the above curves only). The first is conceptually simple (see §4.4 for a short sketch): for primes $p > 2$ of good ordinary reduction, θ becomes a function on the formal group of E, hence a fortiori on $\hat{\mathbf{G}}_m$, to which we may associate a p-adic measure on $\mathbf{Z}_p$ by Cartier duality. The interpolation now follows from the above relation between the Taylor coefficients of θ and central values. The second is along the lines of [15] and might be of independent interest (see §4.1-4.3). We show how we can associate to any quadratic form Q (positive definite, even integral, and of even rank) a measure on the underlying lattice with values in the ring of Katz' p-adic modular forms (extending an idea of B. Perrin-Riou [14]). For Q the norm form of certain imaginary quadratic fields we find that when evaluated at a specific trivialized elliptic curve the two variable measure associated to Q splits as the product of two identical one variable measures. This p-adic version of the factorization formula (25) of [15] then yields the desired interpolation. The precise statements are given in: §4.1 (general theta measures), §4.2 (complex and p-adic factorization formulas), and §4.3 (final p-adic interpolation).

We begin by collecting in §2 various classical results on modular forms and elliptic functions that we need, including a recursion, essentially due to Jacobi and Weierstrass, for the Taylor coefficients of θ. The reader should compare this recursion with the ones in [15].

In a way, the main theme underlying the whole paper is the heat equation, which allows us to pass from the modular variable (z in the upper-half plane) to the elliptic variable ($u \in \mathbf{C}$).

1.1. Acknowledgments. We would like to thanks N. Katz for his help with the theory of p-adic modular forms.

2. Preliminaries

2.1. Classical Modular Forms. We will consider functions f of pairs of complex numbers (ω_1, ω_2) with $\Im(\omega_2/\omega_1) > 0$. We say that such a function has weight $k \in \mathbf{Z}$ if

$$f(\lambda\omega_1, \lambda\omega_2) = \lambda^{-k} f(\omega_1, \omega_2), \qquad \text{for all non-zero } \lambda \in \mathbf{C}.$$

We let $Sl_2(\mathbf{R})$ act on (ω_1, ω_2) via

$$(\omega_1, \omega_2) \mapsto (a\omega_1 + b\omega_2, c\omega_1 + d\omega_2), \qquad \text{for } \begin{pmatrix} a & b \\ c & d \end{pmatrix} \in Sl_2(\mathbf{R}).$$

A function invariant by $Sl_2(\mathbf{Z})$ depends then only on the lattice $L = [\omega_1, \omega_2] \subset \mathbf{C}$.

A C^∞ function $f(\omega_1, \omega_2)$ fixed by $(\omega_1, \omega_2) \mapsto (\omega_1, \omega_2 + \omega_1)$ has a Fourier expansion

$$f(2\pi i, 2\pi i z) = \sum_{n \in \mathbf{Z}} a_n(y) q^n, \qquad \Im(z) > 0, \quad q = e^{2\pi i z}.$$

This is the usual q-expansion (at $i\infty$) if f is holomorphic, in which case the $a_n(y)$'s are simply constants, and if in addition, f is also holomorphic at $i\infty$ then the a_n's are zero for $n < 0$. Any f of weight $k \in \mathbf{Z}$ may be recovered from its expansion via

$$f(\omega_1, \omega_2) = \left(\frac{2\pi i}{\omega_1} \right)^k \sum_{n \in \mathbf{Z}} a_n(y) q^n, \qquad \text{where } z = \omega_2/\omega_1 \text{ and } q = e^{2\pi i z}.$$

In particular, a classical modular form of weight k on a subgroup of $Sl_2(\mathbf{Z})$ gives rise to a holomorphic function $f(\omega_1, \omega_2)$ of weight k invariant by that subgroup.

Given (ω_1, ω_2) with $\Im(\omega_2/\omega_1) > 0$ we let $L = [\omega_1, \omega_2] \subset \mathbf{C}$ be the associated lattice and define

$$(2.1) \qquad A(L) = \frac{1}{\pi} \text{Area}(\mathbf{C}/L) = (\omega_2 \bar{\omega}_1 - \bar{\omega}_2 \omega_1)/2\pi i,$$

We let W be the differential operator acting on C^∞ functions of (ω_1, ω_2) by

$$(2.2) \qquad W = \frac{-1}{A(L)} \left(\bar{\omega}_1 \frac{\partial}{\partial \omega_1} + \bar{\omega}_2 \frac{\partial}{\partial \omega_2} \right).$$

It does not preserve holomorphicity but does preserve the action of $Sl_2(\mathbf{R})$ and sends functions of weight k to functions of weight $k + 2$. Moreover, if $f(\omega_1, \omega_2)$ has weight $k \in \mathbf{Z}$ then $(Wf)(2\pi i, 2\pi i z) = \partial_k(f(2\pi i, 2\pi i z))$, where ∂_k is the differential operator (acting on C^∞ functions of the upper-half plane)

$$(2.3) \qquad \partial = \partial_k = \frac{1}{2\pi i} \frac{d}{dz} - \frac{k}{4\pi y}, \qquad y = \Im(z) > 0,$$

(notice that $A([2\pi i, 2\pi i z]) = 4\pi y$).

2.2. Formulaire Elliptique. We recall various facts from the classical theory of elliptic functions. Our basic reference will be chapter VI of Jordan's book Cours d'Analyse [7].

Given a lattice $L \subset \mathbf{C}$ we have the Weierstrass functions

$$\sigma(u, L) = u \prod_{\omega \in L \setminus \{0\}} (1 - \frac{u}{\omega}) e^{\frac{u}{\omega} + \frac{u^2}{2\omega^2}}, \qquad u \in \mathbf{C}$$

$\zeta = \sigma'/\sigma$, and $\wp = -\zeta'$, where prime indicates d/du. The function σ has weight -1 in the sense that

(2.4) $\qquad \sigma(\lambda u, \lambda L) = \lambda \sigma(u, L), \qquad$ for all non-zero $\lambda \in \mathbf{C}$.

We let

$$s_2(L) = \lim_{s \to 0^+} \sum_{\omega \in L \setminus \{0\}} \frac{1}{\omega^2 \mid \omega \mid^{2s}},$$

$$g_2(L) = 60 \sum_{\omega \in L \setminus \{0\}} \frac{1}{\omega^4}, \qquad and \qquad g_3(L) = 140 \sum_{\omega \in L \setminus \{0\}} \frac{1}{\omega^6},$$

of weights $2, 4$ and 6 respectively. It turns out that s_2 is not holomorphic while g_2 and g_3 clearly are.

Given (ω_1, ω_2) with $\Im(\omega_2/\omega_1) > 0$ we consider the associated lattice $L = [\omega_1, \omega_2]$, and define $\eta_j = \zeta(u + \omega_j, L) - \zeta(u, L)$ for $j = 1, 2$. We let $z = \omega_2/\omega_1 = x + iy$ and $q = e^{2\pi i z}$.

Remark 2.1. The definition of s_2 is due to Hecke [**5**, 468-474] and does not appear in [**7**]. We only need to know that

(2.5) $$s_2(L) = \frac{\eta_1}{\omega_1} + (\frac{2\pi i}{\omega_1})^2 \frac{1}{4\pi y}.$$

We let D be the differential operator acting on C^∞ functions of (ω_1, ω_2) by

(2.6) $$D = -2(\eta_1 \frac{\partial}{\partial \omega_1} + \eta_2 \frac{\partial}{\partial \omega_2}).$$

Notice that D, unlike W, preserves analyticity. It also preserves the action of $Sl_2(\mathbf{R})$ and sends functions of weight k to functions of weight $k + 2$; in fact, if f has weight k then

(2.7) $$W f = \frac{1}{2} D f - k s_2 f.$$

We have

$$D g_2 = 12 g_3, \qquad D g_3 = \frac{2}{3} g_2^2, \qquad D s_2 = 2 s_2^2 + \frac{1}{6} g_2,$$

and

(2.8) $$D\sigma = \sigma'' + \frac{1}{12} g_2 u^2 \sigma.$$

As Jordan shows [**7**, p. 463] this differential equation for σ is essentially equivalent to the heat equation.

Finally, we let

(2.9) $$\theta(u, L) = e^{-\frac{1}{2} s_2(L) u^2} \sigma(u, L).$$

As a function of u, θ is odd, entire, and its only zeroes are simple zeroes at points of L. It satisfies

$$\theta(u + \omega, L) = \pm e^{(\frac{1}{2}|\omega|^2 + u\omega)/A(L)} \theta(u, L) \qquad (+ \text{ if } \omega \in 2L, - \text{ if not}),$$

and therefore $e^{-\frac{1}{2}|u|^2/A(L)}\,|\theta(u,L)|$ is L-periodic.

We have the following expansions

$$(2.10) \qquad s_2(L) \;=\; (\frac{2\pi i}{\omega_1})^2(\frac{1}{4\pi y} + \frac{-1}{12}(1 - 24\sum_{n=1}^{\infty} nq^n/(1-q^n))),$$

$$(2.11) \qquad g_2(L) \;=\; (\frac{2\pi i}{\omega_1})^4\frac{1}{12}(1 + 240\sum_{n=1}^{\infty} n^3 q^n/(1-q^n)),$$

$$(2.12) \qquad g_3(L) \;=\; (\frac{2\pi i}{\omega_1})^6\frac{-1}{216}(1 - 504\sum_{n=1}^{\infty} n^5 q^n/(1-q^n)),$$

$$(2.13) \qquad \theta(u,L) \;=\; \left(\frac{2\pi i}{\omega_1}\right)^{-1} e^{-\frac{1}{2}\frac{1}{4\pi y}v^2}\frac{\vartheta(v,z)}{\eta^3(z)},$$

where

$$(2.14) \qquad \vartheta(v,z) = \sum_{n\in\mathbf{Z}}(-1)^n e^{\pi i(n+\frac{1}{2})^2 z}e^{(n+\frac{1}{2})v}, \qquad v = \frac{2\pi i}{\omega_1}u,$$

and $\eta = e^{\pi i z/12}\prod_{n=1}^{\infty}(1-q^n)$ is Dedekind's eta function.

2.3. The Taylor coefficients of θ. We let $r_n(L)$ be the n-th Taylor coefficient of θ about $u = 0$; precisely,

$$(2.15) \qquad \theta(u,L) = \sum_{n=0}^{\infty} r_n(L)\frac{u^n}{n!}.$$

Notice that r_n is identically zero for even n.

For odd n, the Fourier expansion of r_n is of the form $\sum_{m=0}^{\infty} a_m^{(n)}(y)q^m$, where $a_m^{(n)}$ is a polynomial in $1/4\pi y$ with rational coefficients and degree $(n-1)/2$. Let $a_{m,0}^{(n)}$ be the constant term of this polynomial, then it is not hard to see from (2.13) that

$$(2.16) \quad \sum_{m=0}^{\infty} a_{m,0}^{(n)}q^m = \frac{1}{2^{n-1}}\frac{\sum_{m=0}^{\infty}(-1)^m(2m+1)^n q^{\frac{1}{2}m(m+1)}}{\prod_{m=1}^{\infty}(1-q^m)^3}, \qquad n \text{ odd}$$

(this is, in fact, the p-adic q-expansion of r_n). In particular, $2^{n-1}a_{m,0}^{(n)} \in \mathbf{Z}$.

By (2.4) r_n has weight $n-1$ and the same is true of the n-th Taylor coefficient of σ, which, in contrast to r_n however, is holomorphic everywhere (including $i\infty$), but its q-expansion is only integral outside 2 and 3. It follows that the n-th Taylor coefficient of σ is an isobaric polynomial in g_2 and g_3 of degree n. This together with the differential equation (2.8) allowed Weierstrass [17, V, p. 38] to give a simple recursion for these polynomials. However, we want a recursion for the Taylor coefficients of θ, which are polynomials in s_2, g_2 and g_3. We indicate below one way of doing this (compare with Jacobi [6, II, p.393]).

LEMMA 2.1. *Let c be any C^∞ function of (ω_1, ω_2) independent of u. For (ω_1, ω_2) with $\Im(\omega_2/\omega_1) > 0$ let*

$$f(u, \omega_1, \omega_2) = e^{cu^2} \sigma(u, [\omega_1, \omega_2]) = \sum_{n=0}^{\infty} c_n(\omega_1, \omega_2) \frac{u^{2n+1}}{(2n+1)!}.$$

Then f satisfies the differential equation

$$Df = f'' - 4cuf' + (4c^2 + Dc + \frac{1}{12}g_2)u^2 f - 2cf,$$

where prime means d/du and D is the derivation defined in (2.6). Furthermore, the Taylor coefficients c_n satisfy, and are uniquely determined by, the following recursion

$$c_0 = 1$$
$$c_1 = 6c$$
$$c_{n+1} = Dc_n + 2c(4n+3)c_n - (2n+1)2n(4c^2 + Dc + \frac{1}{12}g_2)c_{n-1}, \text{ for } n > 0.$$

PROOF. It follows from a calculation using the differential equation (2.8) that we leave to the reader. $\square$

PROPOSITION 2.2. *Fix a lattice $L \subset \mathbf{C}$. Let $s_2, g_2, g_3, \theta(u)$ be as in §1.2 and let*

$$\theta(u) = \sum_{n=0}^{\infty} c_n \frac{u^{2n+1}}{(2n+1)!}$$

be the Taylor expansion of θ about $u = 0$. Let $\Delta = g_2^3 - 27g_3^2$, $R_0 = \mathbf{Z}[\frac{1}{6}, s_2, \Delta]$, and $R = R_0[x, y]/(x^3 - 27y^2 - \Delta)$ with x, y indeterminates. Finally, let $\mathcal{D}$ be the derivation of R/R_0 satisfying $\mathcal{D}(x) = 12y$ (hence $\mathcal{D}(y) = \frac{2}{3}x^2$) and let $C_n(x, y) \in R$ be defined recursively by

$$C_0 = 1$$
$$C_1 = -3s_2$$
$$C_{n+1} = \mathcal{D}C_n - s_2(4n+3)C_n - (2n+1)2n(s_2^2 + \frac{x}{12})C_{n-1}, \qquad \text{for } n > 0.$$

Then $c_n = C_n(g_2, g_3)$.

PROOF. A calculation shows that $D(g_2^3 - 27g_3^2) = 0$; now our claim follows from the previous lemma using $c = s_2$ as a constant. $\square$

Remark 2.2. If we take $L = [2\pi i, 2\pi z]$ and let $z \to i\infty$ then $q \to 0$ and by the expansions of §1.2 $s_2 \to -1/12$, $g_2 \to 1/12$, and $g_3 \to -1/216$. Also by (2.16) we have $r_n \to 1/2^{n-1}$ for all $n \geq 0$; therefore, if we run the recursion with $s_2 = -1/12$, and $\Delta = 0$ we should get $C_n(1/12, -1/216) = 1/4^n$ for all $n \geq 0$. This can be easily verified for small n providing a consistency check on our formulas.

3. Taylor Coefficients of θ and Square Roots of Central Values

In this section we reformulate the main theorem of [15] for the simplest cases of class number one, showing that the central value $L(\psi^{2k-1}, k)$ is essentially the square of the $(k - 1)$-st Taylor coefficient of the associated theta function. We start by recalling various facts about the elliptic curves in question and state the theorem in (3.5).

3.1. The Elliptic Curves $A(l)$. For a prime $l \equiv 3 \bmod 4$ let $K = \mathbf{Q}(\sqrt{-l})$ and let $A(l)$ be the elliptic curve studied by Gross in [3], which has CM by the ring of integers of K. We will only consider the cases where $l > 7$ and K has class number 1, namely, $l = 11, 19, 43, 67$, and 163; the curve $A(l)$ is then defined over $\mathbf{Q}$ and $A(l)(\mathbf{Q}) \cong \mathbf{Z}$.

For the sake of completeness we include a table of minimal models taken directly from [3, pp. 82, 86]. We use the standard notation for a generalized Weierstrass model

$$y^2 + a_1 xy + a_3 y = x^3 + a_2 x^2 + a_4 x + a_6.$$

l	a_1	a_2	a_3	a_4	a_6
11	0	-1	1	-7	10
19	0	0	1	-38	90
43	0	0	1	-860	9707
67	0	0	1	-7370	243528
163	0	0	1	-2174420	1234136692

The period lattice of the Neron differential $\omega = dx/(2y + 1)$ is

(3.1) $$L = \Omega_{\mathbf{R}} \mathcal{D}^{-1}, \qquad \text{where } \mathcal{D} = (\sqrt{-l}),$$

and

$$\Omega_{\mathbf{R}} = \int_{A(l)(\mathbf{R})} |\omega|$$

is its positive fundamental real period. This period can be given explicitly by

(3.2) $$\Omega_{\mathbf{R}} = (2\pi)^{(3-l)/4} \prod_{\substack{0 < r < l \\ (\frac{r}{l}) = 1}} \Gamma(r/l) = 2\pi l^{1/4} \mid \eta(\frac{1 + \sqrt{-l}}{2}) \mid^2,$$

where Γ is the gamma function and η is Dedekind's eta function. The first formula is due to Lerch and Chowla-Selberg and the second follows from the fact that the discriminant of ω is $-l^3$.

$A(l)$ has good reduction everywhere except at l, where it has additive reduction of Kodaira type III, and hence its conductor is l^2.

Let now $\mathcal{O}_K$ be the ring of integers of K and ψ the unique Hecke character of K that satisfies

(3.3) $$\psi((\alpha)) = \varepsilon(\alpha)\alpha, \qquad \text{for } \alpha \in \mathcal{O}_K \quad \text{prime to } l,$$

where ε is the quadratic character of K of conductor $\mathcal{D}$. Then the L-series of $A(l)/\mathbf{Q}$ is $L(\psi, s)$.

We refer the reader to [3] for proofs of the above assertions about the curves $A(l)$.

We give below a table of the relevant quantities associated to the lattices (3.1); for s_2, g_2 and g_3 we used the q-expansions of §1.2 choosing $\omega_1 = \Omega_{\mathbf{R}}/\sqrt{-l}$, with $\Omega_{\mathbf{R}}$ as in (3.2), and $z = (1 + \sqrt{-l})/2$; notice that then $A(L) = \Omega_{\mathbf{R}}^2/(2\pi\sqrt{l})$. Naturally, we could also have gotten g_2 and g_3 directly from the minimal model for $A(l)$.

l	s_2	g_2	g_3	$\Omega_{\mathbf{R}}$	$A(L)$
11	-2/3	88/3	-847/27	4.80242132 ...	1.10673668 ...
19	-2	152	-361	4.19055001 ...	0.64118800 ...
43	-12	3440	-38829	2.89054107 ...	0.20278890 ...
67	-38	29480	-974113	2.10882279 ...	0.08646949 ...
163	-724	8697680	-4936546769	0.79364722 ...	0.00785201 ...

Also, if $\theta(u, L) = \sum_{n=0}^{\infty} c_n u^{2n+1}/(2n+1)!$ then, using the recursion of proposition 2.2 with the values of s_2, g_2 and g_3 tabulated above, we get the following values for c_n.

$n \backslash l$	11	19	43	67	163
0	1	1	1	1	1
1	2	6	36	114	2172
2	-8	-16	440	6920	3513800
3	14	-186	-19026	-156282	3347376774
4	304	4176	-8352	-34999056	-238857662304
5	-352	-33984	33708960	3991188960	-3941159174330400

Remark 3.1. Notice that the coefficients c_n appear to be integers when a priori (by (2.16)) we expect them to be, at best, in $\mathbf{Z}[\frac{1}{2}]$; we will prove this fact in §4.2.

3.2. The Formula. Let ψ be the Hecke character of K defined in (3.3). For any positive integer k we may consider the L-series $L(\psi^{2k-1}, s)$; it has a functional equation as $s \mapsto 2k - s$ with root number $(-1)^k$. We will be interested in the central values $L(\psi^{2k-1}, k)$ for even k (by the above considerations these values are zero for odd k).

THEOREM 3.1. *Let* $l = 11, 19, 43, 67,$ *or* 163, $K = \mathbf{Q}(\sqrt{-l})$, *and* $A(l)/\mathbf{Q}$ *the elliptic curve of* §2.1. *Let* ψ *be the Hecke character of* K *associated to* $A(l)$, L *the* *lattice of periods of a Neron differential on $A(l)$, and $\Omega_{\mathbf{R}}$ the positive fundamental real period (these are described in (3.3), (3.1), and (3.2) respectively). Let $A(L)$, $\theta(u, L)$, and $r_n(L)$ be as in (2.1), (2.9), and (2.15) respectively, so that*

$$(3.4) \qquad\qquad \theta(u, L) = \sum_{n=0}^{\infty} r_n(L) \frac{u^n}{n!},$$

is the Taylor expansion of θ about $u = 0$. Then

$$(3.5) \qquad \frac{2(k-1)!}{A(L)^{k-1}\Omega_{\mathbf{R}}} L(\psi^{2k-1}, k) = r_{k-1}(L)^2, \qquad \text{for all integers } k \geq 1.$$

PROOF. Using the expansion for θ in (2.13), it is not hard to see that our claim follows from the main theorem of [15]; we leave the details of the calculation to the reader. $\square$

Remark 3.2. It is not very difficult to extend the theorem to all curves $A(l)$ with $l \equiv 3 \mod 8$ prime (cf. the introduction).

Remark 3.3. Note that the theorem is trivial for odd k since in that case both sides of the equation are zero.

Remark 3.4. We may view this result as a sort of abelian version of Waldspurger's theorem for these particular cases. That is: not only central values of L-series are essentially squares of algebraic numbers in a fixed finite extension of $\mathbf{Q}$, but there are systematic ways of choosing their square roots along certain families so that they are coefficients in the expansion of a particular sort of function. In the case of Waldspurger this family is formed by the quadratic twists of the L-series of a fixed modular form f of integral weight and the systematic choice determined by a form of half integral weight in Shimura correspondance with f .

In our case the family is given by the L-series of odd powers of a fixed Hecke character ψ (which we can think as twisting $L(\psi, s)$ by powers of $\psi/\bar{\psi}$) and the systematic choice determined by a theta function.

Remark 3.5. The relation between $\theta(u, L)$ and values of L-series for CM elliptic curves via the logarithmic derivative is well known and plays a crucial role in the arithmetic of these curves. We are not aware of any previous arithmetic interpretation of the Taylor coefficients of $\theta(u, L)$ itself.

Remark 3.6. Combining the theorem with the fact that $e^{-\frac{1}{2}|u|^2/A(L)} |\theta(u, L)|$ is L-periodic (cf. §2.2) we get the bound $L(\psi^{2k-1}, k) \leq C\sqrt{k}$, with C a constant independent of k. This type of bound is the same one gets by applying convexity arguments to the L-series directly.

4. Interpolation

4.1. *P*-adic Modular Forms and Theta Measures. We start with a brief discussion of theta funtions of positive quadratic forms with coefficients. We refer the reader to [13, ch. VI] for details.

Let (V, Q) be a positive definite quadratic space over $\mathbf{Q}$ of dimension $2k$, i.e: a $\mathbf{Q}$ vector space V of dimension $2k$ together with a positive definite quadratic form $Q : V \longrightarrow \mathbf{Q}$, and let $B(x, y) = \frac{1}{2}(Q(x+y) - Q(x) - Q(y))$ be the associated bilinear form. We consider even integral lattices $L \subset V$ of a rank $2k$; i.e: lattices of full rank for which $Q(m)$ is an even integer for every $m \in L$. Given a basis $v_1, \cdots, v_{2k}$ of L we form the Gram matrix $A = (B(v_i, v_j))$ and define the level of L to be the least positive integer N such that NA^{-1} is integral with even diagonal entries (this does not depend on the choice of basis). Also, we let χ be the character of $\mathbf{Q}(\sqrt{(-1)^k N})/\mathbf{Q}$. Then the theta series

$$\Theta_Q(z) = \sum_{m \in L} q^{\frac{1}{2}Q(m)}, \qquad \Im(z) > 0, \quad q = e^{2\pi i z},$$

is a modular form of weight k on $\Gamma_0(N)$ with character χ.

Given a function $f : L \to \mathbf{C}$, the series

$$\sum_{m \in L} f(m) q^{\frac{1}{2}Q(m)},$$

will not in general be a modular form of any level unless f is of a special type, two typical examples being: (1) f given by congruence conditions (increasing the level but not the weight), and (2) f a harmonic polynomial with respect to Q (increasing the weight but not the level).

In contrast, if we consider the same question p-adically we find that there is no restriction on f as long as it is continuous, since any such f can be approximated by functions of type (1). Before stating the precise result we need to introduce some notation. We refer the reader to [8] and [9] for details on p-adic modular forms and their use in p-adic interpolations.

For the rest of the paper we fix the following: $\bar{\mathbf{Q}}$ an algebraic closure of $\mathbf{Q}$, p a prime number, $\mathbf{C}_p$ the completion of an algebraic closure of $\mathbf{Q}_p$ and two embeddings $\iota_p : \bar{\mathbf{Q}} \mapsto \mathbf{C}_p$, and $\iota_\infty : \bar{\mathbf{Q}} \longrightarrow \mathbf{C}$. In what follows we will tacitly use these embeddings to pass to and from complex and p-adic settings. We let $\mathcal{O}$ be the ring of integers of $\mathbf{C}_p$.

Let $A = \bar{\mathbf{Q}} \cap \iota_p^{-1}(\mathcal{O})$ and for integers k and N, with $N \geq 1$ prime to p, let $M_k(\Gamma_1(Np^\infty), A)$ be the space of true modular forms of weight k on $\Gamma_1(Np^n)$ for some $n \geq 0$, which are defined over A in the sense of Katz [8, 2.4]. Concretely, these correspond (via ι_∞ on q-expansions) to classical modular forms of weight k on $\Gamma_1(Np^n)$ for some $n \geq 0$ whose q-expansion (at $i\infty$) have coefficients in $\iota_\infty(A)$.

There is an inclusion of $M_k(\Gamma_1(Np^\infty), A)$ into V, the full ring of Katz' modular forms on $\Gamma_1(N)$ with coefficients in $\mathcal{O}$, which preserves q-expansions (via ι_p).

We let $M_k(\Gamma_1(Np^\infty), \mathcal{O})$ be the closure of $M_k(\Gamma_1(Np^\infty), A)$ in V (the inherited topology is then that of uniform limits of q-expansions).

We may now turn to the theorem, which was inspired by a similar statement in [14, 2.2.1].

THEOREM 4.1. *Let p, $\mathcal{O}$ and $M_k(\Gamma_1(Np^\infty), \mathcal{O})$ be as above, with $N \geq 1$ prime to p. Let (V, Q) be a positive definite quadratic space over $\mathbf{Q}$ of dimension $2k$, $L \subset V$ a (positive definite) even integral lattice of rank $2k$ and level N. Let $L_p = L \otimes \mathbf{Z}_p$ and $Cont(L_p, \mathcal{O})$ the space of continuous functions on L_p with values in $\mathcal{O}$. Then the following map defines a continuous linear map and hence a measure on L_p with values in $M_k(\Gamma_1(Np^\infty), \mathcal{O})$.*

$$Cont(L_p, \mathcal{O}) \longrightarrow M_k(\Gamma_1(Np^\infty), \mathcal{O})$$

$$f \mapsto \sum_{m \in L} f(m) q^{\frac{1}{2}Q(m)}.$$

PROOF. It suffices to approximate f by locally constant A-valued functions (recall that $A = \bar{\mathbf{Q}} \cap \iota_p^{-1}(\mathcal{O})$) and notice that these give forms in $M_k(\Gamma_1(Np^\infty), A)$. $\square$

We will denote this measure by μ_L and its value on f by $\int_{L_p} f d\mu_L$.

Given a triple $\kappa = (E, \varphi, \beta)$, consisting of an elliptic curve E, an isomorphism of formal groups $\varphi : \hat{E} \longrightarrow \hat{\mathbf{G}}_m$, and an arithmetic $\Gamma_1(N)$ level structure β, all defined over $\mathcal{O}$, we may evaluate at κ to obtain a measure $\mu_{L,\kappa}$ on L_p with values in $\mathcal{O}$.

4.2. P-adic Factorization Formulas. Now we specialize the results of the last section to binary lattices. More specifically, let $l = 11, 19, 43, 67$, or 163 and $K \subset \bar{\mathbf{Q}}$ be the quadratic imaginary field of discriminant $-l$ and $\mathcal{O}_K$ its ring of integers. The level N of §4.1 will now be l, which we assume is not p. The symbol $\sqrt{-l}$ will always denote that square root of $-l$ in K for which $\Im(\iota_\infty(\sqrt{-l})) > 0$. Consider $(K, 2\mathbf{N}_{K/\mathbf{Q}})$ as quadratic space over $\mathbf{Q}$; then $\mathcal{O}_K$ is an even integral lattice of level l and rank 2. Let $\mu_{\mathcal{O}_K}$ be the associated measure given by theorem 4.1.

We want to interpret p-adically the factorization formula (25) of [15] as saying that the measure $\mu_{\mathcal{O}_K}$ evaluated at a particular triple κ splits as the product of two identical measures. Let us start by recalling the factorization formula.

For non-negative even integers h we define

$$(4.1) \qquad \Theta^{(h)} = \sum_{\alpha \in \mathcal{O}_K} \alpha^h q^{\mathbf{N}(\alpha)} \in \mathbf{Z}[[q]].$$

Then $\Theta^{(h)}(z)$ (with $q = e^{2\pi i z}$) is a classical modular form of weight $h + 1$ on $\Gamma_0(l)$ and character $\left(\frac{*}{l}\right)$ (this is a particular case of example (2) of §4.2). Notice that the q-expansion coefficients of $\Theta^{(h)}$ at $i\infty$ are indeed in $\mathbf{Z}$. For odd h we set $\Theta^{(h)} \equiv 0$.

THEOREM 4.2. *Let* $l = 11, 19, 43, 67,$ *or* 163, $K \subset \bar{\mathbf{Q}}$ *the imaginary quadratic field of discriminant* $-l$, *and* L *as in (3.1) the lattice of periods of a Neron differential of the elliptic curve* $A(l)$ *of §2.1. For* $n \geq 0$ *let* $r_n = r_n(L)$ *be the Taylor coefficients of* $\theta(u, L)$ *as in (2.15). Then for all non-negative integers* j *and* h,

$$(4.2) \qquad \left(\frac{2\pi i}{\Omega_{\mathbf{R}}}\right)^{2j+h+1} \partial^j \Theta^{(h)}\left(\frac{l + \sqrt{-l}}{2l}\right) = r_j r_{j+h} \sqrt{-l},$$

where $\Theta^{(h)}, \partial,$ *and* $\Omega_{\mathbf{R}}$ *are defined in (4.1), (2.3) and (3.2) respectively and* $\Im(\sqrt{-l}) > 0$.

PROOF. This is a restatement of (25) of [**15**] applied to these cases. □

COROLLARY 4.3. *With the notation of the theorem* $r_j \in \mathbf{Z}$ *for all* $j \geq 0$.

PROOF. On one hand, the recursion of §2.4 shows that r_j is integral outside 2 and 3. On the other, we can rewrite (4.2) for $j = 1$ in the homogeneous form (see §1.1)

$$W\Theta^{(k)}\left(\Omega_{\mathbf{R}}, \Omega_{\mathbf{R}} \frac{l + \sqrt{-l}}{2l}\right) = r_{k+1}\sqrt{-l},$$

since $r_1 = 1$ in all cases; then the cohomological description of the operator W (see [**8**, IV]) guarantees that r_{k+1} is integral outside l proving that $r_j \in \mathbf{Z}$ as claimed.

Concretely, we may decompose the operator W (in its non-homogeneous form (2.3)) as:

$$\partial_k = \left(\frac{1}{2\pi i}\frac{d}{dz} - k\phi\right) + k\left(\phi - \frac{1}{4\pi y}\right),$$

where $\phi(z) = (rE_2(rz) + E_2(z))/24$, $E_2(z) = 1 - 24\sum_{n=1}^{\infty} nq^n/(1-q^n)$, $q = e^{2\pi i z}$, $y = \Im(z) > 0$, and r is an auxiliary prime (compare with (2.7)). We then verify our claim using the values of $s_2(L)$ of §3.1 and chosing r appropriately. We leave the details to the reader. □

Remark 4.1. We should point out that this corollary is not entirely trivial. There is no a priori reason why the powers of 2 in the denominator of the Fourier expansion of $r_n(L)$ (cf. (2.16)) should cancel out (and indeed they do not for the curve $A(7)$ for example). The reason they do is probably related to the fact that for primes inert in K the valuations of values of modular forms typically grows along families. On the other hand, our proof works because the theorem relates r_j to values of other modular forms, which do have integral q-expansions.

Recall that ω is a Neron differential on $A(l)$ with period lattice L. The choice of basis $\Omega_{\mathbf{R}}[1, (l + \sqrt{-l})/2l]$ of L determines an arithmetic $\Gamma_1(l)$ structure β on $A(l)$, defined over $\mathbf{Z}[1/l, \zeta_l]$, where $\zeta_l \in \bar{\mathbf{Q}}$ is a primitive l-th root of unity (see [**8**, chap. II]).

¿From now on we assume that $A(l)$ has good ordinary reduction at p or, equivalently, that p splits in $K = \mathbf{Q}(\sqrt{-l})$, so that, in particular, $p > 2$. We let $\mathcal{P} = \iota_p^{-1}(p\mathcal{O}) \cap \mathcal{O}_K$ be one of the primes of K above p and $\pi = \psi(\mathcal{P})$ the generator of $\mathcal{P}$ given by the Hecke character ψ of (3.3); the other prime is then $\bar{\mathcal{P}}$, with generator $\bar{\pi} = \psi(\bar{\mathcal{P}})$.

Let $K^\infty \subset \bar{\mathbf{Q}}$ be the union of the ray class fields of K of conductor $l\bar{\mathcal{P}}^n$ for some $n \geq 0$, $K_{\mathcal{P}}^\infty$ be its completion in $\mathbf{C}_p$, and $\mathcal{O}_{\mathcal{P}}^\infty$ be the ring of integers of $K_{\mathcal{P}}^\infty$. Finally, let $\sigma \in Gal(K^\infty/K)$ be the Artin symbol at $\mathcal{P}$; it induces the Frobenius automorphism of $K_{\mathcal{P}}^\infty/\mathbf{Q}_p$.

We choose an isomorphism of formal groups over $\mathcal{O}_{\mathcal{P}}^\infty$

$$\varphi : \hat{A}(l) \longrightarrow \hat{\mathbf{G}}_m,$$

and let c be the associated p-adic period; i.e: $c \in \mathcal{O}_{\mathcal{P}}^{\infty \times}$ is such that the pull-back via φ of the standard differential on $\hat{\mathbf{G}}_m$ is $c\omega$, then

(4.3) $$c^{\sigma-1} = \bar{\pi}.$$

See [**8**, 8.3] for this setup.

We have now a triple $\kappa = (A(l), \varphi, \beta)$ over $\mathcal{O}_{\mathcal{P}}^\infty \subset \mathcal{O}$ at which we can evaluate the measure $\mu_{\mathcal{O}_K}$ obtaining the measure $\mu_{\mathcal{O}_K,\kappa}$. It will be convenient to consider instead the measure $\mu_{\mathcal{O}_K,\kappa}/\sqrt{-l}$ (whose value on a function is that of $\mu_{\mathcal{O}_K,\kappa}$ divided by $\sqrt{-l}$).

Finally, we choose an isomorphism $\nu : \mathcal{O}_K \otimes \mathbf{Z}_p \longrightarrow \mathbf{Z}_p \oplus \mathbf{Z}_p$ so that if x, y are the standard coordinates functions on $\mathbf{Z}_p \oplus \mathbf{Z}_p$, then for all $\alpha \in \mathcal{O}_K$, $x \circ \nu(\alpha) = \iota_p(\bar{\alpha})$ and $y \circ \nu(\alpha) = \iota_p(\alpha)$. To ease the notation we denote by μ^* and μ respectively the measures μ_{OK} and $\mu_{\mathcal{O}_K,\kappa}/\sqrt{-l}$ transported to $\mathbf{Z}_p \oplus \mathbf{Z}_p$ via ν.

We fix the notation and hypothesis above for the rest of the paper. We can now reformulate the previous theorem p-adically.

THEOREM 4.4. *For all non-negative integers j and k we have*

(4.4) $$c^{j+k+1} \int_{\mathbf{Z}_p \oplus \mathbf{Z}_p} x^j y^k d\mu = r_j r_k.$$

PROOF. We may assume that $j \leq k$ since both sides of (4.4) are symmetric in j and k. It is easy to check then that

$$\int_{\mathbf{Z}_p \oplus \mathbf{Z}_p} x^j y^k d\mu^* = (q\frac{d}{dq})^j \Theta^{(k-j)},$$

as modular forms in V, by comparing their q-expansions (via ι_p and ι_∞). Also, in homogeneous form (see §1.1), theorem 4.2 gives

$$W^j \Theta^{(k-j)}(\Omega_{\mathbf{R}}, \Omega_{\mathbf{R}} \frac{l+\sqrt{-l}}{2l}) = r_j r_k \sqrt{-l}.$$

Now the theorem follows from the comparison theorem 8.09 and the formula 5.10.1 of [**8**] by evaluating at the chosen triple κ. $\square$

We immediately get the following corollary.

COROLLARY 4.5. *Let $c_0 \in \mathcal{O}^\times$ be a square root of c. Then there exists an $\mathcal{O}$-valued measure μ_0 on $\mathbf{Z}_p$ such that for all non-negative integers j*

$$(4.5) \qquad\qquad c_0^{2j+1} \int_{\mathbf{Z}_p} x^j d\mu_0 = r_j,$$

and

$$\mu = \mu_0 \bigoplus \mu_0,$$

as measures on $\mathbf{Z}_p \bigoplus \mathbf{Z}_p$.

PROOF. Given a continuous function f on $\mathbf{Z}_p$ with values in $\mathcal{O}$ we define

$$\int_{\mathbf{Z}_p} f \, d\mu_0 = c_0^3 \int_{\mathbf{Z}_p \bigoplus \mathbf{Z}_p} f(x) y \, d\mu.$$

This clearly gives a well defined measure satisfying (4.5) since $r_1 = 1$ in all cases. Now the measures μ and $\mu_0 \bigoplus \mu_0$ agree on all functions $x^j y^k$ and hence, by Mahler's theorem, agree on all f. □

Remark 4.2. The measure is only defined up to ± 1 and we found no way of making a canonical choice.

Remark 4.3. The measure μ^* is not a product measure; what we have shown is that it is when evaluated at a particular κ.

4.3. P-adic Interpolation. We fix a choice of c_0 as in the above corollary hence obtaining a choice of measure μ_0. In the usual manner, we restrict μ_0 to $\mathbf{Z}_p^\times$ (still calling it μ_0) to obtain the p-adic interpolation of suitable variants of r_j.

PROPOSITION 4.6. *Let π and $\bar\pi$ be the generators of the primes of K above p described earlier. Then for all non-negative integers j*

$$(4.6) \qquad\qquad c_0^{2j+1} \int_{\mathbf{Z}_p^\times} x^j d\mu_0 = (1 - \bar\pi^{-1-j}\pi^j) r_j.$$

PROOF. By definition,

$$\int_{p\mathbf{Z}_p} x^j d\mu_0 = c_0^3 \int_{p\mathbf{Z}_p \bigoplus \mathbf{Z}_p} x^j y \, d\mu.$$

On the other hand,

$$\int_{p\mathbf{Z}_p \bigoplus \mathbf{Z}_p} x^j y \, d\mu^* = \bar\pi \pi^j \sum_{\alpha \in \mathcal{O}_K} \bar\alpha^j \alpha \, q^{p\mathbf{N}(\alpha)}$$
$$= \bar\pi \pi^j \, \mathrm{Frob}\Big(\sum_{\alpha \in \mathcal{O}_K} \bar\alpha^j \alpha \, q^{\mathbf{N}(\alpha)} \Big)$$

where Frob is the Frobenius map of V ([**8**, 5.5]). When we evaluate at κ we get $\bar\pi \pi^j c^{-\sigma(j+2)} r_j$ where σ is the Artin map at $\mathcal{P}$ as described in §4.2 ([**8**, 8.3]). A calculation using (4.3) gives our claim. □

Remark 4.4. If we let $a_p = \pi + \bar{\pi}$ then the proposition implies the following congruences

$$r_{j+p-1} \equiv a_p r_j \bmod p, \qquad \text{for all } j \geq 1.$$

Remark 4.5. If for some prime $p > 2$, split in $\mathbf{Q}(\sqrt{-l})$ we find that r_j is not zero modulo p for $j = 1, 3, 5, \ldots, p - 2$ then the above congruences imply that r_j is non-zero for all odd j. In our case we find that we may take e.g. $p = 3, 5, 13, 23, 41$ for $l = 11, 19, 43, 57, 163$ respectively, proving that all $r'_j s$ under consideration are in fact non-zero for odd j.

4.4. *P*-adic θ.

As we mentioned in the introduction there is another way of proving the interpolation properties of r_j, which we describe very briefly now and hope to investigate further in a future publication.

To give an $\mathcal{O}$-valued measure on $\mathbf{Z}_p$ is the same as giving a function on the formal multiplicative group $\hat{\mathbf{G}}_m$ over $\mathcal{O}$; given such a function f, the integral of x^j against its associated measure is the value of $D^j f$ at the identity (where D is the standard derivation on $\hat{\mathbf{G}}_m$), see e.g. [9].

The function θ is not a function on the elliptic curve but it turns out that in the case of good ordinary reduction (and $p > 2$) it is in fact a function on the formal group of the curve (see [11] and also [1], [2], [12] and their references). Moreover, in that case the formal group of the curve is (non-canonically) isomorphic to $\hat{\mathbf{G}}_m$ over $\mathcal{O}$ and we may therefore associate a measure to θ whose moments are the r_j's.

REFERENCES

1. I. Barsotti. Considerazione sulle funzione theta. Symp. Math. **3**, 23-63, (1970).
2. V. Cristante. *p*-adic theta series with integral coefficients. Astérisque **119-120**, 169-182,(1984).
3. B. Gross. Arithmetic of elliptic curves with complex multiplication. Springer Lecture Notes **776**, Springer Verlag, (1979).
4. L. Guo. General Selmer groups ans critical values of Hecke $L-$functions, to appear in Math. Ann.
5. E. Hecke. Math. Werke. Vandenhoeck & Ruprecht, Göttingen, (1959).
6. C. G. J. Jacobi. Gesammelte Werke. Chelsea Publ. Co. (1969).
7. C. Jordan. Cours d'Analyse Chap. VI Gauthier-Vilars, Paris, (1894).
8. N. Katz. *p*-adic interpolation of real analytic Eisenstein series. Annals of Math. **104**, 459-571, (1976).
9. N. Katz. *p*-adic *L*-functions, Serre-Tate local moduli, and ratios of solutions of differential equations. Proc. of the Int. Congress of Mathematicians, Helsinki, 365-371, (1978).
10. N. Koblitz, p-Adic congruences and modular forms of half integer weight. Math. Ann. **274** (1986) 199-220.
11. B. Mazur and J. Tate. *p*-adic sigma function. Duke Math. Journal (3) **62** (3), 663-688, (1991).
12. P. Norman. *P*-adic theta functions. Amer. Journal of Math. **107** (3), 617-661, (1985).
13. A. Ogg. Modular forms and Eisenstein series. NY, Benjamin, (1969).
14. B. Perrin-Riou. Points de Heegner et dérivéées de fonctions *L* *p*-adiques. Invent. Math. **89**, 455-510, (1987).

15. F. Rodriguez Villegas and D. Zagier. Square roots of central values of L-series. Proceedings of the Third Conference of the Canadian Number Theory Assoc., Kingston Ontario, 1991. F. Gouvea and N. Yui eds. Clarendon Press, Oxford, (1993).

16. A. Sofer. p-adic interpolation of square roots of central values of Hecke L-series. Thesis, The Ohio State University, 1993.

17. K. Weierstrass. Math. Werke Berlin, Mayer & Müller, (1915).

DEPARTMENT OF MATHEMATICS, PRINCETON UNIVERSITY, PRINCETON, NEW JERSEY 08544
E-mail address: villegas@math.princeton.edu

Contemporary Mathematics
Volume **174**, 1994

TORSION GROUPS OF ELLIPTIC CURVES OVER CUBIC AND CERTAIN BIQUADRATIC NUMBER FIELDS

HORST G. ZIMMER

ABSTRACT. We report on the determination of all torsion groups of elliptic curves with integral j–invariant over cubic and certain biquadratic number fields. In the cases of torsion groups which occur only for a finite set of elliptic curves over a finite collection of cubic or biquadratic fields, all these curves and fields can be computed and tabulated. Some examples are given.

1. INTRODUCTION

The determination of all torsion groups of elliptic curves E over a number field K of fixed degree $n = [K : \mathbb{Q}]$ over the field $\mathbb{Q}$ of rational numbers is an interesting problem in view of the uniform boundedness conjecture. This conjecture states that there is a bound B_K, depending only on K, such that the order of the torsion group $E_{tors}(K)$ of the Mordell-Weil group $E(K)$ is less than B_K as E varies over all elliptic curves over a fixed, but arbitrary number field K (see [C]). A strong version of the boundedness conjecture asserts that the bound B_K depends only on the degree n of K over $\mathbb{Q}$. The uniform boundedness conjecture was proved by Mazur [M] for $K = \mathbb{Q}$ and by Kamienny [K1], [K2] for quadratic number fields K. Later on Kamienny and Mazur [K3], [KM] extended Kamienny's proof to elliptic curves over number fields K of degree $n = [K : \mathbb{Q}] \leq 8$, and Abramovich [A] modified their method in order to verify the conjecture for number fields K of degree $n \leq 14$.

For $K = \mathbb{Q}$, Mazur [M] proved an assertion which is much stronger than the boundedness conjecture in this case.

1991 *Mathematics Subject Classification.* Primary 11G05, 11G07; Secondary 11Y50.
Work on this topic was supported by DFG grant Zi 329/1-5.
This paper is in final form and no version of it will be submitted for publication elsewhere.

Theorem (Mazur). *Let E be an elliptic curve defined over $\mathbb{Q}$. Then E has one of the torsion groups*

$$E_{tors}(\mathbb{Q}) \cong \left\{ \begin{array}{ll} \mathbb{Z}/m\mathbb{Z} & \text{for } 1 \leq m \leq 12, \ m \neq 11 \\ \mathbb{Z}/2\mathbb{Z} \times \mathbb{Z}/2\mu\mathbb{Z} & \text{for } 1 \leq \mu \leq 4 \end{array} \right\}.$$

Whereas over the rational number field $\mathbb{Q}$, there are no elliptic curves E with a torsion point of order 11, there exist curves E over quadratic number fields K whose torsion group is $E_{tors}(K) \cong \mathbb{Z}/11\mathbb{Z}$. Indeed, Reichert [R] constructed elliptic curves E over quadratic fields K with cyclic torsion groups

$$E_{tors}(K) \cong \mathbb{Z}/m\mathbb{Z} \text{ for } 11 \leq m \leq 18, \ m \neq 17.$$

His construction is based on the calculation of rational points of the modular curves $X_1(m)$ over K. It follows from Kamienny's result [K1], [K2] that there are no elliptic curves E over quadratic fields K such that

$$E_{tors}(K) \cong \mathbb{Z}/17\mathbb{Z}.$$

It appears that on combining Kamienny's boundedness result [K1], [K2] with a meanwhile proved conjecture of Kenku and Momose [KeMo], one obtains the following result.

Theorem (Kamienny, Kenku, Momose). *Let E be an elliptic curve over a quadratic number field K. Then E has one of the torsion groups*

$$E_{tors}(K) \cong \left\{ \begin{array}{ll} \mathbb{Z}/m\mathbb{Z} & \text{for } 1 \leq m \leq 18, \ m \neq 17 \\ \mathbb{Z}/2\mathbb{Z} \times \mathbb{Z}/2\mu\mathbb{Z} & \text{for } 1 \leq \mu \leq 6 \\ \mathbb{Z}/3\nu\mathbb{Z} \times \mathbb{Z}/3\nu\mathbb{Z} & \text{for } 1 \leq \nu \leq 2 \\ \mathbb{Z}/4\mathbb{Z} \times \mathbb{Z}/4\mathbb{Z} & \end{array} \right\}.$$

Of course, the torsion groups

$$\mathbb{Z}/3\nu\mathbb{Z} \times \mathbb{Z}/3\nu\mathbb{Z} \text{ and } \mathbb{Z}/4\mathbb{Z} \times \mathbb{Z}/4\mathbb{Z}$$

can occur only over the fields

$$K = \mathbb{Q}(\sqrt{-3}) \text{ and } K = \mathbb{Q}(\sqrt{-1})$$

of third and fourth roots of unity, respectively.

It is difficult to generalize the theorems of Mazur and Kamienny-Kenku-Momose to elliptic curves over number fields K of higher degree $n = [K : \mathbb{Q}] > 2$. However, by restricting the investigation to elliptic curves with integral j-invariant over certain number fields K of degree $n = [K : \mathbb{Q}] \leq 4$ (say), one obtains only finitely many groups that can occur as torsion groups $E_{tors}(K)$. Moreover, aside from groups of order ≤ 5, all elliptic curves E and all defining fields K with given group $E_{tors}(K)$ can be computed. This program has been carried through recently, and it is our purpose here to report about the results obtained as well as the methods by which they have been derived. The computer algebra system SIMATH (see [SM]) was used to carry out the necessary calculations.

2. Results on Torsion Groups

From now on we shall make the general assumption that E is an elliptic curve with *integral* j-invariant defined over an algebraic number field K of degree

$$n = [K : \mathbb{Q}] \leq 4.$$

Let $\mathcal{O}_K$ denote the ring of integers in K. We consider the following cases of fields K of degree $n \leq 4$.

$\underline{n = 2.}$

In [MSZ1] (see also [MSZ2]) we proved

Theorem 1. *Let E be defined over a quadratic field K and assume that $j \in \mathcal{O}_K$. Then E has one of the torsion groups*

$$E_{tors} \cong \left\{ \begin{array}{ll} \mathbb{Z}/m\mathbb{Z} & \text{for } 1 \leq m \leq 10, \ m \neq 9 \\ \mathbb{Z}/2\mathbb{Z} \times \mathbb{Z}/2\mu\mathbb{Z} & \text{for } 1 \leq \mu \leq 3 \\ \mathbb{Z}/3\mathbb{Z} \times \mathbb{Z}/3\mathbb{Z} \end{array} \right\}.$$

Moreover, up to isomorphism, only finitely many elliptic curves E can have torsion groups

$$E_{tors}(K) \not\cong \mathbb{Z}/2\mathbb{Z}, \ \mathbb{Z}/3\mathbb{Z}, \ \mathbb{Z}/2\mathbb{Z} \times \mathbb{Z}/2\mathbb{Z},$$

and the corresponding defining fields

$$K = \mathbb{Q}(\sqrt{D})$$

have squarefree generators $D \in \mathbb{Z}$ satisfying

$$-31 \leq D \leq 593.$$

In the case of the Kleinian four-group $\mathbb{Z}/2\mathbb{Z} \times \mathbb{Z}/2\mathbb{Z}$, there are infinitely many elliptic curves E over K but they belong to a finite set of j-invariants only.

$\underline{n = 4.}$

What happens if one takes K to be the composite of two quadratic fields? The totally complex case was treated in [H].

Theorem 2. *Let E be defined over the composite K of two complex quadratic fields and assume that $j \in \mathcal{O}_K$. Then*

$$E_{tors}(K) \cong \left\{ \begin{array}{ll} \mathbb{Z}/m\mathbb{Z} & \text{for } 1 \leq m \leq 10 \\ \mathbb{Z}/2\mathbb{Z} \times \mathbb{Z}/2\mu\mathbb{Z} & \text{for } 1 \leq \mu \leq 3 \\ \mathbb{Z}/3\mathbb{Z} \times \mathbb{Z}/3\nu\mathbb{Z} & \text{for } 1 \leq \nu \leq 2 \\ \mathbb{Z}/4\mathbb{Z} \times \mathbb{Z}/4\mathbb{Z} \end{array} \right\}.$$

Moreover, similar finiteness assertions are true as in the quadratic case, but they are to be read off the tables in [H]. Again the infinitely many elliptic curves having the Kleinian four-group as torsion group belong to a finite set of j-invariants. In fact they are twists of finitely many elliptic curves. Note that only the groups

$$\mathbb{Z}/9\mathbb{Z}, \ \mathbb{Z}/3\mathbb{Z} \times \mathbb{Z}/6\mathbb{Z} \ \text{and} \ \mathbb{Z}/4\mathbb{Z} \times \mathbb{Z}/4\mathbb{Z}$$

are to be added to the list of possible torsion groups as one passes from quadratic to totally complex biquadratic fields K.

The totally real case was treated in [St].

Theorem 3. *Let E be defined over the composite K of two real quadratic fields and assume that $j \in \mathcal{O}_K$. Then*

$$E_{tors}(K) \cong \left\{ \begin{array}{ll} \mathbb{Z}/m\mathbb{Z} & \text{for } 1 \leq m \leq 8, \, m \neq 5, \, m = 14 \\ \mathbb{Z}/2\mathbb{Z} \times \mathbb{Z}/2\mu\mathbb{Z} & \text{for } 1 \leq \mu \leq 3 \end{array} \right\}.$$

Again similar finiteness results are to be read off from the tables in [St]. Here the list of possible torsion groups is much smaller than in the totally complex biquadratic case. It is remarkable that, as opposed to the case of totally complex biquadratic ground fields K, the cyclic torsion group

$$E_{tors}(K) \cong \mathbb{Z}/14\mathbb{Z}$$

turns up here. Unfortunately, the case of the cyclic group $\mathbb{Z}/5\mathbb{Z}$ must remain unsettled: on the one hand it could not be shown that $m = 5$ cannot occur and on the other hand, no curves with $m = 5$ while $j \in \mathcal{O}_K$ were found within a reasonable computing time.

$\underline{n = 3.}$

Finally, we consider the class of cubic fields K. At first we shall deal with pure cubic fields, then pass to cyclic cubic fields and eventually treat arbitrary cubic fields.

In [FSWZ] (see also [MSZ2]) the following result is proved.

Theorem 4. *Let E be defined over a pure cubic field K and assume that $j \in \mathcal{O}_K$. Then*

$$E_{tors}(K) \cong \left\{ \begin{array}{ll} \mathbb{Z}/m\mathbb{Z} & \text{for } 1 \leq m \leq 6 \\ \mathbb{Z}/2\mathbb{Z} \times \mathbb{Z}/2\mathbb{Z} & \end{array} \right\}.$$

Moreover, up to isomorphisms or twists, only finitely many elliptic curves E can have torsion groups

$$E_{tors}(K) \not\cong \mathbb{Z}/2\mathbb{Z}, \, \mathbb{Z}/3\mathbb{Z}, \, \mathbb{Z}/2\mathbb{Z} \times \mathbb{Z}/2\mathbb{Z},$$

and the corresponding defining fields

$$K = \mathbb{Q}(\sqrt[3]{ab^2})$$

have squarefree generators

$$a = 2, \, 3, \, 5, \, 31; \quad b = 1.$$

As we see, the list of possible torsion groups is rather small. It grows as one passes to cyclic cubic fields. In [PWZ] (cf. also [W]) we show the following.

Theorem 5. *Let E be defined over a cyclic cubic field K and assume that $j \in \mathcal{O}_K$. Then*

$$E_{tors}(K) \cong \left\{ \begin{array}{ll} \mathbb{Z}/m\mathbb{Z} & \text{for } 1 \leq m \leq 9, \, m = 14 \\ \mathbb{Z}/2\mathbb{Z} \times \mathbb{Z}/2\mu\mathbb{Z} & \text{for } 1 \leq \mu \leq 3 \end{array} \right\}.$$

Moreover, if

$$E_{tors}(K) \not\cong \mathbb{Z}/2\mathbb{Z}, \ \mathbb{Z}/3\mathbb{Z}, \ \mathbb{Z}/4\mathbb{Z}, \ \mathbb{Z}/2\mathbb{Z} \times \mathbb{Z}/2\mathbb{Z},$$

then, up to isomorphism, only finitely many elliptic curves E over finitely many cyclic cubic fields K can have one of the torsion groups listed. Once more, the finitely many fields are to be read off from the tables in [PWZ]. Note that, compared to the pure cubic case, some more groups arise as torsion groups, namely

$$\mathbb{Z}/m\mathbb{Z} \text{ for } m = 7, \ 8, \ 9, \ 14 \text{ and } \mathbb{Z}/2\mathbb{Z} \times \mathbb{Z}/2\mu\mathbb{Z} \text{ for } \mu = 2, \ 3.$$

Remark. Theorem 5 is proved by treating the case of general cubic fields first and separating the cyclic cubic fields from the tables afterwards. The results obtained in [W] are incomplete.

Only the group

$$\mathbb{Z}/10\mathbb{Z}$$

is to be added to the list of torsion groups if one passes to the general cubic case treated in [PWZ].

Theorem 6. *Let E be defined over an arbitrary cubic field K and assume that $j \in \mathcal{O}_K$. Then*

$$E_{tors}(K) \cong \left\{ \begin{array}{ll} \mathbb{Z}/m\mathbb{Z} & \text{for } 1 \le m \le 10, \ m = 14 \\ \mathbb{Z}/2\mathbb{Z} \times \mathbb{Z}/2\mu\mathbb{Z} & \text{for } 1 \le \mu \le 3 \end{array} \right\}.$$

Moreover, similar finiteness results are obtained for the torsion groups

$$E_{tors}(K) \not\cong \mathbb{Z}/2\mathbb{Z}, \ \mathbb{Z}/3\mathbb{Z}, \ \mathbb{Z}/4\mathbb{Z}, \ \mathbb{Z}/2\mathbb{Z} \times \mathbb{Z}/2\mathbb{Z}, \ \mathbb{Z}/5\mathbb{Z},$$

and the finitely many cubic ground fields can be read off from the tables in [PWZ]. It is remarkable that while there are only finitely many elliptic curves E (up to isomorphism) and *cyclic* cubic fields K such that

$$E_{tors}(K) \cong \mathbb{Z}/5\mathbb{Z},$$

there are infinitely many curves E over infinitely many *general* cubic fields having this torsion group. The proof of this fact is nontrivial as will be seen from Propositions 4 and 5 and their Corollaries.

In the finiteness cases arising from Theorems 1 - 6, all elliptic curves and all ground fields have been computed and tabulated. We shall give some examples in Section 4.

3. PROOFS

The proofs of these theorems consist of three components, namely reduction theory, parametrizations and solution of norm equations.

(i) Reduction theory. Let $\mathfrak{p}$ denote a finite place of the number field K, $e_\mathfrak{p}$ and $f_\mathfrak{p}$ the ramification index and residue degree, respectively, and $\tilde{k}_\mathfrak{p}$ the residue field of K with respect to $\mathfrak{p}$. The reduction map, applied to a $\mathfrak{p}$-minimal model of an elliptic curve E over K, yields a cubic curve $\tilde{E}$ defined over the residue field $\tilde{k}_\mathfrak{p}$. In the case of good reduction of E modulo $\mathfrak{p}$, the reduced curve $\tilde{E}$ is an elliptic curve over $\tilde{k}_\mathfrak{p}$, and the order of its Mordell-Weil group $\tilde{E}(\tilde{k}_\mathfrak{p})$ satisfies a fundamental inequality (cf. e.g. [F]):

Proposition 1 ("Riemann hypothesis"). *If E has good reduction modulo $\mathfrak{p}$,*

$$\sharp\tilde{E}(\tilde{k}_\mathfrak{p}) \leq (1 + p^{f_\mathfrak{p}/2})^2,$$

where p is the rational prime of $\mathbb{Q}$ associated with $\mathfrak{p}$.

Reduction theory can be used to get an overview of possible torsion groups of E over K by virtue of the following results (see [F], [MSZ2]):

Proposition 2. *The order of the torsion group of an elliptic curve E over a number field K satisfies the divisibility relations:*

$$\sharp E_{tors}(K) \mid \begin{cases} \sharp\tilde{E}(\tilde{k}_\mathfrak{p}) \cdot p^{2t} & \text{if } E \text{ has good reduction mod } \mathfrak{p}, \\ |v_\mathfrak{p}(j)| \cdot (p^{2f_\mathfrak{p}} - 1) \cdot p^{2t} & \text{if } E \text{ has multiplicative reduction mod } \mathfrak{p}, \\ 12 \cdot p^{2(t+1)} & \text{if } E \text{ has additive reduction mod } \mathfrak{p}, \end{cases}$$

where

$$t := \left\{ \begin{array}{ll} 0 & \text{if } \varphi(p) > e_\mathfrak{p} \\ \max\{\nu \in \mathbb{N};\ \varphi(p^\nu) \leq e_\mathfrak{p}\} & \text{else} \end{array} \right\}$$

with Euler's function φ.

Since we are assuming that E has integral j-invariant, the case of multiplicative reduction is ruled out. Therefore, the orders of possible torsion groups are restricted by this proposition. In fact the orders are bounded by a constant depending only on the field K but not on the curve E. If K has degree $n \leq 4$ over $\mathbb{Q}$, the bound is "small", and we get a pretty good survey of torsion groups that can arise for curves E such that $j \in \mathcal{O}_K$. Actually, already the requirement that E have good or additive reduction at all places $\mathfrak{p}$ lying over 2, 3 or 5 entails a restricted list of torsion groups (cf. Theorems 2 and 2' in [MSZ 2] and Theorem 8 in [H]). As an example, we state the following

Proposition 3. *Let E be defined over a cubic number field K and suppose that E has $\mathfrak{p}$-integral j-invariant at all finite places $\mathfrak{p}$ of K lying over 2, 3 or 5. Then the order of the torsion group of E over K satisfies one of the following divisibility relations:*

$$\sharp E_{tors}(K) \mid 2^2 \cdot 5,\ 2^2 \cdot 7,\ 2 \cdot 11, 2 \cdot 13 \text{ or } 2^\mu 3^\mu$$
$$\text{for } 0 \leq \mu \leq 5,\ 0 \leq \nu \leq 2 \text{ and } (\mu, \nu) \neq (5, 2).$$

(ii) Parametrizations. Having obtained an overview of the possible torsion groups of an elliptic curve E with integral j-invariant over a number field K, we now pass to Kubert's $E(b,c)$-form or Tate's normal form giving a parametrization of curves E over K with a torsion point of order > 3. This normal form, which was used by Kubert [Ku] in his proof of a partial case of the boundedness conjecture (see also [Hu]), is the following (cf. [MSZ 1], [MSZ 2], [FSW]):

$$E: \ Y^2 + (1-c)XY - bY = X^3 - bX^2 \quad (b,c \in K).$$

The *discriminant* is

$$\Delta = b^3(\beta^4 - \beta^3 - 8b\beta^2 + 36b\beta + 16b^2 - 27b)$$

and the j-invariant

$$j = \frac{((\beta - 4b)^2 + 24b\beta)^3}{\Delta},$$

where we use the abbreviating notation

$$\beta := 1 - c.$$

In Tate's normal form,

$$P = (0,0) \in E_{tors}(K)$$

is a point of maximal order $m > 3$. We shall choose the cases

$$m = 5 \text{ and } 7$$

as examples and refer to the papers [MSZ 1], [FSWZ] and [PWZ] for the remaining cases.

If we fix an order m we obtain a specific parametric $E(b,c)$-form of our elliptic curve in which one parameter b or c can be removed. The integrality condition on the j-invariant of E then translates into necessary and (in most cases also) sufficient conditions on the parameter which in turn will be transformed into norm equations. In performing these transformations, we loose the sufficiency of the conditions, but this does not really matter since, in the worst case, we end up producing some elliptic cuves E over K with fractional j-invariant and a torsion point of order m.

Let $\mathcal{O}_K^*$ denote the unit group of K and $v_{\mathfrak{p}}$ be the normalized additive valuation of K corresponding to the prime $\mathfrak{p}$ of K. The degree of K over $\mathbb{Q}$ is again denoted by $n = [K : \mathbb{Q}]$.

A) $E_{tors}(K) \geq \mathbb{Z}/5\mathbb{Z}$, $m = 5$:
The elliptic curve

$$E: \ Y^2 + (1-\alpha)XY - \alpha Y = X^3 - \alpha X^2$$

has parameter

$$b = c =: \alpha \in K \setminus \{0\}$$

and j-invariant

$$j = \frac{((\alpha^2 - 11\alpha - 1)^2 + 5\alpha(2(\alpha^2 - 11\alpha - 1) + \alpha))^3}{\alpha^5(\alpha^2 - 11\alpha - 1)}.$$

Hence we conclude that

$$j \in \mathcal{O}_K \Leftrightarrow \begin{cases} (1) & \alpha \in \mathcal{O}_K^* \\ (2) & 0 \leq v_{\mathfrak{p}}(\alpha^2 - 11\alpha - 1) \leq 3v_{\mathfrak{p}}(5) \text{ for all finite places } \mathfrak{p} \text{ of } K. \end{cases}$$

By taking norms with respect to $K/\mathbb{Q}$, these relations are transformed into the following norm equations:

$$j \in \mathcal{O}_K \Rightarrow \begin{cases} (1') & N_{K/\mathbb{Q}}(\alpha) = \pm 1 \\ (2') & N_{K/\mathbb{Q}}(\alpha^2 - 11\alpha - 1) = \pm 5^\nu \quad \text{for } 0 \leq \nu \leq 3n. \end{cases}$$

B) $E_{tors}(K) \geq \mathbb{Z}/7\mathbb{Z}$, $m = 7$:

The elliptic curve

$$E: \quad Y^2(1 - c)XY - bY = X^3 - bX^2$$

has parameter $\alpha \in K \setminus \{0, 1\}$ such that

$$b = \alpha^2(\alpha - 1), \ c = \alpha(\alpha - 1)$$

and j-invariant

$$j = \frac{(\alpha^8 - 12\alpha^7 + 42\alpha^6 - 56\alpha^5 + 35\alpha^4 - 14\alpha^2 + 4\alpha + 1)^3}{\alpha^7(\alpha - 1)^7(\alpha^3 - 8\alpha^2 + 5\alpha + 1)}.$$

Hence we conclude that

$$j \in \mathcal{O}_K \Leftrightarrow \begin{cases} (1) \ \alpha \in \mathcal{O}_K^* \\ (2) \ \alpha - 1 \in \mathcal{O}_K^* \\ (3) \ 0 \leq v_{\mathfrak{p}}(\alpha^3 - 8\alpha^2 + 5\alpha + 1) \leq 2v_{\mathfrak{p}}(7) \text{ for all finite places } \mathfrak{p} \text{ of } K. \end{cases}$$

Taking norms with respect to $K/\mathbb{Q}$ leads to

$$j \in \mathcal{O}_K \Rightarrow \begin{cases} (\overline{1}) & N_{K/\mathbb{Q}}(\alpha) = \pm 1 \\ (\overline{2}) & N_{K/\mathbb{Q}}(\alpha - 1) = \pm 1 \\ (\overline{3}) & N_{K/\mathbb{Q}}(\alpha^3 - 8\alpha^2 + 5\alpha - 1) = \pm 7^\nu \text{ for } 0 \leq \nu \leq 2n \end{cases}$$

(iii) Solving norm equations. By fixing a degree n and solving the above norm equations (1') - (2') or ($\overline{1}$) - ($\overline{3}$) in all number fields K of degree n over $\mathbb{Q}$, we obtain the complete set of elliptic curves E over finitely many fields K such that E has torsion group

$$E_{tors}(K) \cong \mathbb{Z}/5\mathbb{Z} \text{ or } \mathbb{Z}/7\mathbb{Z}, \text{ respectively.}$$

The methods for solving these norm equations vary, however, depending on whether K is a quadratic, a pure cubic, a cyclic cubic, a general cubic or a biquadratic number field. They also depend on the isomorphism type of the torsion group considered.

We shall treat here only the cases of elliptic curves E over cubic number fields K such that E has cyclic torsion group of order $m = 5$.

In [P2] the following assertion is proved.

Proposition 4. *Let K be a cyclic cubic number field with discriminant D_K. Then*

$$\alpha \in \mathcal{O}_K$$

is a solution of the norm equations

$$(1') \quad N_{K/\mathbb{Q}}(\alpha) = \pm 1$$
$$(2') \quad N_{K/\mathbb{Q}}(\alpha^2 - 11\alpha - 1) = \pm 5^\nu \text{ for } 0 \le \nu \le 9$$

if and only if α generates K and is a zero of one of the following polynomials f with discriminant D_f:

ν	$f(x)$	D_f	D_K
0	$x^3 - 12x^2 + 9x + 1$	$(3^2 \cdot 13)^2$	$(3^2 \cdot 13)^2$
0	$x^3 - 13x^2 + 10x + 1$	139^2	139^2
2	$x^3 - 14x^2 + 11x + 1$	163^2	163^2
3	$x^3 - 9x^2 + 6x + 1$	$(3^2 \cdot 7)^2$	$(3^2 \cdot 7)^2$
4	$x^3 - 12x^2 + 35x + 1$	$(5 \cdot 13)^2$	13^2
4	$x^3 + 3x^2 - 160x + 1$	$(5^2 \cdot 163)^2$	163^2
5	$x^3 + 3x^2 - 10x + 1$	$(5 \cdot 13)^2$	13^2
5	$x^3 - 17x^2 - 25x + 1$	$(2^3 \cdot 5 \cdot 13)^2$	13^2

As a consequence we can state the

Corollary. *There are only finitely many elliptic curves E (up to isomorphism) over finitely many cyclic cubic fields K having torsion group*

$$E_{tors}(K) \cong \mathbb{Z}/5\mathbb{Z}.$$

The proof of Proposition 4 involves Fibonacci and Lucas numbers (see [P2]).

In the general cubic case, the following assertion can be proved ([P1], [P3]).

Proposition 5. *Let K be a cubic number field. If $\alpha \in \mathcal{O}_K$ is a solution of the norm equations*

$$(1'') \quad N_{K/\mathbb{Q}}(\alpha) = n_1$$
$$(2'') \quad N_{K/\mathbb{Q}}(p(\alpha)) = n_2$$

for fixed integers $n_1, n_2 \in \mathbb{Z}$ and a given quadratic polynomial

$$p(x) = x^2 + cx + d \in \mathbb{Z}[x] \text{ with discriminant } D = c^2 - 4d,$$

then there exist integers $w, v, m_1 \in \mathbb{Z}$ such that the equations

$$(3'') \quad w^2 - D(vd + cd - n_1)^2 = 4dn_2,$$
$$(4'') \quad d - \frac{c(vd + cd + n_1) + w}{2d} = m_1$$

hold, and $\alpha \in \mathcal{O}_K$ is a zero of the polynomial

$$q(x) = x^3 - vx^2 + m_1 x - n_1 \in \mathbb{Z}[x].$$

Conversely, if there are integers $w, v, m_1 \in \mathbb{Z}$ satisfying (3") and (4") and if α is a zero of the irreducible polynomial $q(x) \in \mathbb{Z}[x]$, then (1") and (2") have a solution $\alpha \in \mathcal{O}_K$ in the cubic field $K = \mathbb{Q}(\alpha)$.

As a consequence we obtain the opposite assertion to the above corollary.

Corollary. *There are infinitely many (isomorphism classes of) elliptic curves E over infinitely many (noncyclic) cubic number fields K having torsion group*

$$E_{tors}(K) \cong \mathbb{Z}/5\mathbb{Z}.$$

The proof of the Proposition 5 depends on Groebner basis techniques (see [P1]).

The norm equations $(\overline{1})$ - $(\overline{3})$ corresponding to the cyclic torsion group $E_{tors}(K) \cong \mathbb{Z}/7\mathbb{Z}$ are also solved by computing suitable Groebner bases (see [PWZ]).

The solution of norm equations in biquadratic number fields is achieved via relative norms involving quadratic intermediate fields and by referring to the decomposition law in K (see [H], [St]).

4. NUMERICAL EXAMPLES

As we pointed out already, in the finiteness cases all elliptic curves E and number fields K of degree $n = [K : \mathbb{Q}] \le 4$ (with the above-mentioned restrictions in the case of $n = 4$) have been computed. In the tables below we list some examples of curves with torsion groups of orders $m = 5, 7, 8, 10$ and 14. Points of order $m = 10, 14$ are obtained by taking the tables for points of order $5, 7$, respectively, and trying to divide the latter points by 2.

Elliptic curves E over totally complex biquadratic fields K having a torsion group isomorphic to $\mathbb{Z}/5\mathbb{Z}$, $\mathbb{Z}/10\mathbb{Z}$, $\mathbb{Z}/8\mathbb{Z}$, $\mathbb{Z}/7\mathbb{Z}$ or $\mathbb{Z}/14\mathbb{Z}$ turn out to be only those defined already over a quadratic subfield, and hence they are listed already in the tables of [MSZ1], [MSZ2].

Over totally real biquadratic fields K, as pointed out above, we did not find any curves E with integral j-invariant having torsion group isomorphic to $\mathbb{Z}/5\mathbb{Z}$ and, a fortiori, we could not find any curves E over K such that $E_{tors}(K) \cong \mathbb{Z}/10\mathbb{Z}$ either - except, of course, for those curves defined already over a quadratic subfield K and thus listed already in [MSZ1], [MSZ2].

In the tables below, the elliptic curves E over K are given in short Weierstrass form $Y^2 = X^3 + AX + B$. An integral basis (GHB) of K is denoted by $\omega_1, \omega_2, \dots, \omega_n$ for $n = [K : \mathbb{Q}]$.

It would be an interesting task to find a small bound for the order of $E_{tors}(K)$ for number fields K of small degree n and to calculate a complete set of curves E over those fields K with torsion group of order below the bound.

Acknowledgement. The author wishes to thank J. Stein and Th. Weis for setting up the tables taken from [H], [PWZ], [St].

$n = 3, m = 5$:

TABLE 1. $E_{tor}(K) \geq \mathbb{Z}/5\mathbb{Z}$

K	$\mathbb{Q}(\rho)$ with $\rho^3 = 13\rho^2 - 10\rho - 1$
GHB	$\omega_1 = 1$
	$\omega_2 = \rho$
	$\omega_3 = \rho^2$
D_K	$19321 = 139^2$ (cyclic)

α	ρ
j	$275\omega_3 - 3146\omega_2 + 2126$
$E \;:\; A$	$-459\rho^2 - 27\rho$
B	$38718\rho^2 - 25488\rho - 2592$
P	$(\; 3\rho^2 - 18\rho + 3 \;,\; -108\rho \;)$
	is a point of order 5

TABLE 2. $E_{tor}(K) \geq \mathbb{Z}/5\mathbb{Z}$

K	$\mathbb{Q}(\rho)$ with $\rho^3 = 15\rho^2 - 3\rho + 1$
GHB	$\omega_1 = 1$
	$\omega_2 = \rho$
	$\omega_3 = \frac{1}{10}\rho^2 + \frac{4}{5}\rho + \frac{7}{10}$
D_K	$-108 = -1 \cdot 2^2 \cdot 3^3$ (non-cyclic)

α	ρ
j	$12800\omega_3 - 27136\omega_2 + 8448$
$E \;:\; A$	$-1512\rho^2 - 108\rho - 108$
B	$335664\rho^2 - 64800\rho + 22464$
P	$(\; 3\rho^2 - 18\rho + 3 \;,\; -108\rho \;)$
	is a point of order 5

$n = 3, m = 10$:

$$\text{TABLE 3. } E_{tor}(K) \geq \mathbb{Z}/10\mathbb{Z}$$

K	$\mathbb{Q}(\rho)$ with $\rho^3 = 4\rho - 2$
GHB	$\omega_1 = 1$
	$\omega_2 = \rho$
	$\omega_3 = \rho^2$
D_K	$148 = 2^2 \cdot 37$ (non-cyclic)

α	ρ
j	$-4096\omega_3 - 4096\omega_2 + 12288$
$E \ : \ A$	$\frac{770688}{125}\rho^2 + \frac{415152}{125}\rho - \frac{2858544}{125}$
B	$-\frac{184873968}{625}\rho^2 - \frac{99682272}{625}\rho + \frac{685751184}{625}$
P	$\left(-\frac{2424}{25}\rho^2 - \frac{1296}{25}\rho + \frac{9012}{25} \ , \ -\frac{192672}{125}\rho^2 - \frac{103788}{125}\rho + \frac{714636}{125} \right)$
	is a point of order 5
Q	$\left(\frac{192}{5}\rho^2 + \frac{108}{5}\rho - \frac{696}{5} \ , \ \frac{10152}{25}\rho^2 + \frac{5508}{25}\rho - \frac{37476}{25} \right)$
	is a point of order 10

α	$\rho^2 + 2\rho$
j	$10412064\omega_3 + 17440800\omega_2 - 12430736$
$E \ : \ A$	$-\frac{1740852}{125}\rho^2 + \frac{3848742}{125}\rho - \frac{1569024}{125}$
B	$\frac{649681776}{625}\rho^2 - \frac{1438518096}{625}\rho + \frac{586754712}{625}$
P	$\left(\frac{2811}{25}\rho^2 - \frac{6156}{25}\rho + \frac{2532}{25} \ , \ \frac{459648}{125}\rho^2 - \frac{1018008}{125}\rho + \frac{415476}{125} \right)$
	is a point of order 5
Q	$\left(-33\rho^2 + 72\rho - 24 \ , \ -\frac{10908}{25}\rho^2 + \frac{23868}{25}\rho - \frac{9396}{25} \right)$
	is a point of order 10

Recent Titles in This Series

(Continued from the front of this publication)

(See the AMS catalog for earlier titles)